INTERNATIONAL YOUNG PHYSICISTS' TOURNAMENT

Collection of Works for
IYPT Problems
(25th–26th)

INTERNATIONAL YOUNG PHYSICISTS' TOURNAMENT

Collection of Works for
IYPT Problems
(25th–26th)

Editors

Sihui Wang
Wenli Gao

Nanjing University, China

World Scientific

NEW JERSEY · LONDON · SINGAPORE · BEIJING · SHANGHAI · HONG KONG · TAIPEI · CHENNAI

Published by

World Scientific Publishing Co. Pte. Ltd.

5 Toh Tuck Link, Singapore 596224

USA office: 27 Warren Street, Suite 401-402, Hackensack, NJ 07601

UK office: 57 Shelton Street, Covent Garden, London WC2H 9HE

British Library Cataloguing-in-Publication Data
A catalogue record for this book is available from the British Library.

INTERNATIONAL YOUNG PHYSICISTS' TOURNAMENT
Collection of Works for IYPT Problems (25th–26th)

ISBN 978-981-4630-83-2 (pbk)

In-house Editor: Christopher Teo

Printed in Singapore

This book is dedicated to the Fifth Anniversary of NYPT (NJU's Young Physicists' Tournament).

Contents

Nanjing University team won the championship of the 2nd TCPT (Taiwan College-student Physicists' Tournament) (2012, Taipei). The team members (from right) are Cao Tongyi, Zhu Zheyuan (Captain), Li Yaohua, Huang Shan and Xia Qing.

Nanjing University team in 2013 TCPT (Taiwan College-student Physicists' Tournament). The team members (from left) are He Qiying, Du Li (Captain), Qi Jiaan, Ruan Qiyuan and Chen Lan.

Nanjing University team in 2013 CUPT (China Undergraduate Physicists' Tournament). The team members (from left) are Xiong Bo, Zhu Enlin (Captain), Zeng Pei, Zhang Youtian and Qin Zhihang.

Nanjing University team in 2012 CUPT (China Undergraduate Physicists' Tournament). The team members (from left) are Zhu Kejing, Li Xiao, Zhu Zheyuan (Captain), Liu Chang (Leader), Shi Tong, Li Yaohua.

Nanjing University team in 2012 CUPT (China Undergraduate Physicists' Tournament). The team members (from left) are Shi Tong, Zhu Kejing, Li Xiao, Zhu Zheyuan (Captain), Li Yaohua.

Chapter 1

2012 Problem 1: Gaussian Cannon

Qing Xia*, Wenli Gao, Sihui Wang and Huijun Zhou

School of Physics, Nanjing University

Using the theory of elasticity, we establish an accurate collision model and quantitatively explain how Gaussian Cannon gains its most powerful shot under certain experimental parameters. The work done by magnetic force on the steel ball is obtained by measuring the magnetic force. Essential factors to acquire higher ejection speed have been found.

1. Introduction

Problem Statement:

> A sequence of identical steel balls includes a strong magnet and lies in a nonmagnetic channel. Another steel ball is rolled towards them and collides with the end ball. The ball at the opposite end of the sequence is ejected at a surprisingly high velocity. Optimize the magnet's position for the greatest effect.

1.1. *Preliminary Analysis*

Gaussian Cannon is a cool device which can easily be reproduced by yourself. The mechanism of Gaussian Cannon is rather simple–the injected ball is accelerated by the magnetic force and then transfers its energy to the last ball through collision. The last ball overcomes the relatively small energy barrier of the magnetic field and ejects at an incredibly high speed.

In this article, we will investigate how to optimize the magnet's position for the greatest effect, i.e. the highest ejection speed. What intrigue us most are the following questions: Should the magnet be fixed or not if we want to achieve the greatest effect? Which position should the magnet take in the sequence? Will the total number of steel balls influence the

---

*qingxia1234@gmail.com

device's efficiency? If so, how many balls will be most appropriate? We will illustrate the first two questions with experiments and qualitative explanations, while give the last two a numerical calculation and compare it with experimental data.

2. Experimental Setup

Fig. 1.   The wooden channel made by a milling machine.

We made a wooden channel with the milling machine(Fig. 1) and put a sequence of identical steel balls (7.2g,12mm in diameter) including a cylindrical magnet (7.6g,12mm) on it. The speed of the balls is measured by photogates: one photogate is settled 5cm before the magnet to measure the speed of the injected ball, another is put next to the ejected ball. Since the photogates record the period of time the light is blocked, the speed of the ball can be calculated by v=diameter/time (Fig. 2).

In order to understand the phenomenon quantitatively, we measured the magnetic force between each ball, as well as the change of the force's magnitude with distance. This measurement was accomplished by using a jolly balance and an electronic balance. As is shown in Fig. 3, the read of the electronic balance becomes smaller as the distance between the magnet and the steel ball gets smaller.

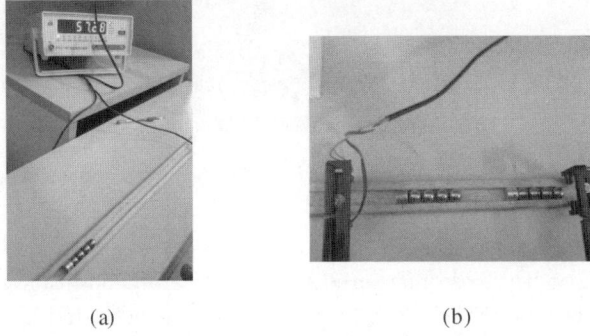

(a) (b)

Fig. 2. Photogates used for measuring the speed of the injected and ejected steel ball. (a) one sequence of balls (b) two sequences of balls. Note, however, that all experimental data shown in this article are measured with only one sequence of balls.

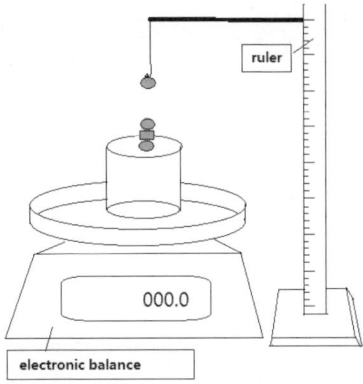

Fig. 3. Setup for measuring magnetic force. The read of the electronic balance becomes smaller as the distance between the magnet and the steel ball gets smaller.

3. Experiment Results

3.1. *Magnet Fixed/Not Fixed*

In the first experiment, we measured the speed of balls while the magnet was fixed by using two small cardboards as is shown in Fig. 4.

The benefit of our fixation method is that the magnet is unable to move backward, ensuring the momentum transfer during collision. The data

Fig. 4.   Magnet fixed by two small cardboards. The advantage of this fixation method is that the magnet is unable to move backward.

are compared with those measured under the same condition except that the magnet was not fixed (Fig. 5). Obviously, properly fixing the magnet produces greater ejection speed. Otherwise, the retreat of the magnet and end balls will reduce the device's efficiency.

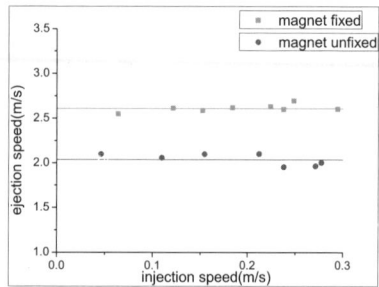

Fig. 5.   The square dots demonstrate the ejected ball's speed when the magnet is fixed, while the round dots demonstrate the speed when the magnet is not fixed. The speed of the ejected ball is higher when the magnet is fixed than not fixed.

Another information we get from Fig. 5 is that the ejected ball's speed almost has nothing to do with the injected ball's speed in our experiment. This is no surprise because the speed of the injected ball was small (less than 0.3m/s) at the spot we measured it. This kinetic energy is much smaller than the energy it gained from the magnetic field afterwards. Therefore, we may assume that the ball is injected at the same speed every time so that the average speed of the ejected ball manifests the device's efficiency.

3.2.   *Influence of the Magnet's Position*

In this experiment, we consider the influence of the magnet's position on the device's efficiency. Schematic picture of the position of the magnet is shown in Fig. 6. Fig. 7 shows the ejection speed is far smaller when the

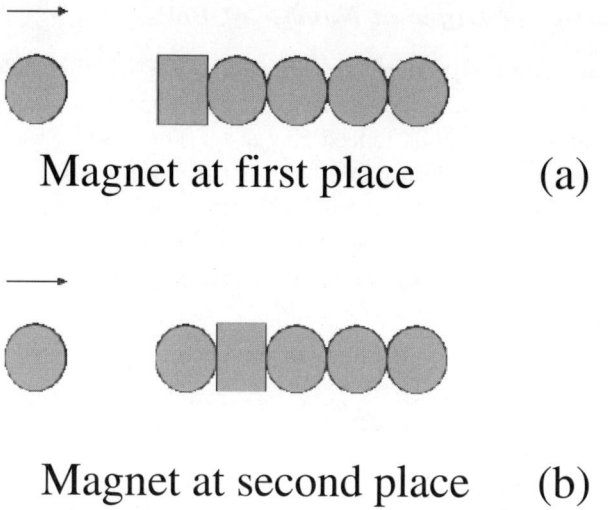

Magnet at first place (a)

Magnet at second place (b)

Fig. 6. Schematic picture of the position of the magnet. (a) magnet at first place (b) magnet at second place.

magnet is put in the second place than that when the magnet is put in the first place. It is obvious that extra balls between the injected ball and the magnet will lower the positive work done on the injected ball by the magnetic force.

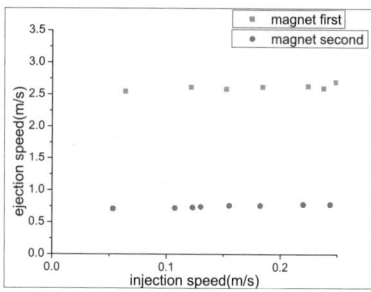

Fig. 7. Speed of the balls under different magnet's position. The speed of the ejected ball is higher when the magnet is put at the first place than when it is put at the second place.

3.3. *Influence of Different Number of Balls*

Another important question we are curious about is whether different number of balls will cause a difference in the average speed of the ejected ball. Fig. 8 indicates that this is indeed the case. The average ejection speed is greatest when there is only one ball on the injection side, namely, the injected ball. Moreover, the average ejection speed is largest when there are 4 end balls in our experiment. This critical point is of peculiar interest so we will address more about the reason of its occurrence in Section 5.

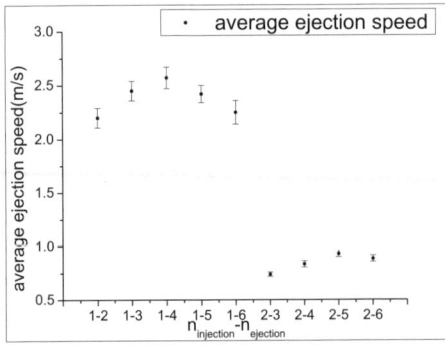

Fig. 8.   Average ejection speed when there are different number of balls on the injection/ejection side. The maximal speed appears when there is one ball on the injection side and four balls on the ejection side.

3.4. *Magnetic Force Measurement*

Fig. 9 shows the change of the magnitude of the magnetic force with distance between the injected ball and the magnet when there is one end ball. The magnetic force-distance relationship is also measured when there are two end balls, three end balls and so forth. It is found that the number of end balls will not make detectable change in the magnetic force.

Using a similar approach, we can measure the magnetic force exerted on the ejected ball when there are different number of end balls (Fig. 10). There are two features about the data: The force on the ejected ball reduces drastically with the increasing number of end balls; the magnetic force on the ejected ball is very small when there are more than 4 end balls.

These data are essential for the calculation in Section 5 when we will give a quantitive description of the whole phenomenon.

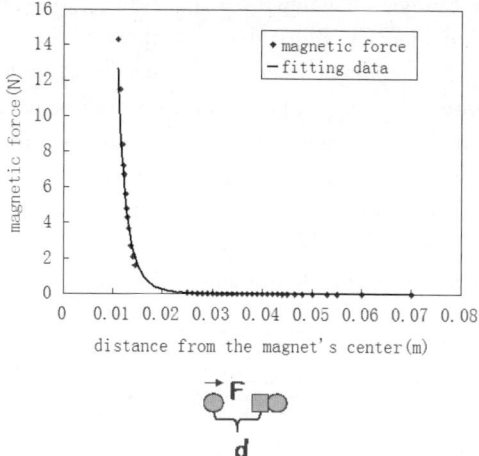

Fig. 9. Magnetic force-distance relationship with one end ball present. These data will later be used when calculating the work done on the steel ball.

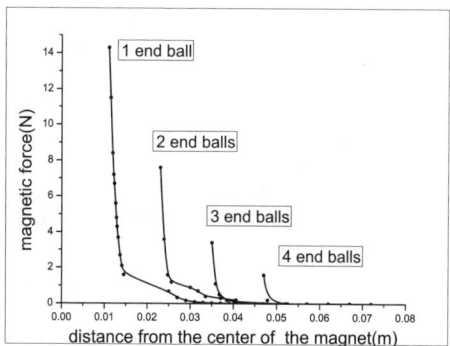

Fig. 10. Magnetic force exerted on the ejected ball when there are different number of end balls.

4. Qualitative Explanation

So far we have made several preliminary discoveries based on the experiment results above: (1) Injection energy is rather small compared with the energy provided by the magnetic field. (2) The magnet should be fixed. (3) Once the total number of steel balls is set, the magnet should be put at the first place, i.e. 1 strikes n balls. We shall also notice that due to energy loss, the last ball can only be ejected when $n_{injection} < n_{ejection}$ (n is the number of balls on injection/ejection side). In our experimental arrangement, the average ejection speed is largest when there are 4 end balls.

Here we propose a simple explanation for the appearance of the maximum point. When there are fewer end balls, there is a stronger magnetic force for the ejected ball to overcome; and when there exist more end balls, although the magnetic force exerted on the ejected ball is smaller, more energy loss is produced during collision.

A quantitative study of the phenomenon is given in the following section.

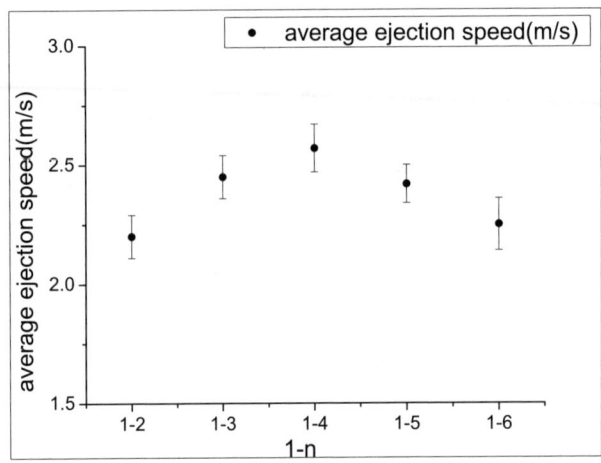

Fig. 11.   1 ball strikes n balls condition. The average ejection speed is the largest when there are 4 end balls in our experiment.

5. Quantitative Model

It is known that sheet steel reaches magnetic saturation easily, so the magnetic moment $\vec{m}$ is approximately a constant when a small external field is applied. Moreover, since the ball in our experiment is made by low carbon steel, its hysteresis effect can be ignored.

5.1. Calculating the Ejected Ball's Speed

5.1.1. General Idea

Fig. 12. Force analysis of the injected ball. The force exerted on the injected ball consists of the magnetic force F and friction f.

Fig. 12 shows the force analysis of the injected ball. At first the injected balls is purely rolling. It keeps doing so provided that

$$a = \beta \cdot R,$$
$$a = \frac{F - f}{m},$$
$$\beta = \frac{f \cdot R}{I}.$$
(1)

where a is the acceleration, β is the angular acceleration of the injected ball, R is the radius and I is the moment of inertia of the steel ball. $F = \frac{7}{2}f$ for a rolling ball.

After collision, the ejected ball slips for a short distance and then does pure rolling. The ejected ball's kinetic energy is calculated by the equation

$$E_{eject} = E_{inject} - E_{collision} - E_{friction} - W_{mag}$$
(2)

where W_{mag} is the work done by the magnet on the steel ball.

Fig. 13 shows the energy transmission during the collision process.

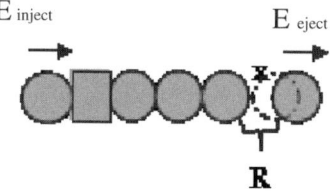

Fig. 13.   Energy transmission during collision. The ejection energy equals the injection energy subtracting the collision loss, friction loss and the negative work done on the ejected ball.

5.1.2. "Touching Ball" Collision Model

The injection energy can be calculated by integrating the magnetic force with respect to distance. As for the calculation of collision and friction loss, we apply here a "touching ball" collision model[1]–Determining the velocities for the case of 1 ball striking "touching balls" is found by modeling the balls as weights with non-traditional springs on their colliding surface. Steel is elastic and follows Hook's force law for springs, F=k·x, but because the area of contact for a sphere increases as the force increases, colliding elastic balls will follow Hertz's adjustment to Hook's law

$$F = k \cdot x^{1.5}. \tag{3}$$

(3) and Newton's law for motion (F=m·a) are applied to each ball, giving several simple but interdependent ("touching") differential equations which can be solved numerically. According to the theory of elasticity[2]

$$x = F^{2/3}[D^2(\frac{1}{R} + \frac{1}{R'})]^{1/3} \tag{4}$$

where $D=\frac{3}{4}\frac{1-\sigma^2}{E}+\frac{1-\sigma'^2}{E'}$. Combining (3) with (4), we get the coefficient k is determined by the Poisson ratio σ, the Young's modulus E of the material and the radius of the touching balls.

$$k = [D^2(\frac{1}{R} + \frac{1}{R'})]^{-\frac{1}{2}} \tag{5}$$

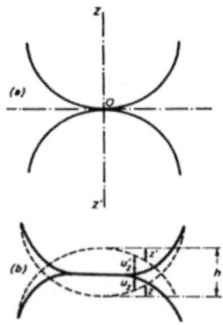

Fig. 14.   Force between touching balls. Hertz's adjustment to Hook's law $F = k \cdot x^{1.5}$.

For cast iron, the Poisson ratio $\sigma=\sigma'=0.3$, the Young's modulus $E = E' = 172 \sim 202$GPa and the radius of the balls in our experiment R=R'=6mm. The value of k is approximately 6.9×10^9-$8.1 \times 10^9 (N/m^{1.5})$.

Taking 1 strikes 3 balls as an example, we now have several interdependent differential motion equations for the balls:

$$m_1\ddot{x}_1 = -k \cdot (x_1 - x_2)^{1.5} + F_{12}$$
$$m_2\ddot{x}_2 = k \cdot (x_1 - x_2)^{1.5} - k \cdot (x_2 - x_3)^{1.5} - F_{21} + F_{23}$$
$$m_3\ddot{x}_3 = k \cdot (x_2 - x_3)^{1.5} - k \cdot (x_3 - x_4)^{1.5} - F_{32} + F_{34} \qquad (6)$$
$$m_4\ddot{x}_4 = k \cdot (x_3 - x_4)^{1.5} - k \cdot (x_4 - x_5)^{1.5} - F_{43} + F_{45}$$
$$m_5\ddot{x}_5 = k \cdot (x_4 - x_5)^{1.5} - F_{54}$$

Where x_i represents the displacement of the i_{th} ball (x_i-x_{i+1}=0 when the two balls are just in contact, i.e. no pressing between the balls), F_{ij} are the magnetic forces between neighboring balls which are measured in experiment and used in our calculation. With initial conditions, these differential equations can be solved numerically.

All we need to do then is to subtract the negative work done on the ejected ball by the magnet. Thus we finally get the theoretical value of the ejected ball's speed at the spot we measured it in the experiment.

5.1.3. *Calculation Results*

The calculated results of the translational kinetic energy before collision, collision loss percentage and the negative work done on the ejected ball are illustrated by Fig. 15 to Fig. 17. We ignore friction loss here because it is negligible according to measurement and calculation.

Fig. 15 shows that the injection energy is insensitive to the number of the end balls n, while the negative work done by the magnet on the end ball decreases rapidly to a stable low value after $n \geq 4$ (see Fig. 16). As the collision loss percentage increases with n (shown in Fig. 17), the ejected ball would get the largest kinetic energy and ejection speed at n=4.

The good consistence between the calculated ejection speed and the experimental data is shown in Fig. 18, indicating the device with one ball striking four end balls is the most effective.

The difference between theoretical calculation and experimental data may come from: (1) Measurement errors (2)The actual magnetic force on a ball in motion is different from the one measured at rest due to the eddy current induced in the moving steel ball.

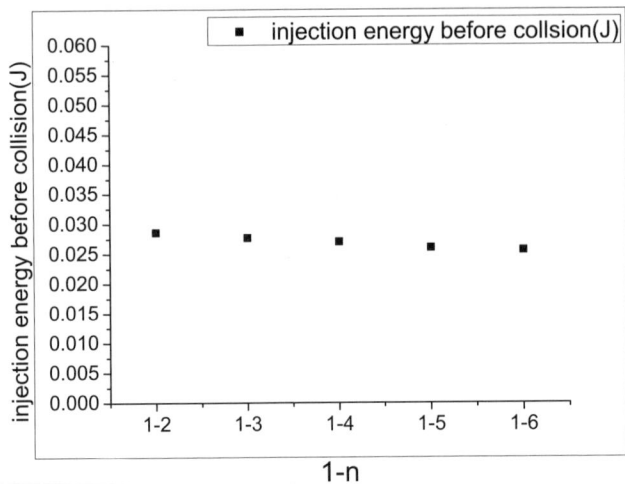

Fig. 15. Translation kinetic energy before collision, calculated by integrating the magnetic force with respect to distance.

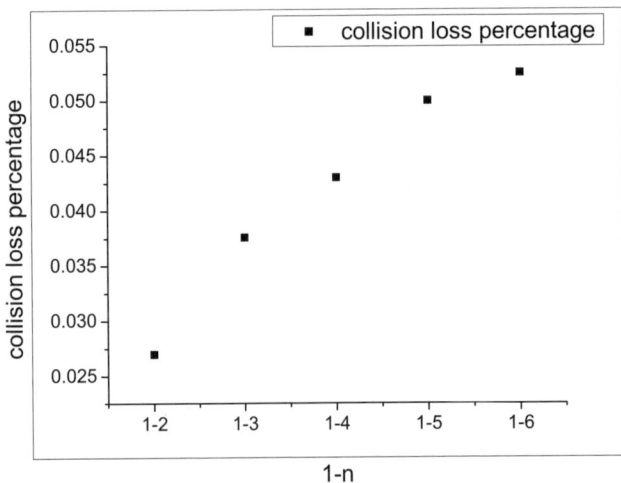

Fig. 16. Collision loss percentage (Energy after collision/Energy before collision), calculated with "Touching ball" collision model.

6. Conclusion

Both the negative work done by magnet W_{mag} and mechanical energy lost in collision affect the kinetic energy of ejected ball. We calculate W_{mag} on

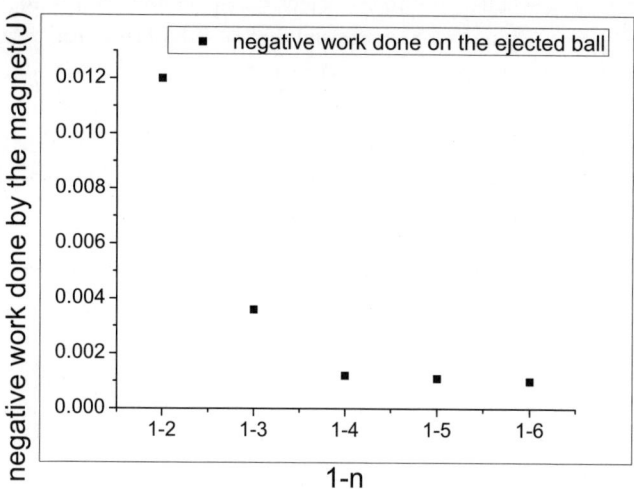

Fig. 17. Negative work done on the ejected ball, calculated by integrating the magnetic force with respect to distance.

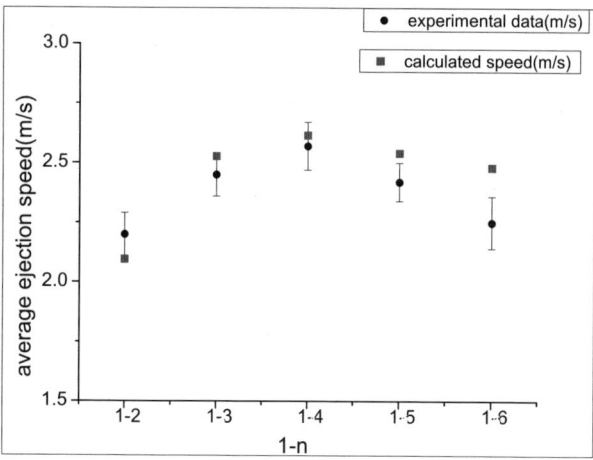

Fig. 18. Comparison between experimental data and calculated ejection speed. Both indicate the ejection speed is highest when there are four end balls in our experiment.

ejected ball quantitatively by measuring the magnetic force. Based on elastic theory, we establish an accurate "touching ball" collision model, whose coefficients are determined by the Young's Module and Poisson ratio, to calculate $E_{collision}$. The calculated ejection speeds fit our experimental data well. Based on our theoretical analysis and experimental results, the following factors are essential to optimize the magnet's position for the greatest effect:

(1) The magnet be fixed properly.

(2) No additional ball between the injected ball and the magnet (1 strikes n balls).

(3) Choose proper number of end balls.

References

1. Wikipedia, Newton's cradle, *Wikipedia, the free encyclopedia* (2012).
2. L. D. Landau and E. Lifshitz, *Theory of elasticity*. Pergamon Press, London (1959).

Chapter 2

2012 Problem 3: String of Beads

Tongyi Cao[*], Zheyuan Zhu[†], Wenli Gao and Sihui Wang

School of Physics, Nanjing University

In this paper, the motion of a string of beads is investigated. By putting forward a continuous model, we build dynamic equations to describe its motion. The numerical solutions to the equations give the dynamics of string and critical velocity of detachment. The results are consistent with the experimental data and give a good explanation for the string to fly from the edge of the beaker at certain moment.

1. Introduction

Problem Statement:

> A long string of beads is released from a beaker by pulling a sufficiently long part of the chain over the edge of the beaker. Due to gravity the speed of the string increases. At a certain moment the string no longer touches the edge of the beaker (Fig. 1). Investigate and explain the phenomenon.

To reproduce this scenario, the most important thing is to put the string into the beaker in sequence to avoid the entanglement that could stop the beads from falling. We try two types of string, one with beads fixed on it and the other unfixed. Only the former flies because the unfixed beads tend to huddle together, which makes the string stiff. Upon reproducing the phenomenon, we see it fly soon after the free drop and then form a shape of β. A high speed camera is used to record the whole process to analyze the motion of the string in detail. It is found that the motion of the string, similar to the take-off of a plane, could be divided into 3 stages: accelerating, taking off and stable flying. The motion of the string happens very fast - a 5m string used in our experiment runs off totally in 4s.

[*]E-mail: caotongyi.is.tc@gmail.com
[†]E-mail: zheyuan.zhu@gmail.com

Fig. 1.   A string of beads flying from the edge of the beaker at certain moment.

2. Dynamics of the String

2.1. *Theoretical Derivation*

The motion of the string is complicated because the motion is affected by a number of factors, such as friction, collision between beads and beaker, drag force by other beads in the beaker and the vibration of the string. In order to simplify the problem, we will ignore the friction and the collision between beads and beaker. During the accelerating stage, the shape of the string will be considered as straight. As the speed of the string is fast and the distance between two beads is small, the string of beads will be regarded as uniform. With these assumptions, a continuous model can be built as follows.

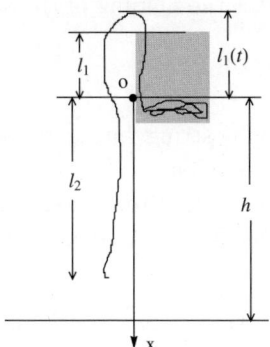

Fig. 2. Model building: x axis vertically downward, with origin at the height of the beads inside the beaker.

(1) Accelerating stage

As shown in Fig. 2, we set the x axis vertically downward, with the origin(O) in the plane of the surface of beads inside the beaker. Although the surface of the beads will descend as dropping, it is ignorable if a beaker with a large diameter is used. We set the vertical distance between origin and the ground as h, the brim of beaker as l_1, the peak of string as $l_1(t)$. The string of beads has an average density of λ.

The motion of the string is hampered by a force to accelerate new static beads to the same speed as that of the moving string instantly. Assume T is the dragging force applying to the running string (with a velocity of $\dot{x}$) by the static ones, and consider a small segment of string lying still in the beaker. After a short time duration dt, the static segment is pulled up by the moving string and also has a velocity of $\dot{x}$. The mass of this small segment is:

$$dm = \lambda\dot{x}dt \tag{1}$$

Using the law of momentum on this small segment of string, we have:

$$Tdt = \dot{x}dm \tag{2}$$

The force T can be easily derived as $\lambda\dot{x}^2$. Consider the whole moving string, along it we have:

$$\lambda l_2 g - \lambda\dot{x}^2 = \lambda(2l_1 + l_2)\ddot{x} \tag{3}$$

where $l_2 = x(t)$, if $x(t) \leq h$; and $l_2 = h$, if $x(t) > h$, corresponding to the cases before and after the landing of the string respectively.

(2) Detachment (Taking off)

Investigate the whole flying string during a time interval Δt, the change of vertical momentum contains two parts: one come from the velocity change of a small segment $\dot{x}\Delta t$ (from upward $\dot{x}$ to downward $\dot{x}$); while the other comes from the velocity change of the remaining string, with l_1 part cancels out.(Fig. 3):

$$\Delta p = \lambda(\Delta\dot{x}l_2 + 2\dot{x}^2\Delta t) \qquad (4)$$

$$\frac{\Delta p}{\Delta t} = \lambda(\ddot{x}l_2 + 2\dot{x}^2) \qquad (5)$$

The change of momentum is provided by the gravity force and the pull by the static beads in the beaker. According to the change of

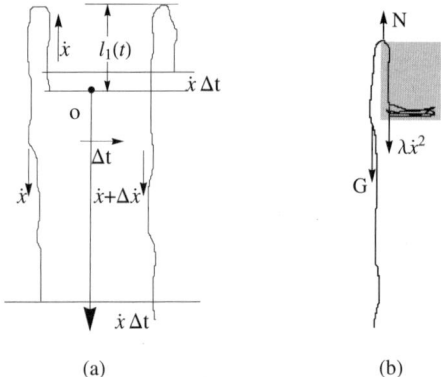

(a)   (b)

Fig. 3.   (a) The momentum change can be derived by investigating the change in a small time duration Δt. (b) Force analysis of a string of falling of beads.

momentum and force analysis (Fig. 3), the theorem of momentum in the vertical direction goes as:

$$\lambda\dot{x}^2 + \lambda(l_2 + 2l_1)g - N = \lambda l_2\ddot{x} + 2\lambda\dot{x}^2 \qquad (6)$$

Where N is the normal force given by the brim of beaker. The critical velocity of detachment can be derived by setting N=0.

(3) **Ascending to stable flying**

After the detachment, l_1, instead of being a constant, becomes a function of time $l_1(t)$.

$$\lambda \dot{x}^2 + \lambda(l_2 + 2l_1(t))g = 2\lambda \dot{x}^2 + \lambda l_2 \ddot{x} \tag{7}$$

By solving the differential equations above, we can achieve the dynamics of the string during the entire process, which will be done in the next section. But before running into that, we try to get the physics essence of the process in the perspective of energy. The stable flying stage observed in experiment can be explain by the following analysis: the string gains kinetic energy from the work done by gravity is proportional to $\dot{x}$, but lose some of them in bringing up new beads into moving, which is proportional to $\dot{x}^3$. The dissipating power in collision to the ground of $0.5\lambda\dot{x}^2$. At the start, the driving power is larger than the resisting terms. Then as speed increases, the resistance increases with $\dot{x}^3$ till an equilibrium is obtained.

2.2. *Numerical Solution*

Setting parameters in accordance with our experiment apparatus we solve the differential equations mentioned in section 2.1 and connect them according to boundary condition. We obtain dynamic graph and normal force of the falling string (Fig. 4, Fig. 5).

3. **Experiment Verification**

3.1. *Experiment Design*

A digital camera (with 480fps high speed filming), a 5m string with fixed beads ($\lambda = 0.026kg/m$) and a 1000ml baker are used in the experiment.

The method of measuring the displacement is as follow. We count the number of frames once each bead or several beads go out of screen and convert the number of frames into time. In this way we role out the influence of vibration and only take the vertical displacement into account.

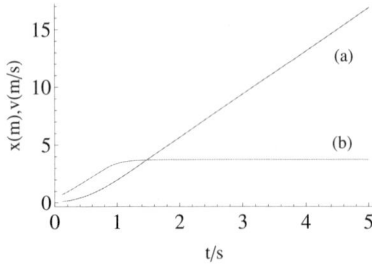

Fig. 4. Numerical solution of (a) displacement and (b) velocity.

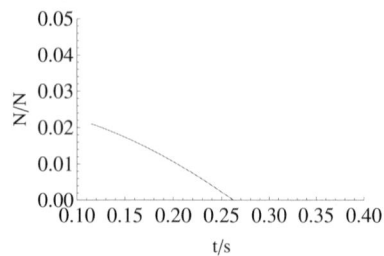

Fig. 5. Numerical solution of the normal force of the falling part N, the calculation starts from the landing moment $t = 0.1s$. Notice that when N reduces to zero, the string of beads detaches from the beaker.

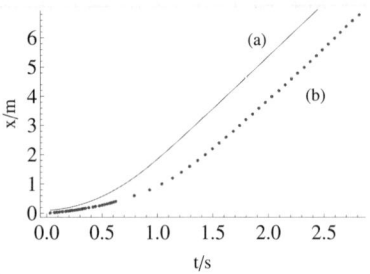

Fig. 6.   Comparison between (a) numerical solution and (b) experimental data.

3.2. *Comparison*

We find our numerical solution have the same trend but diverge slightly to the experiment data (Fig. 6). It is the friction caused by collision at the brim of the beaker that slows down the acceleration process. And this friction depends on the normal force N, which can be varied by setting different l_1. The shorter the segment hanging inside the beaker, the smaller the friction will be in the acceleration process. We do another experiment with beads placed on the brim of beaker (Fig. 7) to reduce the normal force N and the friction, then a good match (Fig. 8) can be obtained.

Another approach to this problem is to write a dynamic equation including friction and fit it to the experiment data. As we can calculate the normal force, we can define a friction index as a simplification for the collisions on the brim of the beaker. The friction index μ varies from 0.14 to 0.27 in our experiments.

Fig. 7. Setting l_1=0.01 by raising the still beads in the beaker.

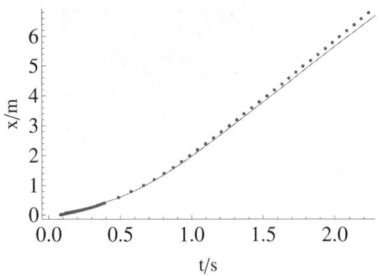

Fig. 8. The comparison between experiment data(dots) and theoretical result(line) .

We also find the critical velocities predicted v_{pre} by setting $N = 0$ which is in good match with the experimental results v_{exp}.(Table 1)

Table 1. Comparison of critical velocity between experiment and prediction.

l_1 (m)	v_{exp} (m/s)	v_{pre}(m/s)
0.01	1.50	1.46
0.05	1.54	1.53
0.12	1.80	1.78

4. Vibration

We also detect wave-like vibration on the string. The string vibrates because the static beads being dragged up do not lie exactly under the up-going segment of the string, but distribute around it. At a moment the new beads are on the left, a horizonal force to the left is exerted on the string when dragging them up. The next moment new beads are lying on the right, the horizonal force will be to the right. In this way waves are produced. But, as the beads are distributed randomly in the beaker, it is hard to study. So we try a more regular initial state (Fig. 9).

With a regular initialization we produce a sinuous wave. If the initial string is placed vertical to the line connecting the falling part and the static part of the string, a vertical vibration is observed; for a parallel initialization a parallel vibration is observed. In both cases the speed of the wave should be the sum of the speed of the wave propagating on the string and the

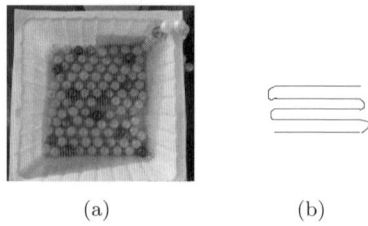

(a)   (b)

Fig. 9.   Beads' initial state. String of beads is placed regularly in a rectangular box (a), the sequence of the string of beads (b).

speed of the string itself.

$$v'_w = \sqrt{\frac{T}{\lambda}} = \sqrt{\frac{\lambda \cdot \dot{x}^2}{\lambda}} = \dot{x} \tag{8}$$

$$v_w = \dot{x} + v'_w = 2\dot{x} \tag{9}$$

Eq. (9) indicates that the speed of the wave should be twice the speed of the string. In experiment the speed of wave we measured is 7m/s while the falling speed is 3.6m/s, which is consistent with the theory.

5. Conclusion

A continuous model is built to answer the phenomenon described in the problem. The dynamic equations of the string are derived and numerical solution is given concerning the acceleration process and critical velocity of detachment. The condition of detachment and other results are compared with experimental data and they are in good matches. Further discussion on the vibration of the string is given. The theoretical prediction on the velocity of the wave on the string is verified.

Chapter 3

2012 Problem 5: Bright Waves

Zheyuan Zhu[*], Kejing Zhu, Wenli Gao and Huijun Zhou

School of Physics, Nanjing University

In this paper we study the relation between waves on the water surface and the corresponding patterns casted on the bottom of water tank illuminated by a light source from above. We establish a model based on the fundamental mechanism of this phenomenon — refraction, and provide an approach to calculate the pattern formed under arbitrary light source shining above any water surface. Using this model we also investigate several peripheral factors such as depth of water, direction of incident light etc. that could influence the properties of the patterns.

1. Introduction

Problem Statement:

When there are waves on the water surface, you can see bright and dark patterns on the bottom of the tank. Study the relation between the waves and the pattern.

1.1. *Preliminary Experiments and Analysis*

This problem asks us to find the relation between the shape of water surface and the patterns casted on the bottom of the tank. In the following sections, we will first reproduce this phenomenon, then introduce a straightforward explanation and computation on the patterns based on refraction. Factors like depth, incident direction etc. that could influence the pattern are considered. The patterns are calculated and compared with what we have observed from experiments.

---

[*]E-mail: zheyuan.zhu@gmail.com

1.1.1. *Experimental Setup*

In order to reproduce the bright waves phenomenon, we set up the experiment shown in Fig. 1. In the half-filled transparent tank, waves are generated by a burette dripping droplets into the tank with a moderate rate. Just above the bottom of the tank lays a piece of cloth which is designed to enhance the image and minimize unwanted specular reflection of light on the bottom of tank. The whole tank is illuminated by parallel incident light and a point light source in Fig. 1. The bright and dark strips are recorded through the side of transparent tank to avoid distortion when the light from the pattern undergoes second refraction as it comes out of water.

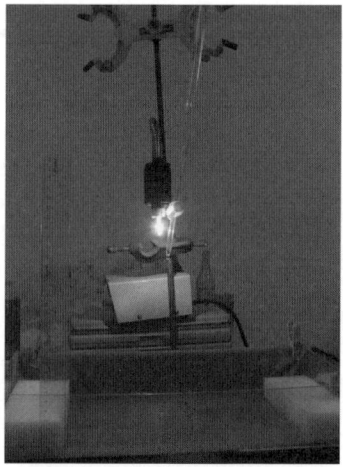

(a) Illumination from a parallel source   (b) Illumination from a point source

Fig. 1. Experiment setup to observe "bright wave" phenomenon. Waves are generated by falling droplets onto the surface. The tank is illuminated from above with two kinds of lamps (parallel source in (a) and point source in(b)) and patterns are recorded using a camera from the transparent side of the tank to avoid distortion.

It is observed that a ripple on the surface will create a bright stripe on the bottom if the whole apparatus is illuminated by parallel light from above. If a point light source is used, the light source, ripple on the surface and the corresponding bright pattern lies roughly in a line, and the whole image is almost the projection of water surface onto the bottom.

1.2. *Background and Our Approach*

This phenomenon has long been named as "caustics", which is defined as the envelop of light either reflected or refracted by an object.[1] Before the wide application of computer, theoretical investigations on caustic have been conducted using geometric optics in an attempt to find analytical solutions for the shape of caustic patterns. Due to the complexity of water surface, analytical solution is only limited to some specific patterns. Lock *et al.* investigated analytical pattern of the Shadow-sausage[2] effect for a tree branch extending diagonally out of shallow water. It was further pointed out that fine structures involving diffraction pattern[3,4] will occur at steep junctions on the surface, since these junctions, which are almost singular, represent the smallest ripples on water surface that might be comparable to the wavelength of light.

Computer simulations have been made available based on the principle of geometric optics that governs the formation of caustic. Several methods have been proposed to calculate the shape and intensity of caustic. "Backward beam tracing"[5] traces the path of each beam from the source until it reaches the receiver, or screen, where the image is casted. In this method, the surface of water is meshed into small polygons, typically triangles. Incident light on each polygon is simplified to include only rays connecting the point light source and vertexes of polygon. Snell's Law is applied to these incident rays and produces the direction of refracted rays, which forms the profile of this polygon (called "caustic polygon"[5]) casted on the surface of receiver. The conservation of light flux is used to compute the intensity of caustic polygon contributed by the corresponding surface polygon, and final intensity of this area segment is the accumulation of all intensity from surface polygons that could cast a profile in this area segment. Several efficient algorithms for computing volumetric caustics,[6,7] which could cast light onto a 3D receivers, have also been implemented in an effort to reproduce real-time computer-generated image, yet still based on the same fundamental principle "ray tracing".

Our approach is generally more direct. Light beams are considered to be composed of "light rays". We trace each ray and "bend" it according to Snell's Law when it hits the surface. The direction of refracted ray can be found so that the position of the spot on the bottom due to this ray is thus directly computed. With sufficient sampling points picked on the water surfaces, the resulting spots on the bottom could reveal the shape of the pattern as well as its intensity.

2. Theoretical Analysis

2.1. *Light Ray Tracing*

2.1.1. *2D Case*

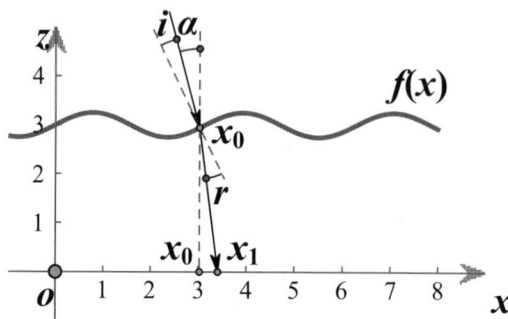

Fig. 2.   2D illustration of refraction on the water surface using geometric optics. α indicates the angle between incident ray and vertical direction. i and r are the incident and refracted angle respectively.

We begin with a simple 2D case in Fig. 2, which could imitate the pattern created by parallel waves on the surface. Shape of water surface is denoted as a continuous function $z = f(x)$. The trajectory of a ray that incidents onto the surface at an angle α against vertical axis is calculated according to the geometric relations shown in Fig. 2. The normal direction of surface at the incident point x_0 makes an angle $\arctan(f'(x_0))$ with respect to vertical direction, where $f'(x_0)$ denotes the derivative of f at x_0. Assuming the incident angle is i, we could determine the incident angle by

$$\tan(\alpha + i) = f'(x_0). \tag{1}$$

According to Snell's Law

$$\sin i = n \sin r, \tag{2}$$

where r is the refracted angle and n is the refractive index of water. Then we have the position of spot on the bottom x_1 according to the geometric relation

$$x_1 = x_0 + f(x_0) \tan(\alpha + i - r). \tag{3}$$

This intuitive approach above using geometric construction is not extendible to include a more general 3D case, where the angle of incident ray against the normal direction of surface at that incident point might not be as obvious as that in 2D case. This calls for a generalized analytical expression in vector form,[8] which could be processed with software like *Mathematica* automatically.

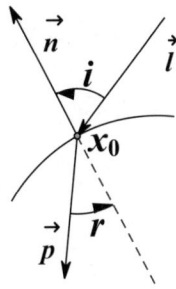

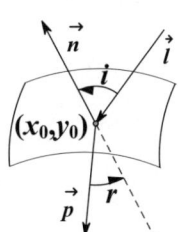

(a) Vector illustration of incident $\vec{l}$, normal $\vec{n}$, and refracted $\vec{p}$ direction in 2D case

(b) Vector illustration of incident $\vec{l}$, normal $\vec{n}$, and refracted $\vec{p}$ direction in 3D case

Fig. 3. Expressing Snell's Law in vector form for both 2D and 3D cases. In both two cases the angle of incidence and refraction is always positive and direction of refracted ray $\vec{p}$ is determined automatically through the superposition coefficients in front of $\vec{l}$ and $\vec{n}$

In our vector expression shown in Fig. 3, we try to avoid any intermediate variables such as incident angle and refracted angle, and try to express solely in terms of incident ray and normal direction at the point on the surface. The tangential vector to the surface at point x_0 could be expressed as $(1, f'(x_0))$. Normal vector at this point is simply $(-f'(x_0), 1)$, where we pick the one pointing outward of the surface. After normalization, the normal direction is

$$\vec{n} = \frac{(-f'(x_0), 1)}{\sqrt{1 + f'^2(x_0)}}. \qquad (4)$$

The incident angle i is defined as the angle between incident ray and normal direction of surface, and can be determined from

$$\cos i = -\vec{n} \cdot \vec{l} \qquad (5)$$

where $\vec{l}$ is the normalized incident ray vector pointing from light source to the point on water surface. Next we resort to the Snell's Law as appeared in Equation 2 to determine the angle of refraction r. Since $\sin r = \sin i/n$, we could obtain the expression for $\cos r$

$$\cos r = \sqrt{1 - \sin^2 r} = \sqrt{1 - \frac{1}{n^2}(1 - \cos^2 i)} \tag{6}$$

The last but crucial step is to determine the vector for refracted ray $\vec{p}$, which could always be expressed as the superposition of two linearly independent[†] vectors $\vec{n}$ (the normal direction of surface) and $\vec{l}$ (the direction for incident ray): $\vec{p} = a\vec{l} + b\vec{n}$. If we require $\vec{p}$ normalized, and provided the angle between vectors $\vec{p}$ and $\vec{n}$ is $\pi - r$ (refer to Fig. 3(a) for details), we have

$$\sqrt{a^2 + b^2 + 2ab\vec{l} \cdot \vec{n}} = 1 \tag{7}$$

$$(a\vec{l} + b\vec{n}) \cdot \vec{n} = -\cos r \tag{8}$$

which could yield two sets of solutions for coefficients a and b. The positive choice of a should be accepted as the final solution since the projection of refracted ray on the incident direction is positive, which means that the refracted ray is generally travelling forward but with slight deviation from incident direction $\vec{l}$, instead of travelling backward. The solution is therefore

$$a = \frac{\sin r}{\sin i} \tag{9}$$

$$b = \cos i \frac{\sin r}{\sin i} - \cos r \tag{10}$$

Substitute $\sin r/\sin i$ with $1/n$, $\cos i$ with $-\vec{l} \cdot \vec{n}$ and $\cos r$ with $\sqrt{1 - \frac{1}{n^2}(1 - (\vec{l} \cdot \vec{n})^2)}$ (refer to Equation 6), the final solution for the normalized refraction vector is

$$\vec{p} = \frac{1}{n}\vec{l} - \left(\frac{1}{n}\vec{l} \cdot \vec{n} + \sqrt{1 - \frac{1}{n^2}(1 - (\vec{l} \cdot \vec{n})^2)}\right)\vec{n} \tag{11}$$

2.1.2. *3D Case*

Calculation in the case of 3D surface has been made easier thanks to the vector relation presented in Sec. 2.1.1. Suppose the surface of water is represented by a function $z = f(x, y)$. In the final expression for refracted ray

---

[†]This is always true unless the incident angle is 0, which means the incident ray is aligned parallel to the normal direction of surface, and in this case the vector for refracted ray $\vec{p}$ is also parallel to that of incident ray l

vector $\vec{p}$ (indicated in Fig. 3(b)), we simply need to replace $\vec{n} = (-f'(x_0), 1)$ in 2D case with

$$\vec{n} = \frac{(-\frac{\partial f}{\partial x}\big|_{(x_0,y_0)}, -\frac{\partial f}{\partial y}\big|_{(x_0,y_0)}, 1)}{\sqrt{1 + \left(\frac{\partial f}{\partial x}\big|_{(x_0,y_0)}\right)^2 + \left(\frac{\partial f}{\partial y}\big|_{(x_0,y_0)}\right)^2}} \tag{12}$$

which indicates the normal direction on the curved surface at point (x_0, y_0).

2.2. *Formation of the Pattern*

Vector form of expression is preferred in both 2D and 3D calculation on the pattern. In the case of 2D, setting the source as parallel light and inputting common forms of waves (i.e. sin curve) as f in the above algorithm, we calculate the pattern on the bottom in Fig. 4(a). This output of simulation reveals that the pattern possesses the same feature as that of the surface: Each stripe in the pattern arises from a ripple on the surface. This can be ascribed to the focusing effect of convex surface, which concentrates light energy distributed uniformly along curve $a \rightarrow b$ into a smaller line segment $a' \rightarrow b'$ with relatively higher intensity of light. Conversely, uniformly distributed energy over a concave curve $b \rightarrow c$ will diverge this amount of energy onto a larger line segment $b' \rightarrow c'$, resulting in a lowered intensity. A comparison with experimental results under similar condition is presented in Fig. 4(b).

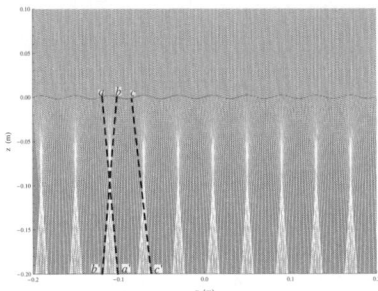

(a) Computation result for the light rays and pattern under parallel light source.

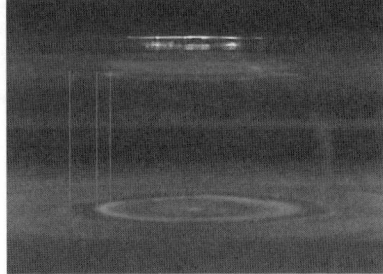

(b) Experimental result for the pattern under parallel light.

Fig. 4. Computation result and experimental observation for the pattern created under parallel light source. Each bright stripe in the pattern corresponds to a ripple on the surface.

If we switch from parallel to point light source, the calculated pattern is shown in Fig. 5(a). In this 2D case, bright strips do not appear directly beneath the ripples on the surface. Instead, they appear roughly at the projection of each ripple on the bottom. This result fits with experimental results shown in Fig. 5(b).

(a) Computation result for the light rays and pattern under parallel light source.

(b) Experimental result for the pattern under parallel light.

Fig. 5.   Computation results and experimental observation for the pattern created under point light source. Each bright stripe in the pattern is roughly the projection of ripple onto the bottom under the point light source.

3. Features of the Pattern: Predictions and Experimental Verifications

3.1. *Shape of Waves Dominates the Pattern*

3.1.1. *Curtate Cycloid Wave*

From the model proposed above we could notice that the shape of wave $f(x)$ is the dominant factor on the pattern. In Sec 2.2, we plug in sinusoidal function in f to represent water wave. However, it is known that real water wave resembles an curtate cycloid[‡9] rather than sine curve, especially for deep water. Theoretical calculation suggests that curtate cycloid would create a pattern (in Fig. 6(a)) that shares the same characters with pattern created by sine curve (in Fig. 6(b)). Therefore, it is reasonable to use sine

---

[‡]This is valid for water waves with low amplitude and whose wavelength is small compared to the depth of water.

curve in the model as it would preserve all the features that the pattern under curtate cycloid would possess, with relatively easier calculation.

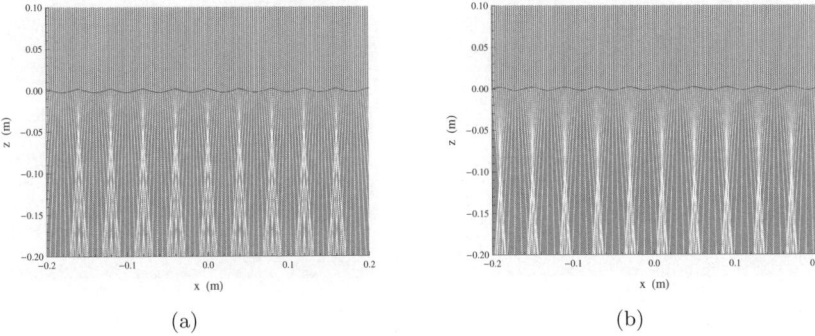

(a)   (b)

Fig. 6.   Comparison between calculated patterns that appear under (a) curtate cycloid and (b) sine curve of same wavelength and amplitude reveals similar characters of patterns under these two types of curves.

3.1.2. *Superposition of Two Waves*

In our model it is surprising to find that superposition of two plane waves on the surface does not directly lead to the superposition of patterns. Instead, it would create a node beneath the intersecting area of two waves (Fig. 7).

3.2. *Influence of Depth on the Pattern*

In the 2D case with sine waves as an example in Fig. 4(a), it can be clearly seen that the direction of refracted ray is always confined in a "dense" range of angles. We define the "deviation angle" δ as the angular difference between refracted and incident ray. In 2D case the deviation angle takes the form:

$$\delta = i - r \tag{13}$$

where i and r denote the incident and refracted angle respectively in Fig. 2. Refracted ray is confined in a certain angular range because the derivative of f (which describes the shape of surface), thus the angle of refraction r and the deviation angle of incident light δ are bounded as long as f is continuous.

The bounded angular range ensures that the envelop of refracted light is almost linear when the water is deeper than the focus of the surface.

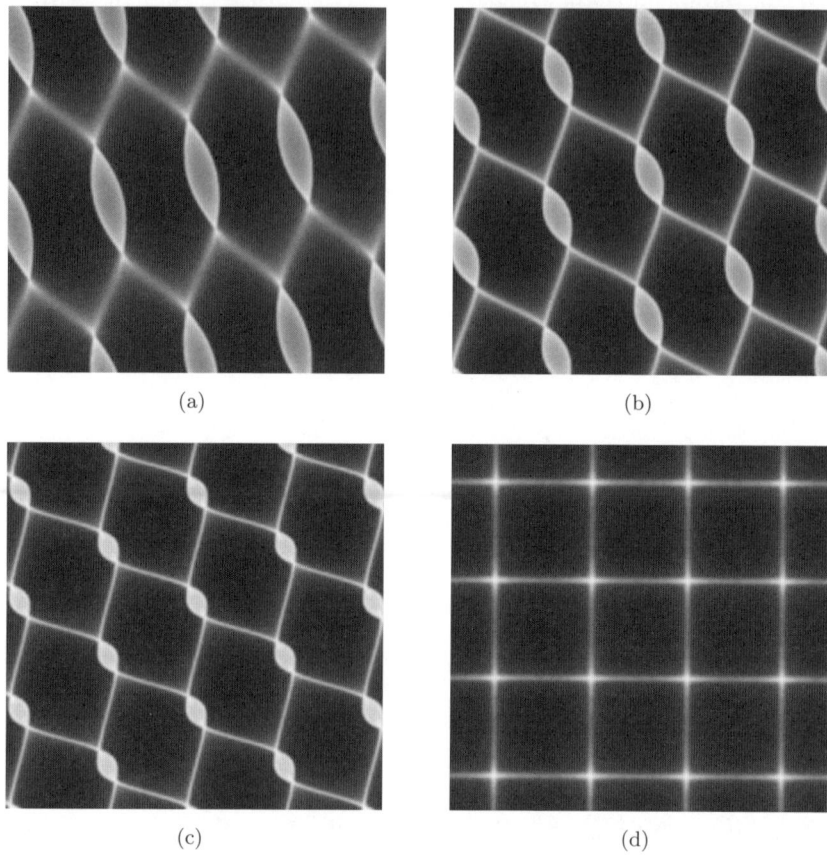

Fig. 7. Patterns casted on the bottom of tank by superposition of two plane water waves travelling at (a) 30°, (b) 45°, (c) 60°, (d) 90° against each other.

This linear boundary implies that the width of bright stripe would increase linearly with the depth of water. Fig. 8 is the width of stripe measured in experiments. Camera and tank are both fixed and width is measured by counting the pixels that each stripe covers, and then convert to millimeter by comparing with a reference object in each photo. Depth is adjusted by injecting water into the tank and the distance from the burette to the surface is relatively fixed to ensure almost same wave on the surface. A linear relation is clearly revealed in this diagram.

If the depth of water is further increased, two neighbouring bright strips might overlap or even merge, thus forming a second-order bright stripe as

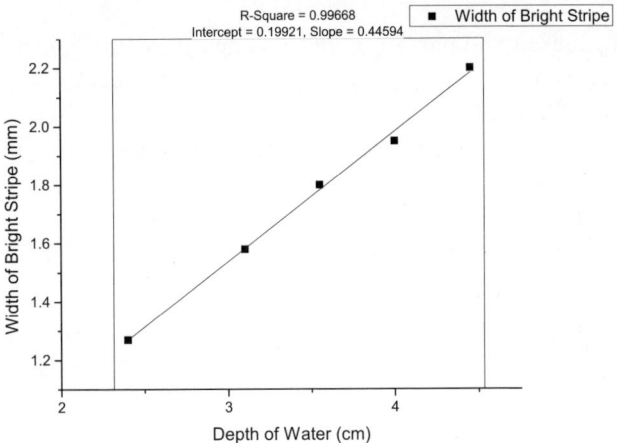

Fig. 8.   Width of bright strip versus depth of water as measured in experiments.

indicated in Fig. 9. Although this second-order pattern is predicted to exist theoretically, it could not be clearly distinguished in experiments due to the diffusion when light travels a sufficiently long distance in water.

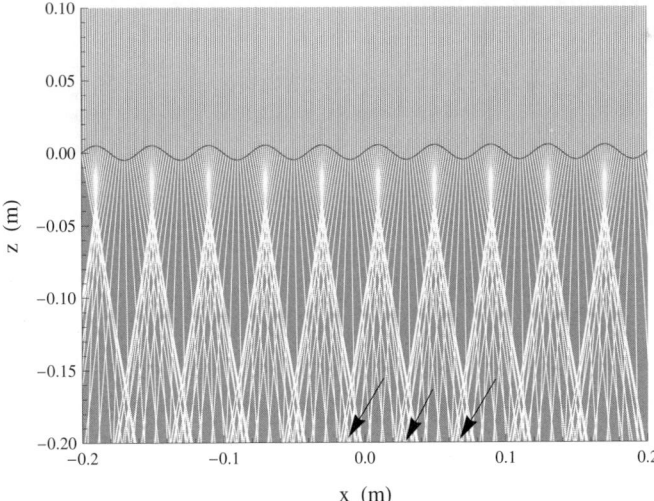

Fig. 9.   The prediction on second-order bright pattern due to the merging of two neighboring bright stripes. Several second-order stripes are marked with arrow in this diagram.

3.3. *Influence of Incident Direction on the Pattern*

Computation on the pattern under inclined incident light in Fig. 10 indicates that the pattern shifts to the same side as inclination of incident light. This calculation could also explain the character of pattern formed under point light source: beams shining onto each ripple make different angles against vertical direction. If each ripple is casted onto the bottom in the direction of beam shining onto it, the composite effect would be an enlarged pattern analogous to projecting water surface onto the bottom.

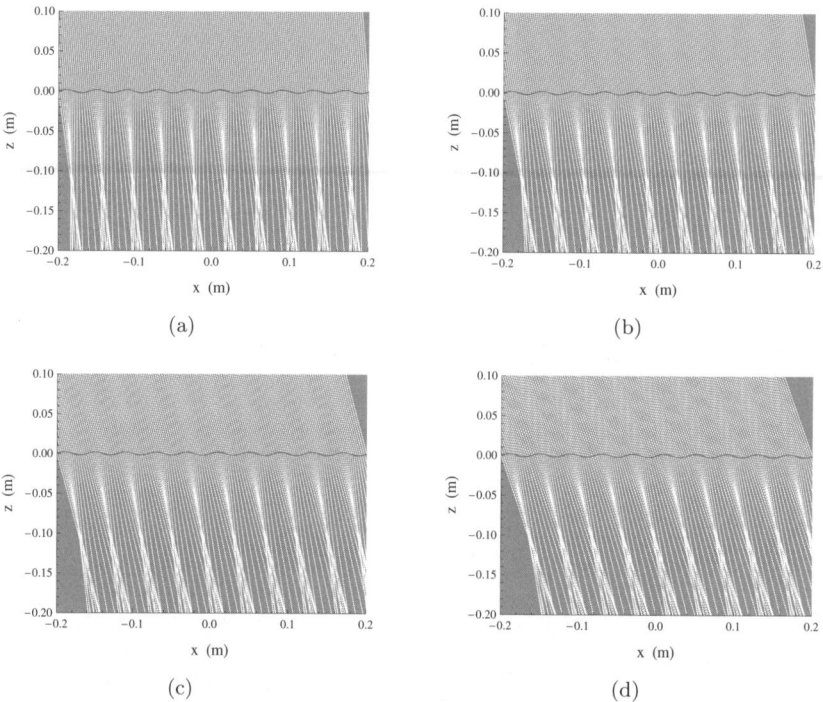

Fig. 10. Simulation results showing the position and width of pattern formed under inclined incident light at different angles (a) 5°, (b) 10°, (c) 15° and (d) 20° against vertical direction.

3.4. *Discussions on the Results*

Previous studies on the pattern[3,4] have pointed out that diffraction contributes to the fine structure of the pattern, especially at steep junctions on the surface. In our experiment we produced neither steep junctions nor waves with extremely small wavelength, and thus diffraction is not included in our model.

Dispersion plays a minor role in this phenomenon, and could only be observed if the water is sufficiently deep for polychromatic light to separate into different colors. Consider a simple 2D case in Fig. 11. The angular difference between red and blue light could reach 2.8×10^{-3}rad[§].[10] Under shallow water with depth of 20cm in our experiments, this narrow colored stripe could hardly be distinguished. In a large swimming pool with depth of 1m, colored range becomes approximately 2.8mm wide, which might be perceivable. This may be the right reason why sometimes we could find colored strips on the bottom of swimming pool.

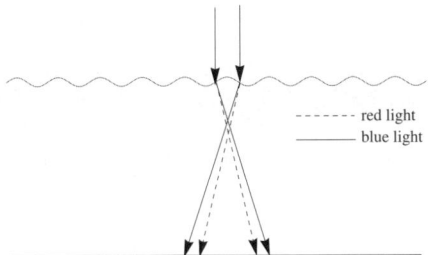

Fig. 11. A comparison between deviation angle and angular ranges of red and blue light. Solid and dashed lines illustrate the blue and red light respectively.

4. Conclusion

We apply the ray-tracing method to calculate the pattern formed under different water surfaces. Shape of water wave determines the pattern. Depth of water will influence the width of bright strips and even induce a second-order pattern. The superposition of two waves forming different angle against each other will induce nodes on the pattern beneath the intersecting area of waves intersect. Direction of illumination will result in patterns shifted from vertical position. The peripheral factors could not alter its correspondence to the ripples on surface.

[§]This calculation is based on the different refractive indexes of water: n=1.331 for red light (wavelength 680nm) and n=1.337 for blue light (wavelength 450nm). Both of red and blue light are shined vertically onto the most steep point on the water surface, where the slop is 1, corresponding to an incident angle of 45 degrees. These angles of refraction are calculated and then compared.

References

1. J. Nye, *Natural Focusing and Fine Structure of Light: Caustics and Wave Dislocations.* Taylor & Francis (1999). ISBN 9780750306102. URL `http:// books.google.com.hk/books?id=L5jOjl7GeGYC`.
2. J. A. Lock, C. L. Adler, D. Ekelman, J. Mulholland, and B. Keating, Analysis of the shadow-sausage effect caustic, *Applied optics.* **42**(3), 418–428 (2003).
3. M. Berry and J. Nye, Fine structure in caustic junctions, *Nature.* **267**(5606), 34–36 (1977).
4. C. Upstill, Light caustics from rippling water, *Proceedings of the Royal Society of London. A. Mathematical and Physical Sciences.* **365**(1720), 95–104 (1979).
5. M. Watt, Light-water interaction using backward beam tracing, *ACM SIGGRAPH Computer Graphics.* **24**(4), 377–385 (1990).
6. M. A. Shah, J. Konttinen, and S. Pattanaik, Caustics mapping: An image-space technique for real-time caustics, *Visualization and Computer Graphics, IEEE Transactions on.* **13**(2), 272–280 (2007).
7. T. Nishita and E. Nakamae. Method of displaying optical effects within water using accumulation buffer. In *Proceedings of the 21st annual conference on Computer graphics and interactive techniques*, pp. 373–379, ACM (1994).
8. A. Glassner, *An Introduction to Ray Tracing.* Academic Press, Academic Press (1989). ISBN 9780122861604. URL `http://books.google.com.hk/books?id=YPblYyLqBM4C`.
9. Z. Zhang and G. Cui, *Fluid Dynamics.* Tsing Hua University Press (1999). ISBN 9787302031680. URL `http://books.google.com/books?id=sfbGmtOsawOC`.
10. G. M. Hale and M. R. Querry, Optical constants of water in the 200-nm to 200-μm wavelength region, *Applied optics.* **12**(3), 555–563 (1973).

Chapter 4

2012 Problem 6: Woodpecker Toy

Zheyuan Zhu*, Wenli Gao, Sihui Wang and Huijun Zhou

Nanjing University, School of Physics

This paper discusses the motion of a woodpecker toy under various conditions. Equations of motion are established to investigate the motion in detail, including the oscillation of woodpecker and sliding of sleeve. We also propose some approaches to measure the relevant parameters such as rotational inertia of woodpecker and angular stiffness of spring. In addition, solutions under different initial conditions are discussed. The phase diagram is presented to reveal the typical dynamical process of a woodpecker motion.

1. Introduction

Problem Statement:

A woodpecker toy exhibits an oscillatory motion. Investigate and explain the motion of the toy.

1.1. *Preliminary Analysis*

Woodpecker is a traditional toy which is fun to play with. If you bend the spring and release it, the woodpecker will oscillate and descend along the pole, which imitates the behavior of a real woodpecker in the forest digging for worms from the bark. Shortly after being excited, the oscillatory motion seems unstable. After a couple of cycles the woodpecker enters a stable motion, marked by uniform sliding along the pole and fixed period between collisions against the pole. The task of this problem is to investigate its oscillatory motion. Since the oscillation is accompanied with sliding, and considering the nonlinearity brought by friction, it is almost impossible to figure out its period directly from the mass of woodpecker and stiffness

---

*E-mail: zheyuan.zhu@gmail.com

of spring. Therefore, we resort to the most fundamental Newton's Law to study the motion of woodpecker. The solution to the motion will be compared with experiment, and some aspects of oscillation (such as period and phase diagram) will be analyzed using the results from the solution.

1.2. *Background and Our Approach*

It was pointed out that periodic impact and friction between the sleeve and pole would regulate its oscillation behavior and turns the whole system into a nonlinear dynamic system,[1] where the motion can no longer be perceived as a simple superposition of oscillation and sliding.

Studies on the motion of this toy applied analytical mechanics[2] which is less efficient in dealing with dissipative forces such as friction than Newtonian Mechanics. Therefore, the "impact with friction" model[3] is proposed to in which the impact between sleeve and pole is treated as an instantaneous collision process, and the impulse of friction is introduced into the system as μ (friction coefficient) times that of the normal force between two contact surfaces. The essence of this model is to decompose the velocity $\vec{u}$ before collision into a normal component $\vec{u}_n$ subject to normal force between contact surfaces, and a tangential component $\vec{u}_t$ subject to the friction force. Given restitution coefficient e between two surfaces, the change of momentum in the normal direction can be obtained as $m(e-1)\vec{u}_n$[a], and it is exactly the impulse of normal force. Hence the impulse from friction and thus the change of momentum in the tangential direction is derived easily without the necessity to calculate the magnitude of normal force or friction force. Note that this model relies on the assumption that the sleeve bounces back and loses contact with the pole immediately after collision, so as to guarantee a frictionless motion except for the collision process. However, as observed in high-speed video on the motion of woodpecker and sleeve, the collision between sleeve and pole is nearly non-elastic. In addition to the vanish of normal velocity after the collision, the sleeve remains in contact with the pole and slides, which would induce friction impeding the motion of sleeve. Moreover, in some cases it would first slide along the pole for a while before getting stuck to the pole, or sometimes when the woodpecker is inclined upward, it does not stop sliding at all. Based on these phenomena, it is necessary to include friction during the whole process of motion. Therefore, depending on the state of woodpecker and sleeve, we try to set

---

[a]Here m represents generalized mass (i.e. if $\vec{u}$ represents angular velocity, then m corresponds to moment of inertia).

up equations of motion determining their evolution and observe how the system will behave under these equations.

Fig. 1 illustrates a typical woodpecker toy. The motion of this toy generally appears as a coupling between oscillation and sliding. However, further observation under high speed camera reveals that instead of a simple superposition of oscillation and sliding, the system performs 3 distinct types of motion within each period — *oscillation*(Type 1), *oscillation with sliding*(Type 2) and *sliding*(Type 3). We would first extract a model according to the behavior and then analyze these 3 types of motions in the model respectively.

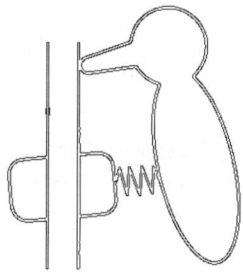

Fig. 1.   Sample of the Woodpecker Toy.

2. Dynamic Model of Woodpecker Toy

The following assumptions were made in our model in an attempt to simplify its motion shown in Fig. 2.

First of all, we assume the spring only exerts restoring torque $\vec{\tau}$ on both the woodpecker and sleeve which is proportional to the deflected angle of spring β: $|\vec{\tau}| = k\beta$, where k is called angular stiffness of spring. The deflected angle β of the spring in this case equals the difference between the inclination of woodpecker and sleeve, which were denoted by ϕ and θ respectively. θ_m, a positive value, is used to denote the maximum inclination of sleeve.

In addition, the collision between sleeve and pole is supposed to be non-elastic since as observed from the high-speed video, the sleeve never bounces back after hitting the pole.

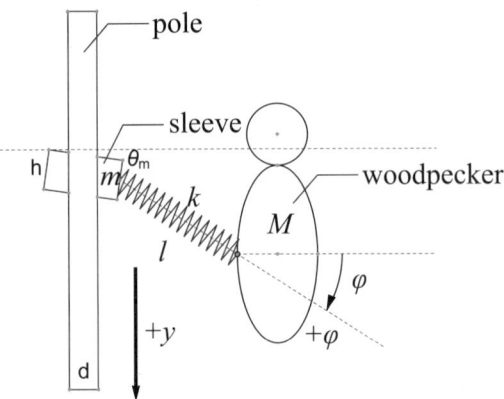

Fig. 2.   Model of woodpecker toy. Two degrees of freedom for the system, y and ϕ, and their positive directions, are indicated in this figure.

Finally, for simplicity we assume that the friction coefficient μ between sleeve and pole is identical everywhere on the pole, and kinetic and static friction coefficient are also identical.

We need to rely on another simplification that the woodpecker does not revolve around the pole while it is descending. Revolving phenomenon can rarely be seen as long as the beak of woodpecker hits on the center of the pole. Without revolving, the system has two degrees of freedom as shown in Fig. 2, the inclination of woodpecker ϕ and the vertical displacement of sleeve y.

The woodpecker motion is classified as *oscillation, oscillation with sliding* and *free sliding*. *Oscillation* means the sleeve is locked on the pole under static friction while the woodpecker oscillates around the sleeve under the restoring torque from spring. When static friction fails to hold up the sleeve, it begins to slide along the pole while the woodpecker continues to oscillate around the sleeve. This type of motion is named *oscillation with sliding*. *Pure sliding* only occurs during a tiny interval when the sleeve detaches from the pole. In this type of motion, woodpecker rotates and slides together with sleeve and the spring hardly deforms to provide internal torque on the system.

2.1. Equations of Motion

2.1.1. Oscillation

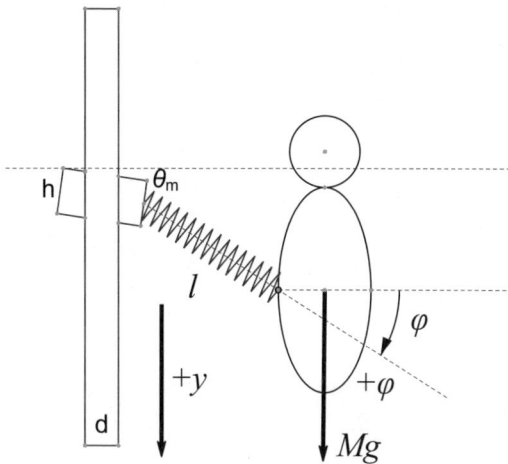

Fig. 3. Free-body diagram for *Oscillation*. The woodpecker is subject to the torque from gravity $Mgl\cos\phi$ and restoring force from spring $-k(\phi - \theta_m)$.

In this type of motion the sleeve stands still on the pole as shown in the free-body diagram (Fig. 3). The equation of rotation describing a woodpecker oscillates around the sleeve under the torque from gravity and spring can be expressed as

$$Mgl\cos\phi - k(\phi - \theta_m) = J\ddot{\phi} \qquad (1)$$

where the 2nd-order derivative of inclination is the angular acceleration, and J means the moment of inertia of woodpecker pivoting on the center of sleeve.

Since the sleeve at this stage is stuck to the pole under friction, there is no change on y:

$$\dot{y} = 0 \qquad (2)$$

2.1.2. Oscillation with Sliding

When the sleeve slides along the pole, and the spring still bears deformation, the woodpecker would oscillate around the sleeve, which is coupled with the descending of sleeve. According to the free-body diagram (Fig. 4),

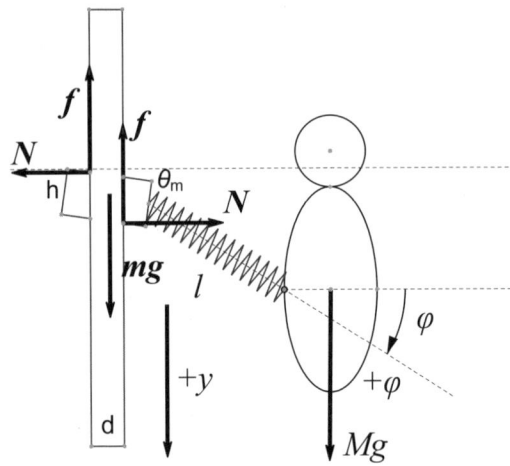

Fig. 4.   Free-body diagram for *Oscillation with Sliding*.

we could calculate the normal force on the sleeve from the equilibrium of torque:

$$\frac{Nh}{\cos\theta_m} = k|\phi - \theta_m| \tag{3}$$

The friction acting on the sleeve is μ times of normal force. The dynamic equation of the "woodpecker + sleeve" system along vertical direction is now expressed in the form:

$$(M + m)g - 2f = M(l\ddot{\phi}\cos\phi + \ddot{y}) + m\ddot{y} \tag{4}$$

The left side describes the net force on the system along vertical direction. $(M+m)g$ is the gravity acting on the sleeve and the woodpecker. f denotes kinetic friction impeding the sliding of sleeve and is obtained from $f = \mu N$,[b] where N is calculated from Equation 3. On the right side $(l\ddot{\phi}\cos\phi + \ddot{y})$ indicates the vertical acceleration of woodpecker while $\ddot{y}$ denotes the acceleration of sleeve.

The rotational equation of woodpecker is affected by the acceleration of sleeve. Since we take the center of sleeve as reference point, apart from the

---

[b]This is only valid when the maximum static friction cannot hold up the sleeve, leaving it sliding. If the sleeve does not slide, Equation 1 and 2 are applicable. The critical condition to divide the two types of motion will be discussed in section 2.2.1.

torque of gravity and that of spring, we need to take the torque of inertia force into account. So the rotational equation is modified as:

$$(Mgl - M\ddot{y}l)\cos\phi - k(\phi - \theta_m) = J\ddot{\phi} \tag{5}$$

2.1.3. *Pure Sliding*

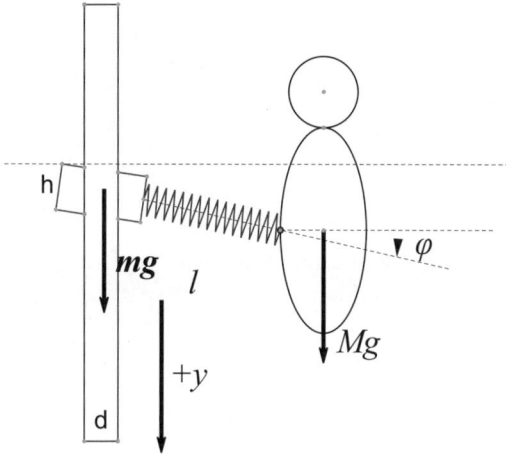

Fig. 5. Diagram for *Pure Sliding*. The whole system bears only two gravities on the sleeve and woodpecker respectively.

When the sleeve detaches from the pole, the woodpecker moves and rotates together with the sleeve (Fig. 5), so that we could neglect the deformation on the spring. The translational and rotational equations are respectively:

$$(M + m)g = M(l\ddot{\phi}\cos\phi + \ddot{y}) + m\ddot{y} \tag{6}$$

$$(Mgl - M\ddot{y}l)\cos\phi = J\ddot{\phi} \tag{7}$$

They resemble Equation (4) and (5) but without a friction term and a term for restoring torque.

2.2. *Additional Conditions*

These 3 types of motion discussed above are connected by additional conditions discussed below:

2.2.1. Sliding Occurs

For each moment of sleeve standing on the pole, we need to compare the maximum static friction available with that required to hold the sleeve on the pole. Here we assume equal static and kinetic friction coefficient $\mu_s = \mu_k = \mu$. The friction required to hold the sleeve could be solved via the equation of "woodpecker + sleeve" system (Refer to Fig. 4 for free-body diagram):

$$(M + m)g - 2f = Ml\ddot{\phi}\cos\phi \tag{8}$$

The normal force on the sleeve, calculated from equilibrium of torque on the sleeve (Equation 3), contributes to a corresponding maximum static friction

$$f_{max} = \mu\frac{k|\phi - \theta_m|\cos\theta_m}{h} \tag{9}$$

As long as the sleeve is not sliding, we need to check whether f has exceeded the maximum f_{max}, which determines whether it would slide or not in the next moment.

2.2.2. Detachment of Sleeve from the Pole

If the woodpecker reaches an inclination $\phi = \theta_m$ (identical to the maximum inclination of sleeve) during the relief of spring, and passes either $\phi = \theta_m$ from above or $\phi = -\theta_m$ from below, it would drive the sleeve to the same angular velocity with itself so that they could rotate together with no deflection on the spring subsequently. Angular moment of woodpecker and sleeve in this "driving process" (Fig. 6) is conserved, since the impulse of gravity is negligible in this short interval. Therefore we have the following equality:

$$J\dot{\phi} = (J + j)\dot{\phi}' \tag{10}$$

where j means the moment of inertia of the sleeve with respect to the center of sleeve.

2.2.3. Collision

Since the collision between sleeve and pole are assumed to be inelastic, the sleeve would not bounce back but keep contact with the pole instead. It happens when the woodpecker switches from free sliding to oscillation with sliding. We consider the angular velocity of woodpecker unchanged before and after the collision, with the inclination of sleeve stays at θ_m or $-\theta_m$.

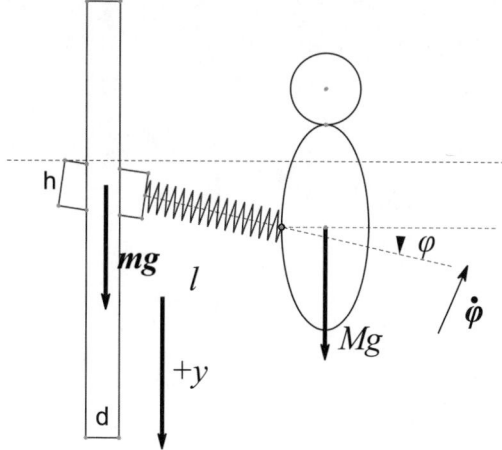

Fig. 6. "Driving" process when the woodpecker drives the sleeve to rotate together at same angular velocity.

2.3. *Numerical Calculation*

If a woodpecker is inclined upward, simply replace θ_m in previous equations with $-\theta_m$ (Fig. 7) and we obtain similar equations and additional conditions applicable for every stage of its motion.

We used computer to solve those equations of motion and test for additional conditions. The program is compiled based on the following flow chart (Fig. 8) and written with Mathematica.

3. Comparison with Experiments

3.1. *Measuring Relevant Parameters*

3.1.1. *Moment of Inertia J*

Due to the irregular shape of woodpecker, the moment of inertia of woodpecker has to be measured by experiment rather than by calculation. We hang it under a string to form a compound pendulum(Fig. 9). The period of the pendulum under small vibration is pertinent to its moment of inertia and the length of string. According to parallel-axis theorem, the moment of inertia about a parallel axis separated r away from the one that passes through center of mass is given by $J = J_0 + mr^2$, where J_0 is the moment

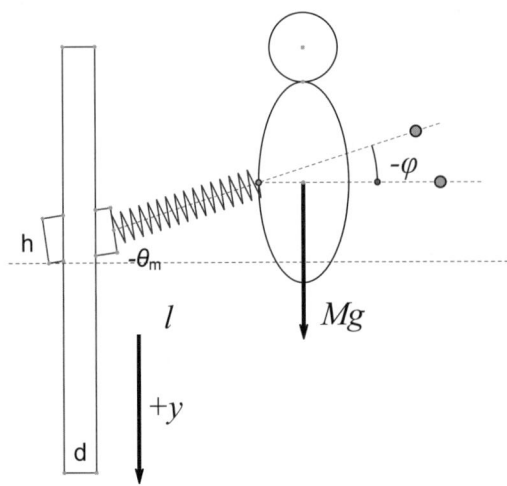

Fig. 7.   Diagram for a woodpecker inclined upward.

of inertia of woodpecker with respect to its center of mass. The period of a compound pendulum under small amplitude is $T = 2\pi\sqrt{\frac{J}{mgr}}$ and in this case it becomes

$$T = 2\pi\sqrt{\frac{J_0 + mr^2}{mgr}} \tag{11}$$

If we rewrite this equation, we could find a linearity $T^2mgr = 4\pi^2(J_0 + mr^2)$. We measured the period under different hanging length r and plot the diagram concerning T^2mgr vs mr^2(Shown in Fig. 10 respectively), whose interception on y axis equals $4\pi^2 J_0$. In this way J_0 is measured as $7.6 \times 10^{-7}\text{kg} \cdot \text{m}^2$.

3.1.2. Angular Stiffness k

For the measurement on angular stiffness of the spring, we hang a certain amount of mass on one side of a spring and measure its corresponding angular deflection — "deflected angle β" (Fig. 11) under different hanging mass. After converting the gravity to the torque exerted on one end of the spring, we plot the diagram comparing torque vs. deflected angle β (Fig. 11). The linearity confirms our assumption that the restoring torque

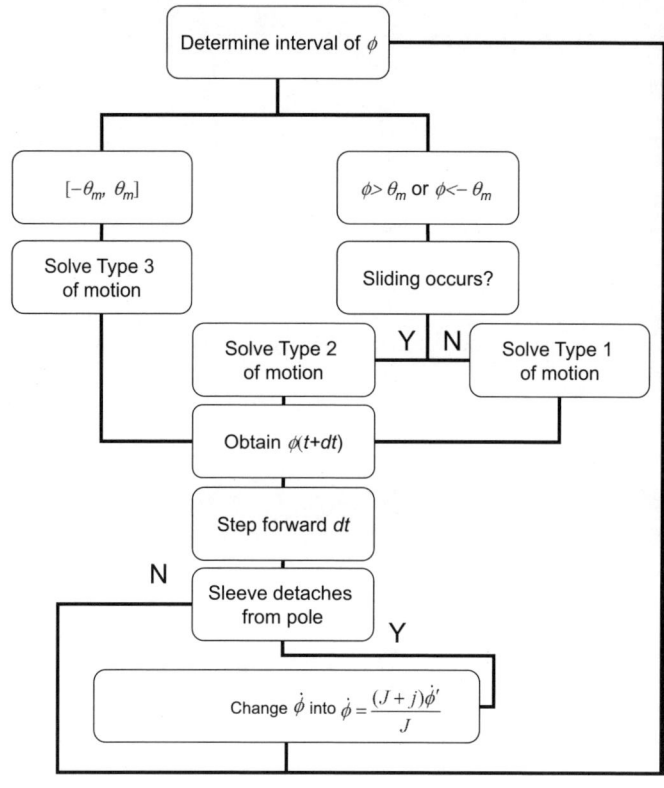

Fig. 8.   Flowchart for the simulation in each time step.

is proportional to its deflection, and its slope $k = 1.2 \times 10^{-2} \text{N} \cdot \text{m/rad}$ gives the angular stiffness of the spring.

3.1.3. *Friction Coefficient Between Sleeve and Pole*

Friction coefficient determines when the sleeve starts to slide on the pole. It is measured by gradually increasing the inclination of pole to a critical angle at which the sleeve starts to slide. The critical angle is measured as $\theta = 23°$ and therefore $\mu = \tan \theta = 0.42$.

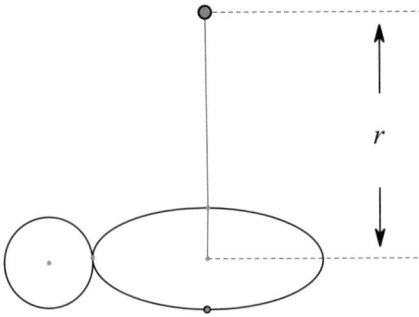

Fig. 9.   Illustration of measurement on the moment of inertia for woodpecker.

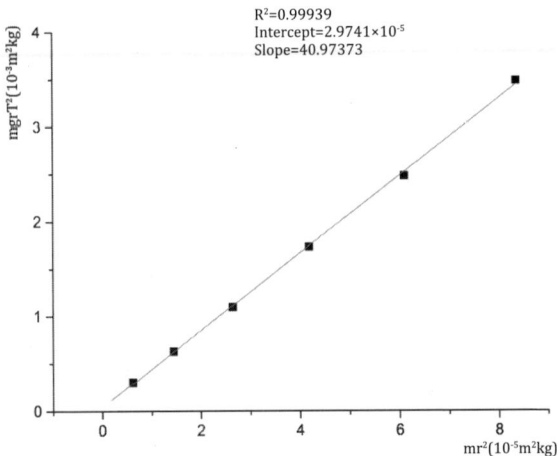

Fig. 10.   Measuring moment of inertia J from the period T of compound pendulum vs. its hanging length r. J can be determined from the intersection of the linear fitting curve.

3.1.4. *Geometric Parameters of Sleeve and Pole*

Geometric parameters of the sleeve and pole are shown in Fig. 12. Based on simple geometric relations, we can find the maximum inclination of sleeve θ_m, which must satisfies

$$h \tan \theta_m = D - \frac{d}{\cos \theta_m} \qquad (12)$$

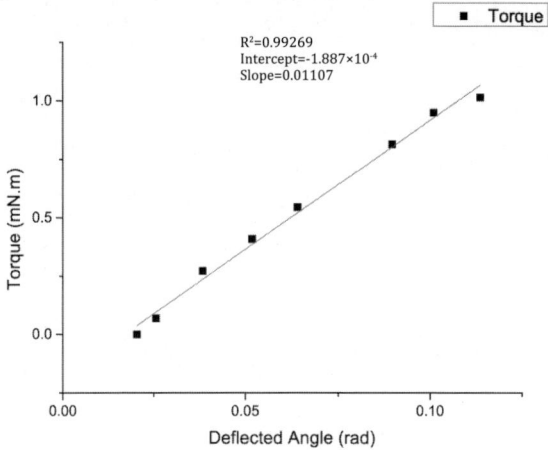

Fig. 11. Measuring angular stiffness of spring k from the torque created by the hanging mass vs. the deflected angle β. k can be determined from the slope of the linear fitting curve on (b).

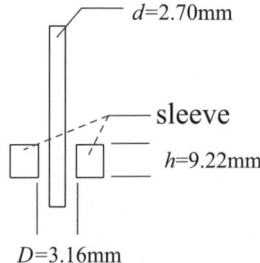

Fig. 12. Geometric parameters of sleeve and pole used in experiments.

3.2. Measuring the Period of Motion

For a more accurate measurement on its oscillation period, we record the sound emitted when the sleeve hits the pole. Since in every period, the sleeve hits the pole twice, so we count the neighboring 2 peaks in the sound wave as shown in Fig. 13, and take the average as $T = 146$ms.

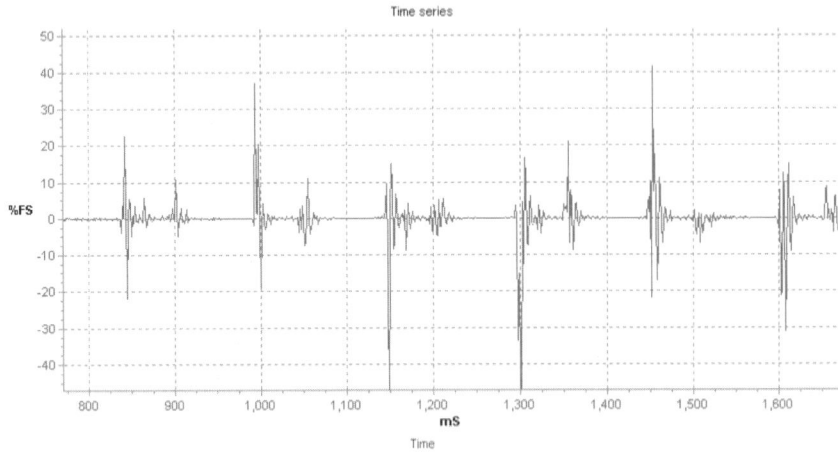

Fig. 13.   Sound record of sleeve colliding with the pole. The horizontal axis represents time in ms and the vertical axis is the relative sound intensity.

3.3. *Results of Simulation and Discussions*

According to experimental results, parameters of calculation are listed in Table 1. The calculation results indicate how inclination of woodpecker ϕ and velocity of sliding $\dot{y}$ evolve over time as shown in Fig. 14.

Table 1.   List of parameters used in simulation.

Parameter	Description	Value
J	Moment of inertia of woodpecker	$7.6 \times 10^{-7} kg \cdot m^2$
μ	friction coefficient between sleeve and pole	0.42
k	Angular stiffness of spring	$0.012 N \cdot m/rad$
ϕ_i	Initial inclination of woodpecker	$15°$

It is worthwhile to notice from these diagrams that after the first several unstable periods, both the oscillation and sliding of woodpecker would eventually reach a steady-state with a stable oscillation period of 0.145s, which matches the experimental result of 146ms fairly well. Also, the system is insensitive to its initial condition only if the motion is excited.

However, if we input a smaller initial inclination of woodpecker, insufficient energy is initially injected into the system to overcome the friction, which would always hold up the sleeve, and the system would never meet the critical condition for sliding as described in section 2.2.1. Fig. 15 illustrates how friction is always large enough to withhold the sleeve under an initial inclination of $\phi = 5°$, by comparing the maximum static friction

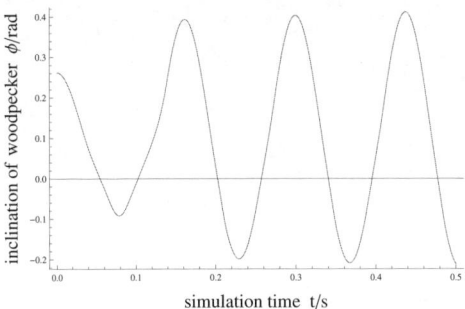

(a) Inclination of woodpecker ϕ vs. time

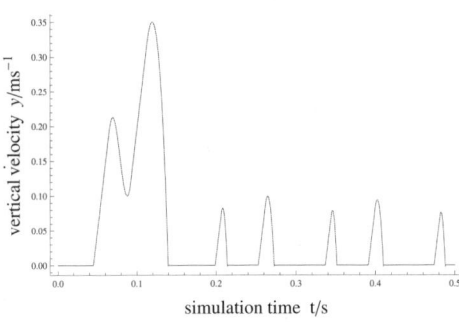

(b) Vertical velocity $\dot{y}$ vs. time

Fig. 14. Results from simulation. (a) indicates the inclination of woodpecker ϕ changing over time, while (b) is the sliding velocity along the pole.

and the friction needed to hole the sleeve. The maximum static friction available between the sleeve and pole is calculated from the the friction coefficient μ times the normal force between them as suggested in Equation 3. The friction needed is computed from Equation 4 by setting vertical acceleration $\ddot{y}$ to 0.

We also plot the phase diagram that reveals the basic mechanical process of the phenomenon, see Fig. 16.

Three types of motion in steady-state are clearly shown in sequence on the phase diagram. The trace on the phase diagram is composed of several distinctive curves corresponding to each type of motion conjuncted together. For example, the curves $i \to a \to b$, $e \to f$ correspond to *oscillation*; $b \to c$, $d \to e, f \to g$ and $h \to i$ correspond to *Oscillation with Sliding*; $c \to d$ and $g \to h$ correspond to *Pure sliding*. This phase diagram also covers the condition of small amplitude, when the woodpecker could

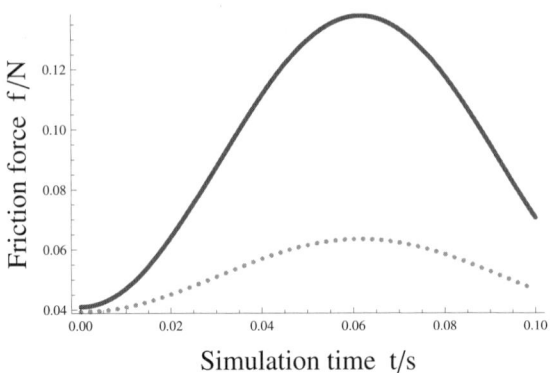

Fig. 15. Evolution of friction over time when the initial energy injected into the system is too small to overcome the friction. The dashed curve indicates friction needed to hold the sleeve on the pole while the solid curve indicates maximum static friction available.

not overcome the friction, so it sticks to the pole and undergoes Type 1 of Motion — *Oscillation* only, illustrated by a small closed loops at the center of the phase diagram. When a sufficient amount of energy is injected to excite the system, it undergoes 3 types of motion in sequence, and gains in amplitude in the first few periods to reach steady-state, as indicated by the peripheral, open but gradually convergent loop. All the mechanical energy comes from work done by gravity. Regulated by the periodic motion, gravity channels a certain amount of energy into this system in every period to drive the oscillation and overcomes those consumed by friction or lost during collision. Upon reaching steady-state, energy gains from gravity balances the lost in collision and friction.

4. Conclusion

In conclusion, we apply Newtonian Mechanics to establish the equations that can determine the motion of woodpecker under different stages. Its motion can be described by three basic types — *Oscillation, Oscillation with Sliding* and *Free Sliding. Oscillation* is a major type of motion dominating a stable period. Parameters are obtained from experiments. The results of calculation have been presented to illustrate the woodpecker motion. The phase diagram has been made to reveal the a typical woodpecker motion. In addition, a certain amount of energy is needed to excite the woodpecker before it slides down. From the perspective of energy, the energy originates

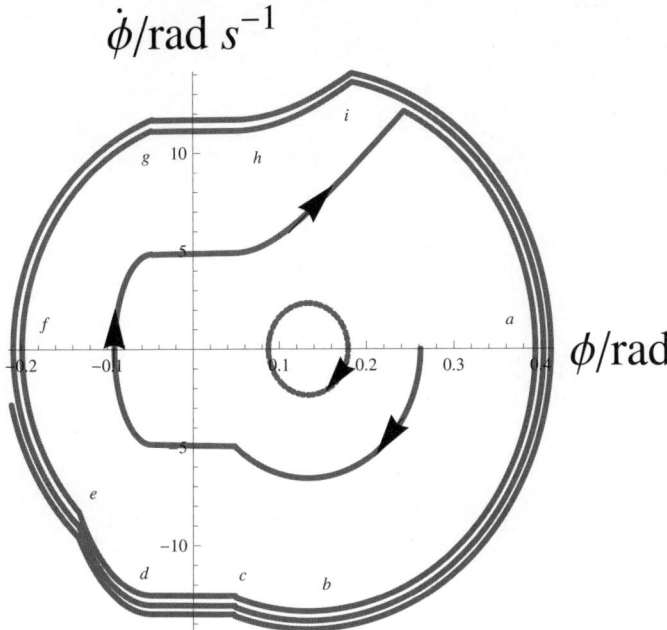

Fig. 16. Phase diagram of the motion by comparing $\dot{\phi}$ vs. ϕ. Three types of motion are jointed by distinctive curves on the loop. The central, dashed curve indicates the case when initial deflection of spring is too small to excite the system. The continuous curve outside corresponds to the case that the initial inclination is large enough to drive the descending motion of woodpecker.

from gravity and is consumed by friction and collision. The balance of these energy leads to a steady-state.

References

1. D. H. v. C. R.I.Leine, C.Glocker. Nonlinear dynamics of the woodpecker toy. In *ASME 2001 Design Engineering Technical Conference and Computers and Information in Engineering Conference*, ASME, Pittsburgh, PA (September, 2001).
2. C. Glocker and C. Studer, The woodpecker toy, *Proceedings of the IMES Center for Mechanics, ETH Zurich* (2003).
3. C. H. G. R. I. Leine, D. H. Van Campen, Nonlinear dynamics and modeling of various wooden toys with impact and friction, *Journal of Vibration and Control.* **9**, 25–78 (2003).

Chapter 5

2012 Problem 8: Bubbles

Kejing Zhu, Qing Xia, Sihui Wang and Huijun Zhou

The School of Physics, Nanjing Univerity

When a large number of bubbles exist in the water, an object may float on the surface or sink. The assumption of equivalent density is proposed in this article to explain the concrete example. According to the assumption, an object is floatable only if its density is less than the equivalent density of the water-bubble mixture. This conclusion is supported by the floating experiment and by measuring the pressure underwater to a satisfactory approximation.

Keywords: equivalent density, buoyancy, pressure

1. Introduction

Problem Statement:

Is it possible to float on water when there are a large number of bubbles present? Study how the buoyancy of an object depends on the presence of bubbles.

The problem asks if it is possible to float on water when there are a large number of bubbles present. A preliminary experiment has been done to see whether it happens.

The bubble producer is made by putting air stones connected to an air pump in a container. Put a piece of paraffin block in the water which can float on the surface. Once the air pump is turned on, bubbles go up from the bottom and the water level begins to rise. When the bubbles reach the surface, the paraffin block starts sinking. The whole process is shown in figure 1.

It's easy to reproduce this phenomenon in lab, which shows a floating object would sink when there exist a large number of bubbles. After submerging, the piece of paraffin block begins to move randomly in the water, and this can be explained by the irregular motion of water flow and

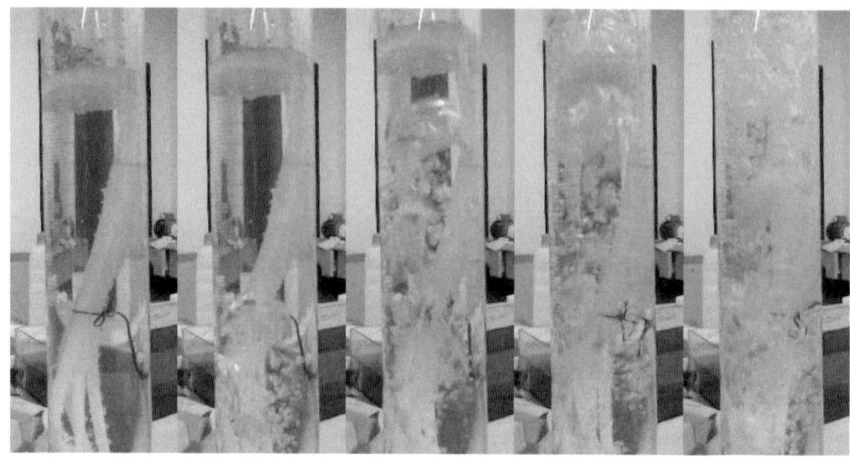

Fig. 1.   The floating states of a paraffin after turning on the air pump.

bubbles. What we are interested in is the reason why the paraffin block sinks, provided that the density of paraffin block is less than that of water.

2. Assumption of Equivalent Density

One might believe that the density of the mixture of air and water is less than that of sole water. But this intuitive idea doesn't hold when it comes to the following counter-example: A sunken object will not rise even if one makes a mixture of water and stones. Evidently, giving a reasonable explanation to the change in density requires more careful examination.

2.1. *Buoyancy with "Static Bubbles" Present*

Imagine that there is a kind of invisible rope which can hold the bubbles, shown as figure 2(a). Suppose that the ith bubble has a volume V_i, so the buoyancy of this bubble is $B_i = \rho g V_i$. Bubbles are suspended in the water, and the total force pulling upward by the ropes on the bottom of the container is

$$T = \sum_{i=1}^{n} \rho g V_i = \rho g V, \tag{1}$$

where V is the total volume of the bubbles, ρ is the density of water, g is the gravitational acceleration.

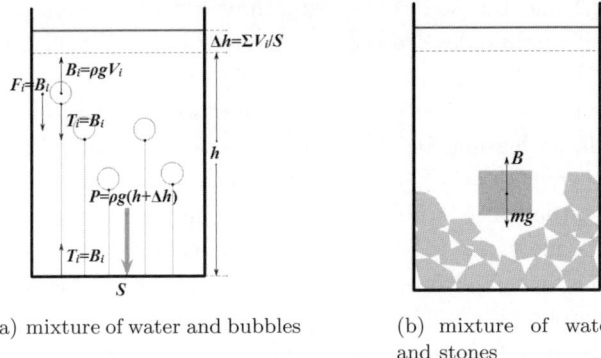

(a) mixture of water and bubbles

(b) mixture of water and stones

Fig. 2. Force analysis.

With the extra volume of bubbles, the height of water surface rises Δh from h, where $\Delta h = V/S$, and the cross-sectional area of the container is S. So the pressure on the bottom of the container is $P = \rho g(h + \Delta h)$ while the total force is $F = \rho g S(h + \Delta h)$. But what is interesting is that the gravity of water is $G = \rho g S h$ while that of air bubbles can be ignored. Amazingly, we find that the pressure at the bottom is greater than that caused by the gravity of water inside the container. What is more, such abnormity of pressure exists on any cross section within the water.

To solve the riddle of pressure, we need to take the tension on the rope into account. As the tension of rope is $T = \rho g V = \rho g S \Delta h$, then the resultant force on the bottom is $F - T = \rho g S h$, which equals the gravity of the water.

A bubble that experiences a buoyancy of $B_i = \rho g V_i$ then exerts a reactive force on the water of the same magnitude as $B_i = \rho g V_i$, seemingly acquire the density of water in producing pressure. This is also true if the bubbles are replaced by stones, shown as figure 2(b). In other words, static objects under water will not change the water pressure. The only difference they make is an extra force exerting on the bottom– positive for the stones which press the container's floor, and negative for static bubbles which lift it.

2.2. *Buoyancy with Rising Bubbles*

In section 2.1 a prediction is made that a floating object would not sink if the water is mixed with static bubbles. Then how to explain the preliminary experiment? Notice that the difference between the assumption in section 2.1 and the preliminary experiment is the motion of bubbles, it seems that if the bubbles are rising, a floating object would get less buoyancy. Why?

Rising bubbles experience two stages of motion: the acceleration stage, very soon after releasing, and the uniform motion at terminal speed. As the acceleration stage is very short, we only consider the situation of uniform motion. Like the static bubbles, bubbles in a uniform motion are also force balanced, the buoyancy is balanced by the viscous force f_i , similar to the action of the "invisible rope". The reactive viscous force is applied to the water surrounding the bubbles in an upward direction but not to the bottom, and the force is of the same magnitude as the buoyancy. On average, a minus pressure is applied to the water body just canceling the positive one, the counter-force of the buoyancy. If the bubbles are small enough and uniformly distributed in the water, it makes the density and pressure of water decrease in proportion to the percentage of bubbles' volume, which results in the notion of equivalent density.

3. Equivalent Density-Experimental Verification

To check the validity of equivalent density, we will investigate the floating states of objects with various densities under water-bubble mixture of different ratio. Then measure the pressure under water and make a comparison to that given by the equivalent density.

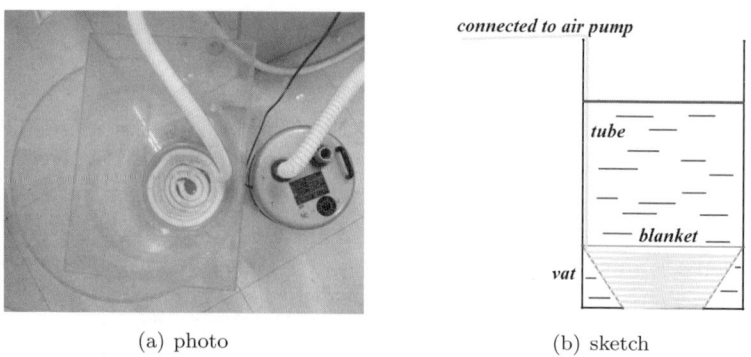

(a) photo   (b) sketch

Fig. 3.   The experimental setup.

To avoid boundary effect, we use a large transparent vat, 29cm in diameter, 50cm in height. To produce uniform distributed bubbles, a holey circular truncated cone basket is stick to the bottom of the vat in which coiled a pierced tube, shown as figure 3(b). we will ignore the boundary effect if the object is far from the boundary.

3.1. Floating Objects with the Presence of Bubbles

(a) photo   (b) sketch

Fig. 4.   Paraffin samples of different densities.

The equivalent density of water ρ' is controlled by the air flow valve of the pump. The ratio of equivalent density ρ' to the density of water ρ is inversely proportional to their volume, so that ρ' can be obtained by the change in the height of water level δh. In computing the volume, the region beside the basket should not be taken into account, because no bubbles can be produced in this area.

Table 1.   Mass, volume and density of the ten pieces of paraffin block.

No.	m(g)	V(cm^3)	ρ(g/cm^3)
1	10.7	12.98	0.83
2	13.1	14.85	0.88
3	13.9	15.39	0.90
4	13.8	15.20	0.91
5	14.8	15.95	0.93
6	13.7	14.52	0.94
7	12.9	13.18	0.98
8	13.7	13.70	1.00
9	14.0	13.65	1.03
10	12.9	12.40	1.04

To test the floating states of objects with different densities, we prepare a group of samples by mixing foams or steel inside a piece of paraffin as shown in figure 4(a). The parameters of the samples are listed in Table 1. Samples 8-10, whose densities are equal or larger than that of water, will not float at all. So, sample 1-7 are used in the floating experiment. In the test, we place a sample on the water surface and adjust the air flow valve gengtly. With more and more bubbles produced in the water, the sample starts sinking. The water level at this moment is recorded to calculate the equivalent density ρ'. If the assumption of equivalent density is right, the sample's density should be equal to the equivalent density of water ρ', which is shown as the straight line in figure 5. In the figure, the discrete points are experimental data. We noticed that the they are a bit lower than the predicted value of equivalent density. The reason is probably that the equivalent density assumption overestimates the effect of bubbles by ignoring the bubbles' initial acceleration motion.

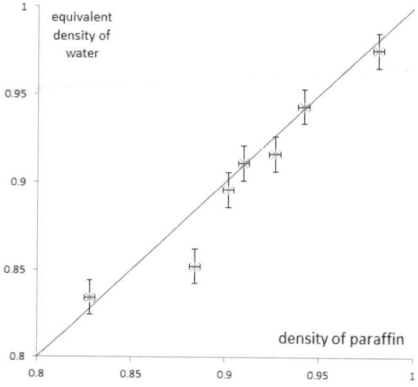

Fig. 5.   Density of the samples vs the equivalent density.

3.2. *Pressure Measurement*

A deduction of equivalent density assumption is that when bubbles exist, the pressure under water is modified as $P' = \rho'g(H + \delta h - h)$ rather than $P = \rho g(H - h)$, see figure 6.

Here ρ' is a function of δh expressed as

$$\rho' = \rho\frac{V_0 + SH}{V_0 + SH + S\delta h}. \tag{2}$$

where V_0 is the volume of basket. The pressure change ΔP can be calculated as

$$\Delta P = P' - P = \rho g\delta h\frac{V_0 + Sh}{V_0 + SH + S\delta h}. \tag{3}$$

and compare to experimental data. A piezometer is used to measure the pressure under water. The result is shown in figure 7. In the figure, the solid

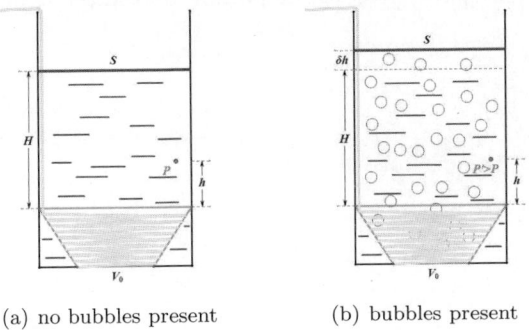

(a) no bubbles present   (b) bubbles present

Fig. 6.   Pressure with and without bubbles.

line is the result of equation(3), and the discrete points are the experimental data. We see that most of the points are a bit higher than the theoretical line. The reason might also be, as mentioned before, that the equivalent density assumption overestimates the effect of bubbles, giving a pressure lower than actual value.

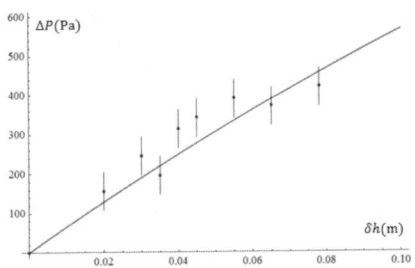

Fig. 7.   Pressure changes with the presence of bubbles.

4. Motion After Sinking

What the object's motion is like after sinking? Will it sink to the bottom or not? From our observation, the answer is NO! With the presence of turbulent flow, it's unlikely for the object to stay steady at the bottom. But it's really hard to trace the paraffin block inside the vat with a large number of bubbles.

In order to trace the motion of an object, we made a new paraffin block sample with LED and batteries embedded as shown in figure 8(a), and 8(b).

So we can follow the track of the paraffin block, see figure 9.

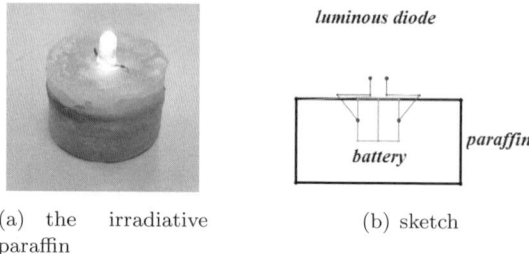

(a) the irradiative paraffin   (b) sketch

Fig. 8.   Irradiative paraffin sample that can float on the water.

Fig. 9.   Trace the paraffin under water.

As we can see, the paraffin block moves up and down inside the vat without sinking to the bottom. Maybe the shape of the object would influence the motion, but actually, it doesn't matter here. We need to concentrate to the cause of the irregular motion, which can be explained by the turbulent flow. When a bubble is rising, the water on the path is driven aside and reoccupies the space taken by the bubble, in this way, a local water cycle is formed as figure10. Since the cycles are not steady, the turbulent flow makes the paraffin block move randomly as observed from the experiment.

5. Summary

In summary, when a large number of bubbles exist in the water, an object may float on the surface or sink. The assumption of equivalent density is proposed. According to the assumption, an object is floatable only if

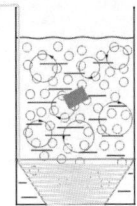

Fig. 10.   Irregular local cycle of the water.

its density is less than the equivalent density of the water-bubble mixture, which is consistent with Denardo el's work.[1] The buoyancy is also corrected by the equivalent density. This conclusion is supported by our floating experiment and by measuring the pressure underwater to a satisfactory approximation.

References

1. C. D. M. M. Bruce Denardo, Leonard Pringle, When do bubbles cause a floating body to sink, *Am J Phys.* **69**(10), 1064–1072 (October, 2001).

Chapter 6

2012 Problem 10: Rocking Bottle

Yaohua Li*, Wenli Gao, Sihui Wang and Huijun Zhou

Nanjing University, School of Physics

In this paper, the motion of a bottle partly filled with water is investigated. Two stages of motion showing different kinetic properties, named as "moving stage" and "rocking stage", can be clearly identified in the experiment. In the moving stage, the bottle moves forward with a short period vibration, while in the rocking stage, the bottle oscillates with a significantly longer period around a certain spot. Theoretical and numerical methods are employed to explain these phenomena. By simplifying the system into a rigid body model, it is found that in the moving stage, classical mechanical method gives results that fit our experiment well. And the rocking stage is thought to be the result of the asymmetric torque generated by the gravity of a liquid layer adhered to the inside wall of the bottle.

1. Introduction

Problem Statement:

Fill a bottle with some liquid. Place it on a horizontal surface and give it a push. The bottle may first move forward and then oscillate before it comes to rest. Investigate the bottle's motion.

The motion of a liquid-filled container is rather common in daily life, but a precise dynamic solution to this problem is not easy to reach. Given a push, the bottle firstly moves forward while vibrating periodically, then before it comes to rest, it moves back and forth around the ending spot. Shallow water theory[1] and Nervier-Stokes equation[2] are generally applied in previous studies on the motion of a cylinder filled with liquid.[3]

---

*E-mail: liyaohuanju@gmail.com

For the moving stage, we will present a dynamic solution describing the bottle and water's motion by adopting the rigid-body model, which is found in good agreement with experiment.

For the rocking stage, however, the rigid-body model cannot give a satisfactory explanation. Contrary to the intuitive expectation that the bottle would slow down gradually to rest, the bottle performs a large scale oscillation, whose period and amplitude are significantly different from those of the previous stage. A reasonable explanation to this phenomenon is not easy to find. The symmetry defect of the bottle's shape and the minute curvature of the bottle's course can both interfere with our observation. Actually this kind of interesting motion occurs even in a perfectly cylindrical bottle on a perfect plane, given that a fluid with non-zero viscosity is filled in the bottle.

2. Experiments

In the experiment, the motion of the bottle and the water is recorded by using a Casio EX-FC150 high speed camera.[a] The bottles used are metal cylindrical bottles with a transparent top, and are placed on a horizontally adjusted plane. As the initial push is hard to reproduce, instead of trying to quantitatively control the initial condition, we record the initial velocity v of the bottle, the angular position θ and angular velocity ω of the "liquid bulk". Tap water is used as the liquid for most experiments, and viscous liquid is also studied later for comparison.

According to our experimental observation, we may assume that bottle rolls without slipping. A black line drawn on the transparent top of the bottle helps determine the bottle's rolling angle $\triangle\phi$. In this way, we obtain the bottle's displacement $x = R\triangle\phi$ by recording the angular displacement of the bottle $\triangle\phi$, as shown in Fig. 1. Meanwhile, this method allows us to observe the motion of liquid rather clearly and to figure out the relation between the motion of the bottle and that of the liquid, i.e. phase difference of the oscillation.

A typical motion is described by the v-t graph (Fig. 2) and x-t graph (Fig. 3) of the bottle by analyzing data from the video.

Two stages of motion can be seen clearly from the diagrams. At the first stage, named as "moving stage", the bottle moves forward with an oscillatory motion with a short period. At the second stage, named as "rocking stage", the bottle rocks with a longer period around a fixed point

---

[a]30fps is sufficient to record the motion of bottle, while recording the motion of water requires 120fps.

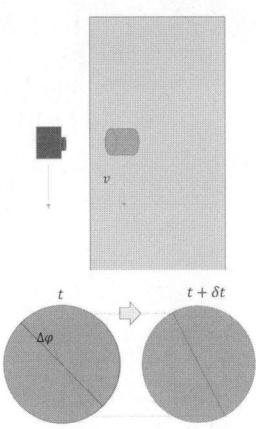

(a) Experiment setup

(b) Schematic of quantitative measuring

Fig. 1.   Illustration of the method of tracking the motion of the bottle and liquid. 1(a) A black line drawn on the transparent top of the bottle helps determine the bottle-rolling angle $\triangle\phi$. 1(b) A camera follows the bottle along with the moving bottle.

without moving forward. We will give explanations to this phenomenon in the following section.

3. Theoretical Analysis

3.1. *Theory for the First Stage*

The forces acting on this system (bottle and water) are body forces, liquid friction and rolling friction. Body forces include gravity, inertia force and pressure between the bottle and the water. In this problem, body forces are the dominant forces, while internal friction is the cause for dissipation of oscillatory energy and rolling friction decelerates the translation. Before we go further, two assumptions need to be made:

- The water surface is a plane, and the vortex inside the water is neglected.
- The gradient of velocity only exists in a thin layer (δ) between the water and the bottle, as shown in Fig. 4

Under these assumptions, a force analysis of the moving stage can be performed. When the "water block" swings to the left, the torque generated from gravity pushes the system leftwards, and the vice versa when the water is on the right. Thus the liquid-filled system performs an oscillating motion under the alternating gravity torque.

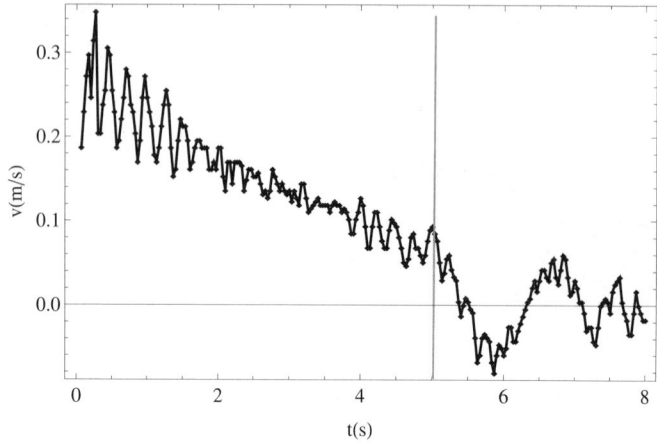

Fig. 2.   Velocity diagram of a bottle with 40% of its volume filled with water. Two stages are divided by the vertical line at $t \simeq 5$s. The first stage, whose velocity is above 0, can be identified with a shorter oscillating period of around 0.3 seconds; the latter can be clearly identified by the large peaks and valleys.

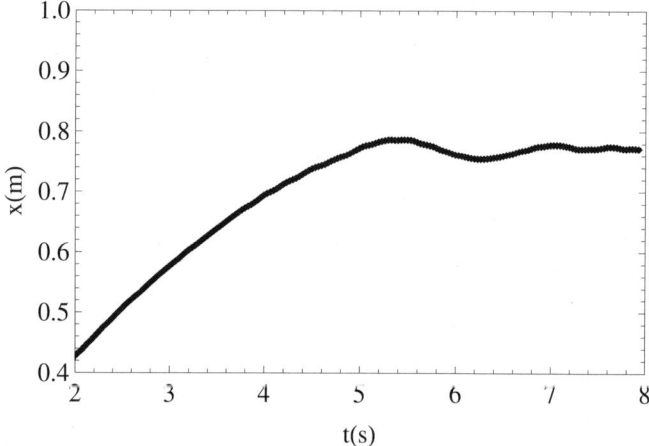

Fig. 3.   Displacement diagram of the same bottle in Fig. 2. To illustrate the rocking motion more clearly, the diagram does not start from 0s.

Hitherto we have drawn a simple picture of this system qualitatively. As we haven't introduced any dissipation, the bottle would just move on without stop. Now let's make it closer to reality. With the second assumption we made above and the formula of internal friction:[4]

$$F_{drag} = -\eta \cdot A \cdot \mathrm{d}u/\mathrm{d}r \tag{1}$$

Table 1. Important variables used in the figures.

Variable	Meaning
m	Mass of contained liquid
M	Mass of the bottle
r	Distance between center of mass (liquid) and center of circle
δ	Assumed thickness of the layer where velocity gradient exists
N_{lb}	Normal force exerted by liquid to the bottle
N_{bl}	Normal force exerted by bottle to the liquid
F_{drag}	Internal friction exerted by the liquid to the bottle, vice versa for F'_{drag}
p	the portion of in-bottle space filled with liquid

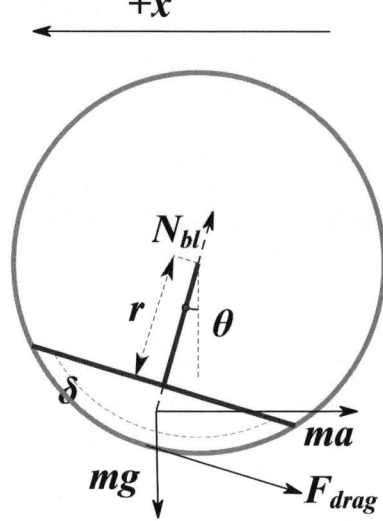

Fig. 4. Force analysis of the liquid with respect to the moving bottle.

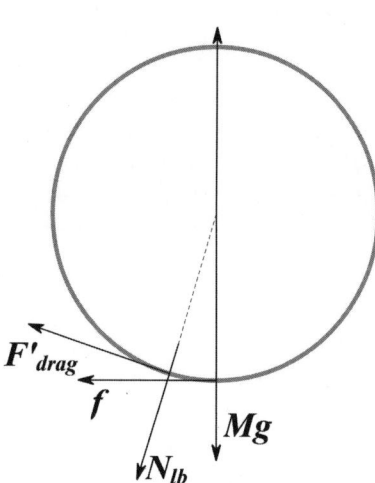

Fig. 5. Force analysis of the bottle with respect to the ground.

where η is the coefficient of viscosity, u is the velocity of liquid in the reference frame of the bottle, and A is the area of the contacting layers.

For simplicity, we convert F_{drag} into its torque M_{drag} by introducing a parameter k:

$$M_{drag} = -k(\dot{\theta} + \frac{\dot{x}}{R}). \tag{2}$$

Here the effect of viscosity is included in k, which is derived experimentally by fixing the bottle and measuring the decay of the water wave's amplitude.

With the above analysis, following equations can be derived:

$$-mgr\sin\theta - m\ddot{x}r\cos\theta - k(\dot{\theta} + \tfrac{\dot{x}}{R}) = I\ddot{\theta} \tag{3}$$

$$fR + k(\dot{\theta} + \tfrac{\dot{x}}{R}) = -I_b\tfrac{\ddot{x}}{R} \tag{4}$$

$$f = (M+m)\ddot{x} - m\dot{\theta}^2 r\sin\theta + m\ddot{\theta}r\cos\theta \tag{5}$$

$$N = (m+M)g + m\dot{\theta}^2 r\cos\theta + m\ddot{\theta}\sin\theta \tag{6}$$

where I is the liquid block's moment of inertia. Under small angle approximation, i.e. when θ and $\dot{\theta}$ are small, the following equation can be derived by canceling f and $\ddot{x}$ with Equation 3, 4, 5, and neglecting those dissipation terms ($k(\dot{\theta} + \dot{x}/R)$ and $F_r R$):

$$mgr\theta + (\frac{m^2 r^2}{2M+m} - I)\ddot{\theta} = 0 \tag{7}$$

Hence we get a simple harmonic oscillation with period T without considering the dissipation

$$T = 2\pi\sqrt{\frac{I}{mgr} - \frac{mr}{(2M+m)g}} \tag{8}$$

We can make a comparison between Equation 8 and our experiments. Given $m = 0.12$ kg, $p = V_{liquid}/V_{container} = 0.3$, we obtain the calculated value T=0.292s,[b] while the corresponding measured value is $T = 0.285$s.

In order to take dissipation into account, we employ numerical methods to solve Equations (3)~ (6). The calculation is done with the help of Mathematica. After we figure out x and v as a function of time, we can also get the period T as a function of liquid volume ratio (represented by p), as shown in Fig. 6. The curve connecting the squares is the theoretical result, the dots indicate the experimental data. The period of the moving bottle is measured as the time interval between two adjacent local maximums or minimums of its velocity.

In the first half of the curve in Fig. 6, where liquid volume ratio p is below 0.4, the theoretical curve fits with experimental data well, with an error less than 5%. However, the error increases quickly as liquid volume increases. Also, one can notice that the theoretical prediction is systematically above the experimental value. This is because in reality, the "liquid block" is not a rigid body. Velocities of different parts of it varies when the liquid swings, forming vortex inside the body. Deformation of water level also exist changing the effective moment of inertia. Such effect increases

[b]It might take some time for readers to verify this result, as the moment of inertia of water is done by numerical integration of the circular segment.

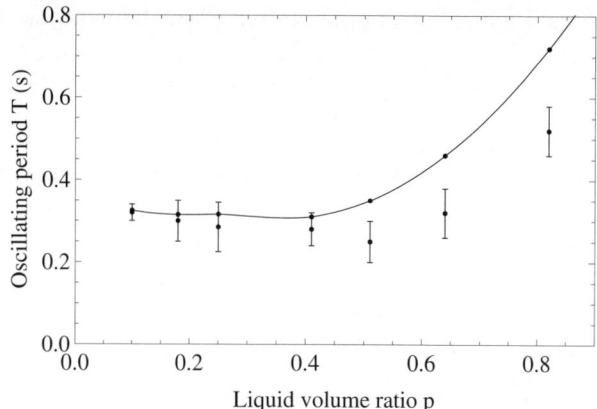

Fig. 6.   Oscillation period as a function of liquid volume ratio p. The curve connecting the squares is the theoretical result, the dots indicate the experimental data.

with the volume, or depth of the liquid. Observation supports this qualitative analysis: when the liquid is shallow ($p < 0.4$), the liquid surface is basically plain without any observable curvature; but if more liquid is filled into the bottle, the liquid surface bends and even splashes upon violent push.

3.2. Analysis for the Rocking Stage

In this part we will explain the bottle's motion in the second stage, in which the bottle rolls back a little and preforms a rocking motion without moving forward. Several factors may account for the rocking phenomenon, such as the shift of mass center from the symmetry axis, the curvature or defect of horizontal surface as well as the behavior of the liquid.

3.2.1. Influence of the Mass Center Shift

Suppose the center of mass deviates a distance of ϵ from the axis of the bottle, gravity of the bottle would cause an additional restoring moment to the system.

$$Mg\epsilon sin\theta = -(M + m)\ddot{x} \cdot R \qquad (9)$$

With $\ddot{x} = R\ddot{\theta}$ and under the approximation of small angles, we have

$$T = 2\pi\sqrt{\frac{(M+m)R^2}{Mg\epsilon}} \tag{10}$$

This equation of motion is derived assuming that the vibration of liquid has disappeared, so that it remains at the bottom of the bottle, maintaining the same speed with the bottle wall. However, to create a vibration with observed period (1s in our experiment), a shift of $\epsilon = 2$cm is needed for the bottle in the our experiment. Then we measured the distance of the shift in different bottles by measuring the period of their rocking without liquid, shown in table 2.

Ordinary kinds of bottle	Shift of mass center ϵ (mm)
Beer bottle (glass)	2.0±0.3
Plastic bottle	< 1.0
Metal bottle	< 1.0

What's more, if this is the cause for the rocking, the additional torque should be acting on the bottle throughout the process, thus there will not be two stages, the rocking (which is 3 times slower than that in the first stage) should also occur in the first stage, which contradicts with our experiments. Therefore, this can not be the main reason for the rocking motion. But it may be the right answer for a glass bottle or some other asymmetrical bottle. In examining the explanation of defects of the surface, we calculate the curvature needed to create such a rocking motion and also found that it requires a much higher roughness than that of our horizontal plane. In conclusion, after careful calculation and scrutiny of experimental condition, for our case the mass center shift and surface defect of the track only account for some disturbance to the final stage of bottle motion. The true reason of the rocking motion lies in the liquid adhered to the inner wall of the bottle, as will be discussed below.

3.2.2. *Influence of the Adhered Liquid*

If the wall of bottle brings up liquid as it rolls, the adhered liquid would cause an additional torque. This hypothesis can explain well why the bottle has two different types of motion. When the bottle is moving fast, adhered water covers the whole upper surface (See Fig. 1), no net moment is created. When it slows down, the water shrinks to one side (see Fig. 2), creating a

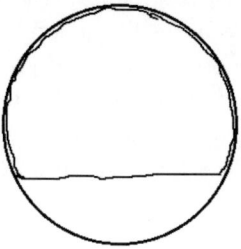

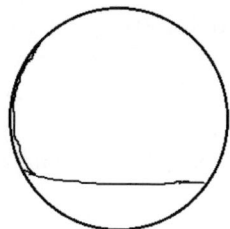

Fig. 7. Illustration of liquid in
the moving stage.

Fig. 8. Illustration of liquid in
the rocking stage.

force opposite to the direction of motion. When moving direction changes,
so does the direction of force. This adhered water hypothesis is consistent
with our experimental results using extreme viscous liquid like glycerin-
water solution. When the viscosity is large enough, almost all the liquid
would adhere to the wall of the bottle, and thus the rocking stage becomes
more evident, or even overwhelms the moving stage.

3.3. *An Interesting Discovery*

In order to study the influence of viscosity to rocking stage we try to fill
glycerin-water solution with different concentration into the bottle. The
liquid-filled system preforms 3 types of motion depending on the viscosity
of the liquid.

Low viscosity occasion (viscosity coefficient η far below 1pa·s): the bottle moves forward and rocks–just as we have discussed in section 3.1.

With η around 1pa·s, the bottle behaves like a lazy man, resisting any
push. This is a unique kind of motion that only occurs in a specific range of
viscosity. It happens because the characteristic flowing time (time needed
to flow to the bottom from around the wall) matches with the oscillating
period of the system. Therefore the internal friction always serves to con-
sume kinetic energy of the system greatly, converting the energy to heat.
This phenomenon may be useful if one wants to stop the rolling of a round
object. Surprisingly our equations proposed in section 3.1 produce very
similar motion to our observations in this situation (high viscosity occasion), although the rigid body model looks nothing like a layer of sticky
glycerin used in our experiment. The computed $v - t$ diagrams are shown
in Fig. 3 and 4.

In very high viscosity occasions (i.e. glue, cold honey, whose viscosity η is much greater than 1 pa·s), the liquid behaves like a solid, because it doesn't have enough time to flow down the bottle wall. The system is more like a rigid body now, with the liquid sticking to the bottle. However, the deformations caused by shear stress do cause energy loss to the system, but much more insignificant compared to the "lazy man" situation above.

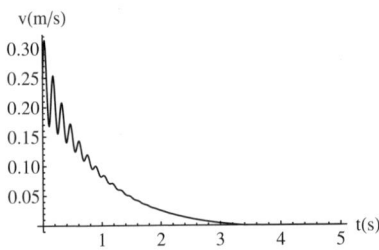

Fig. 9. Computed $v - t$ diagram of a bottle filled with thin glycerin-solution.

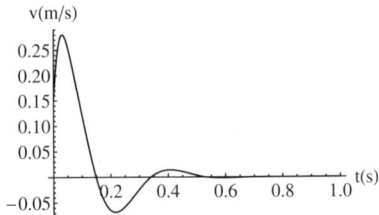

Fig. 10. $v-t$ diagram for thick glycerin-solution. Corresponding to a 97% glycerin solution under $25°$C.

4. Conclusion

In this study of the problem "Rocking Bottle", we identify the motion of the bottle into two stages, namely the "moving stage" and the "rocking stage". With a rigid body model, we find the vibration of the bottle's velocity due to coupled oscillation of liquid and bottle, reaching a theoretical period that matches with our experiment especially in the shallow liquid (small p) domain. As for the second stage, we discuss several potential factors that can cause the large-scale rocking motion, and find out that the origin of this motion is the viscosity of the liquid. Viscosity causes the liquid to adhere to the wall of the bottle, causing asymmetric gravitation torque that drives the bottle to rock.

References

1. A.-g. e. a. Yokomichi, Nagakute, Self-excited vibration analysis of a rotating cylinder partially filled with liquid, *Journal of System Design and Dynamics*. 5(2), 372–387 (2011).

2. Y. L. ZL. Wang, *Dynamics of Liquid Filled System.* Science Press, China (2002).
3. W. P. Harold R. Vaughn, William L. Oberkampf, Fluid motion inside a spinning nutating cylinder, *Journal of Fluid Mechanics.* **150**, 121–138 (1985).
4. Y. Qin, *Thermal Physics.* Higher Education Press (2011).

Chapter 7

2012 Problem 13: Misty Glass

Shan Huang*, Xiao Li, Wenli Gao and Huijun Zhou

School of Physics, Nanjing University

Based on diffraction theory, we propose a model to explain the formation of colorful rings created by a misty glass. The model is verified by examining the relation between the size of the ring and size of the droplets.

1. Introduction

Problem Statement:

Breathe on a cold glass surface so that water vapor condenses on it. Look at a white lamp through the misted glass and you will see colored rings appear outside a central fuzzy white spot. Explain the phenomenon.

White light can be separated into components of different wavelengths (different colors). The most familiar example is probably a rainbow which is explained by dispersion. However, interference and diffraction can also cause similar phenomenon. N. J. Bridge[1] studied colorful rings on a mirror covered by water droplets or other small particles (e.g. dust). He ascribed the formation of the colorful rings to the interference of two beams of light, one of which changed its direction by small particles and the other didn't. This was based on J. Walker's study on monochromatic rings on dusted mirror in 1981.[2] It is clear that their explanations need incident light and reflected light to create a proper phase difference, which is crucial to interference. But in the present problem, we look "through" a glass and there's no reflected light in the system. So we need to develop another model to explain the phenomenon.

This paper is organized as follows: Firstly, three preliminary experiments are introduced to study the behavior of the rings and determine

---

*E-mail: b111120051@nju.edu.cn

the origin of the phenomenon (i.e. refraction, diffraction or interference). Secondly, we propose a model to explain the formation of colorful rings. Finally, we reproduced the colored rings on an artificial screen according to the model. We also estimate the angular width of the rings.

2. Preliminary Experiment

The first step is to reproduce the phenomenon that the problem describes. Instead of breathing on the glass, we use hot water to produce more uniform water droplets on a glass plate. Looking through the misty glass to a white paralleled light source, a set of concentric rings appear. The typical pattern of colored rings has following features: The sequence of the colored rings is (from inside to outside) from blue to green to red (Fig. 1). Sometimes more than one order of the rings can be observed. Last but not least, only small droplets can produce colored rings, when the average size of the droplets grows to some extent, the rings totally disappear.

Fig. 1. The phenomenon described in the problem: Looking through a misty glass to a white paralleled light source, two orders of concentric rings appear. The sequence of the colored rings is from blue to red from inside to outside.

The second preliminary experiment is to measure the size of the droplets. A piece of glass is placed above the hot water for different time to control droplets's size. Then the size of the droplets is measured immediately by a reading microscope. The temperature of hot water is about $70°C$ and the time of vaporing varies from 2 to 10s. It is hard to measure the sizes precisely as the droplets keep evaporating during the measurement. However, a critical size of the droplets about $50\mu m$ is still found. Droplets with diameters less than $50\mu m$ can produce colored rings. Furthermore, we also find that when droplets are small, their shapes are close to circle, with nearly uniform sizes, while the larger ones are more irregular (Fig. 1).

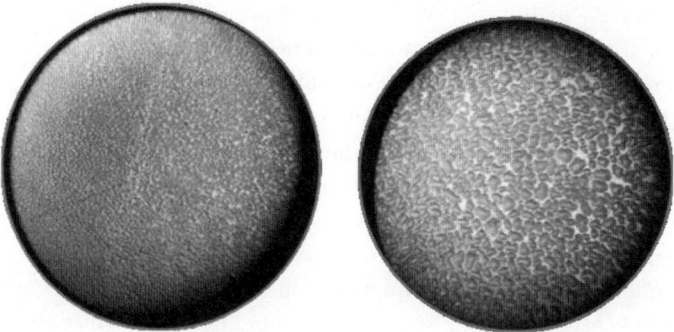

Fig. 2. Water droplets under the reading microscope: on the left, the diameter of droplets is smaller than 10 μm; on the right, the diameter is about $50\mu m$.

In the third preliminary experiment(Fig. 3), we use an expanded beam laser (He-Ne, $632nm$) to illuminate the misty glass. A number of bright and dark rings with different centers appear on the screen. Diffraction patterns on the screen, instead of the uniform pattern caused by refraction, indicate that wave optics effect is dominant in the phenomenon.

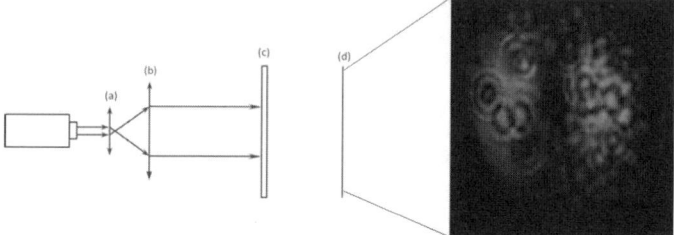

Fig. 3. Equipment and phenomenon of preliminary experiment 3: (a) and (b) are convex lenses to expand the laser. (c) is the misty glass. (d) is the screen. A number of bright and dark rings with different centers appear on the screen.

3. Theoretical Analysis

3.1. *Formation of Colorful Rings*

As shown in the third pre-experiment, water droplets can create diffraction patterns on the screen. But it is also shown that what camera has taken through the misty glass in Fig. 1 is quite different from what appears on the bare screen in Fig. 3. What causes the difference between these two phenomena? And what is the mechanism for the colorful rings to appear? The existence of lens in a camera or human eyes may be a clue.

Human eyes have a very complex structure. But generally, it can be simplified into three major parts: the pupil, the lens and the retina. The retina acts as the screen. Lens can focus the paralleled light. Based on this fact, we develop the model as follows:

Suppose that parallel monochromatic light illustrates the water droplets, and assume that the water droplets are of the same size and shape, which is reasonable according to Fig. 2. The parallel incident light with the same diffraction angle θ_1 corresponding to the first order maximum intensity, will be focused on the retina as a single bright ring of first order. See figure 4. So does the second order bright ring. Thus what appears in human eye is a set of concentric bright and dark rings.

When we change the monochromatic light into white light, using the model, we can easily find the solution to the problem. The colorful rings result from diffraction effect of light with different wavelengths. According to theory of Fraunhofer diffraction, with the small-angle approximation, the angular width of the bright rings is proportional to wavelength of the light but reciprocal to diameter of the object:

$$\theta \propto \frac{\lambda}{a} \tag{1}$$

Here, if the diameter a remains unchanged, the angular width θ only depends on wavelength. So, as shown in Fig. 5 and Fig. 1, red color is located outside and blue is inside. Since the blue light has shorter wavelength than red light, the first order bright ring of its diffraction pattern has smaller angular width, and focuses on inside place of retina.

To sum up, an imaging model has been established to explain how the multiple patterns on the screen transfer to one set of concentric rings through human eyes or camera.

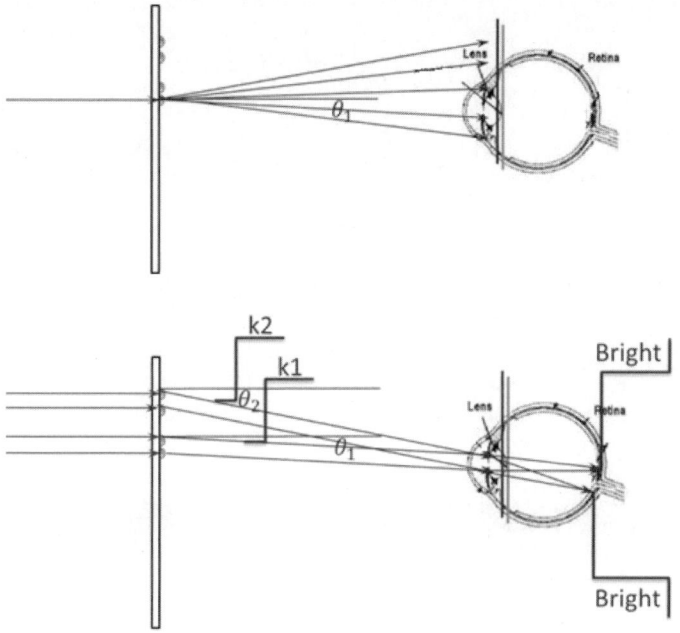

Fig. 4. Mechanism for a single circular diffraction pattern to form in human eyes (monochromatic case): k_i indicates the ith bright ring, θ_i refers to the diffraction angle. The convex lens in the eye can focalize the light of the same diffraction angle, so different order of bright rings are separated.

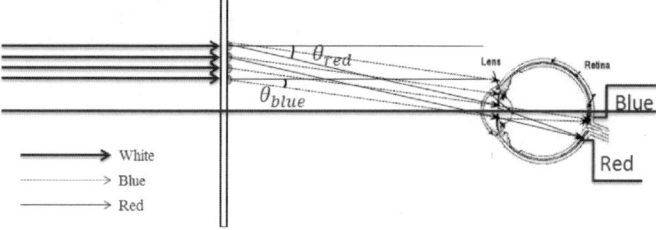

Fig. 5. Schematic picture for the formation of concentric diffraction pattern in human eyes (polychromatic case): The first order bright ring of red and blue light will focus on different part of the screen so that different colors are separated.

3.2. Angular Width of the Rings

According to theory of diffraction,[3] the intensity of the pattern at certain angle θ should be:

$$I(\theta) = I_0 [(\frac{2J_1(x)}{x})]^2 \tag{2}$$

where

$$x = \frac{2\pi a}{\lambda} \sin \theta \tag{3}$$

I_0 is the intensity of incident paralleled light. $J_1(x)$ is the first order Bessel Function. a is the radius of object. The graph of factor $\frac{2J_1(x)}{x}^2$ is shown in Fig. 6.

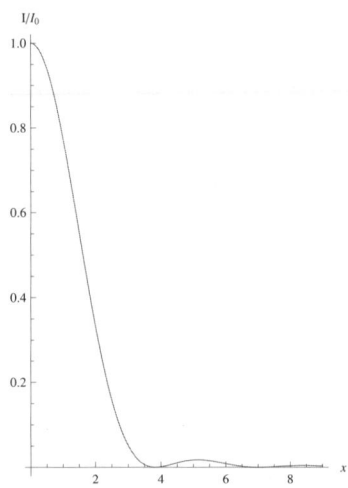

Fig. 6.   The graph of diffraction factor: the first and second maximum point occurs at $x_1 = 5.136$ and $x_2 = 8.417$.

Table 1.   Angular position of diffraction maxima and minima.

I/I_0	1	0	0.0175	0	0.0042	0
diffraction order	0		1		2	
θ(400nm) (rad)	0	0.024	0.033	0.044	0.054	0.065
θ(700nm) (rad)	0	0.043	0.057	0.078	0.094	0.114

Take the diameter of the object to be 10μm according to our observation, see Fig. 2. Table 1 shows the θ values which correspond to the maximum and minimum intensity of light of first and second order. From the table we can approximately estimate that:

$$\frac{\theta_{red}}{\theta_{blue}} \approx \frac{5}{3} \tag{4}$$

We also calculate the relationship between the angular width of the first order bright ring and the size of the droplet.

$$2\theta \approx 2\sin\theta = \frac{\lambda x}{\pi a} \tag{5}$$

It is clear that $\theta(a)$ is a decreases with a, indicating that the rings will shrink with the increasing diameter of droplets.

4. Experiment Verification

4.1. *The Necessity of Convex Lens*

In our model, human eyes are needed for concentric rings. The function of lens is to form a real image and separate light of different colors on the screen.

We use a convex lens and a screen to imitate the structure of human eye, expecting that we can reproduce the concentric colorful rings on the screens directly. We also change the type of the lens to test the necessity of convex lens. Fig. 7 shows our set up for this experiment.

Besides convex lens, we also use concave lens and plate lens to repeat the experiment. It is found that only convex lens can produce the same colorful rings as we have observed on the misty glass with eyes (Fig. 8).

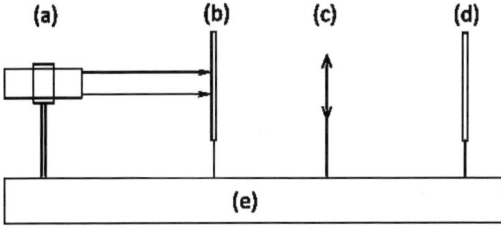

Fig. 7. The equipment of experiment in section 4.1: (a) paralleled light source, (b) misty glass, (c) lens (replaceable), (d) screen. In the experiment we have used convex lens, concave lens and plate lens respectively.

(a)   (b)   (c)

Fig. 8.   The result of experiment in 4.1: (a), (b), (c) is respectively the case where we use convex lens, plate lens and concave lens. It's obvious that only convex lens is capable of producing a set of concentric colorful rings.

4.2. *Angular Width*

Using the same set of equipment in Sec. 4.1, we can measure the angular width of the first order rings. As the size of the droplet is changing very fast (especially at the edge of the mist) during measuring, we can only examine the relationship between angular width of the ring and the size of the droplet qualitatively. As predicted in Equation (1), θ is reciprocal to a.

When we breathe on the glass, the size of the droplet on the glass is obviously increasing. We recorded the change of the rings during the vapor condensation. The results (shown in Fig. 9) demonstrate that as the size of the droplets increases, the diameter of the rings decreases. This is coherent with the prediction.

5. Conclusion

It is found that small droplets condensed on a glass plate can produce colored rings while large droplets can't because of the diffraction effect. Our model shows that a convex lens can converge light with same diffraction angle on the same place of the screen (or retina, in human-eye case). By building an artificial "eye" and comparing the effect of convex lens and other lenses, it is found that only convex lens plays a significant role in transferring multiple irregular diffraction patterns into single set of concentric rings described in the problem. Experiments also verify that the angular

(a) $t = 0.2s$ (b) $t = 0.4s$ (c) $t = 0.6s$

Fig. 9. The result of experiment in section 4.2: From (a) to (c) we recorded the change of the rings during the vapor condensation. During this process, the size of the droplet increases while the angular width of the ring decreases.

width decreases with increasing size of droplets, which is consistent with the model.

References

1. N. J. Bridge, A novel effect of scattered-light interference in misted mirrors, *Phys. Educ.* **40**(359) (2005).
2. J. Walker. Interference patterns made by motes on dusty mirrors. `http://jesseenterprises.net/amsci/1981/08/1981-08-fs.html` (August, 1981).
3. M. Born, E. Wolf, and A. Bhatia, *Principles of Optics: Electromagnetic Theory of Propagation, Interference and Diffraction of Light.* Cambridge University Press (1999). ISBN 9780521642224. URL `http://books.google.com.hk/books?id=aoX0gYLuENoC`.

Chapter 8

2012 Problem 15: Frustrating Golf Ball

Shan Huang*, Zheyuan Zhu, Wenli Gao and Sihui Wang

School of Physics, Nanjing University

This paper studies the condition for a golf ball to escape from a hole. The two determining factors are the the ball's initial velocity v_0 and its deviation from the center of the hole d. There is a critical escaping velocity v_c for every deviation d. The ball's motion is analyzed by calculating the change of velocity whenever the ball collides with the hole. The critical conditions predicted by our theory are verified through experiment.

1. Introduction

Problem Statement:

It often happens that a golf ball escapes from the hole an instant after it has been putted into it. Explain this phenomenon and investigate the conditions under which it can be observed.

Golf ball, as a popular sport worldwide, has a meticulous international standard which specifies the size, material and shape of both the ball and the hole (USGA standard).[1] This allows us to reduce the number of parameters to two key factors: the initial velocity of the ball v_0 and the deviation of the motion trail from the center of the hole d, as shown in figure 1.

In our model and simulation, we use following parameters according to USGA (United States Golf Association). The radius of hole is $R = 5.4cm$, the diameter of ball is $42.5mm$, and the mass of ball $m = 43.7g$.

In this paper we first qualitatively explain the phenomenon, then present a model from rigid body dynamics perspective. After that, we numerically simulate the model to demonstrate the conditions. The verification experiment is given at the end of the paper.

---

*b111120051@nju.edu.cn

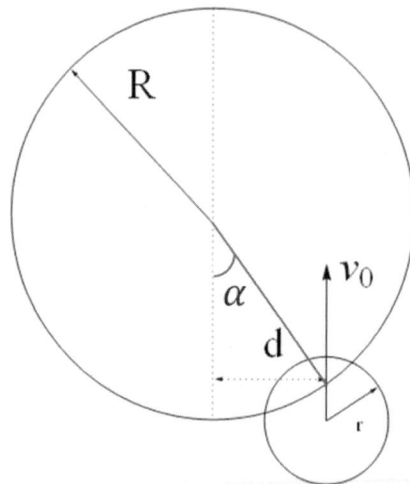

Fig. 1. A top view of the ball and the hole: R is the radius of the hole, r is the radius of the ball. Two key factors d and v_0 are as shown.

2. Theoretical Analysis

2.1. *Overall Analysis of Ball's Motion*

For simplicity, we assume the ground is flat and the ball rolls smoothly, thus when it reaches the hole, there will be no vertical velocity. Then the ball flies over the hole, collides and interacts with the other side of the hole. After the interactions, the ball gains another velocity, which may cause it to escape or fall into the hole, depending on different conditions.

When the ball is rolling, it combines both translation and rotation. When it collides with the opposite rim, the direction of rotation will result in an upward friction. If the ball initially moves faster, the collision position will be higher and the reaction force exerted on the ball will be larger. It makes it easier for the ball to escape. On the other hand, the deviation d will affect the flying distance of the ball. The bigger the deviation is, the easier for the ball to escape. But when d is large enough, the collision will result in an apparent change in direction of velocity, causing more complex situations, e.g. multiple collisions.

According to the relative position of the ball's center to the holes'rim, we divide the collision into two types: below-rim collision and above-rim collision. For a certain d, we could find a critical initial velocity v_{0c} to determine the collision type of the ball. From Fig. 1 we get following equations.

$$R \sin \alpha = d \tag{1}$$

$$R^2 + s^2 - 2Rs \cos \alpha = (R - r)^2 \tag{2}$$

$$v_{0c} t = s \tag{3}$$

$$\frac{1}{2} g t^2 = r \tag{4}$$

Combining Equation (1), (2) and (3), the diagram of the $v_{0c} - d$ relation is solved as in Fig. 2. When $v_0 > v_{0c}$, above-rim collision occurs, when $v_0 < v_{0c}$, below-rim collision occurs.

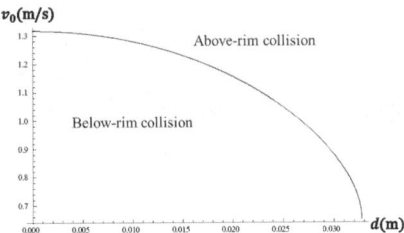

Fig. 2. $v_{0c} - d$ image: v_{0c} is the critical velocity that determines whether an above-rim collision or a below-rim collision will occur.

2.2. Below-Rim Collision

Since process of collision is very short, we neglect the impulse of gravity. With the force analysis in figure 3 we can derive the dynamic equations.

Based on the law of momentum in x and y direction respectively we obtain

$$-\int N dt = m(v_x - v_0 \cos \beta) \tag{5}$$

$$\int f dt = m(v_y + gt) \tag{6}$$

From definition of dynamical friction, we have

$$f = \mu N \tag{7}$$

From definition of restitution coefficient, we get

$$v_x = -e v_0 \cos \beta \tag{8}$$

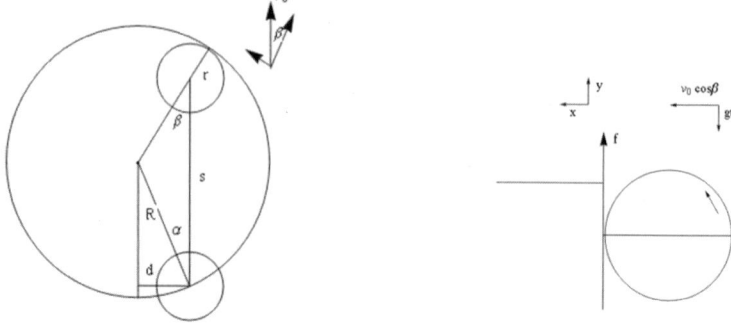

(a) the top view of the hole   (b) the side view of rim and the ball

Fig. 3.   Force analysis to a below rim collision when $d \neq 0$.

We use Mathemetica to solve Equation 5 to 8 to get v_x and v_y after the collision. Then consider such a projectile process to see whether it is possible for the ball to escape.

As an example, we demonstrate situation when $d = 0$, i.e. $\cos \beta = 1$, the four equations become:

$$-\int N dt = m(v_x - v_0) \tag{9}$$

$$\int f dt = m(v_y + gt) \tag{10}$$

$$f = \mu N \tag{11}$$

$$v_x = -e v_0. \tag{12}$$

Let v_0 be v_{0c} and change μ and e to investigate the behavior of the ball. The calculated results show that the bigger μ and e are, the higher the final position will be. This is shown in Fig. 4. Theoretically, we find that when $\mu > 0.8$ and $e > 0.7$, the ball could escape from the hole in below-rim case. In our experiment where $\mu = 0.4$ and $e = 0.4$, the ball couldn't escape from the hole in below-rim collision case. This is coherent with the model.

2.3. *Above-Rim Collision*

First of all, we analyze the change of velocity.

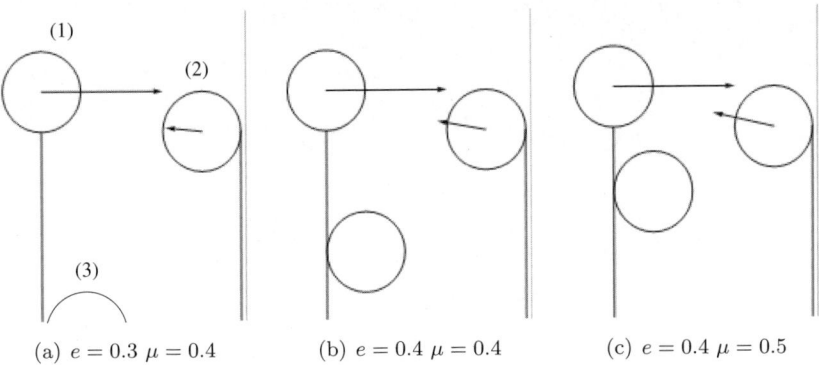

(a) $e = 0.3$ $\mu = 0.4$ (b) $e = 0.4$ $\mu = 0.4$ (c) $e = 0.4$ $\mu = 0.5$

Fig. 4. Simulation of below-rim collision: (1) the start position, (2) position of collision, and (3) the end position are shown respectively.

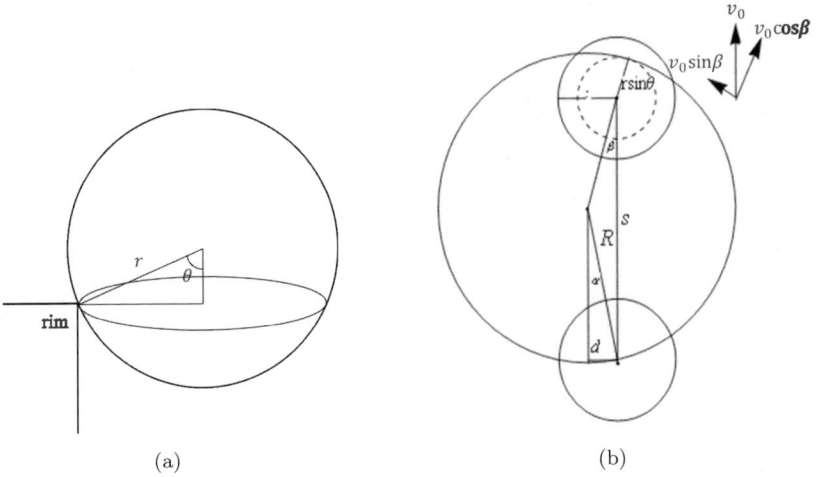

(a)   (b)

Fig. 5. Above- rim collision (a) Side view (b) Top view.

With Fig. 5 we get the following geometrical relations:

$$R^2 + s^2 - 2Rs \cos \alpha = (R - r \sin \theta)^2 \tag{13}$$

$$d = (R - r \sin \theta) \sin \beta \tag{14}$$

$$\sin \alpha = \frac{d}{R}. \tag{15}$$

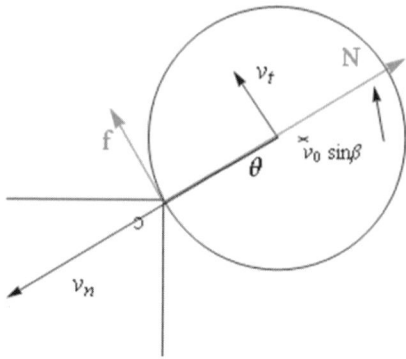

Fig. 6.   Force analysis for above-rim collision, where O is the position of collision.

From figure 5 we also have following kinetic equations

$$v_0 t = s \tag{16}$$

$$\frac{1}{2}gt^2 = r(1 - \cos\theta). \tag{17}$$

The normal and tangential velocities of the golf ball an instant before the collision are

$$v_{t0} = v_0 \cos\beta \cos\theta - gt\sin\theta \tag{18}$$

$$v_{n0} = -v_0 \cos\beta \sin\theta - gt\cos\theta. \tag{19}$$

The change of the velocities will be determined by following equations,

$$v_{nf} = -ev_{n0} \tag{20}$$

$$-\int N dt = m(v_{nf} - v_{n0}) \tag{21}$$

$$\int f dt = m(v_{tf} - vt0) \tag{22}$$

$$f = \mu N. \tag{23}$$

Now, we need to figure out the escaping condition. In the discussion above, we have analyzed the change of velocity, i.e. for every given deviation d and initial velocity v_0, we can calculate the velocity after collision, from which we can predict whether the ball will escape. If the velocity and position of the ball satisfy the following two conditions, an escape will occur.

Condition 1: After collision, the radial velocity of the ball is pointing outwards.

Condition 2: After collision, the effective kinetic energy overcomes gravitational potential energy difference between the rim and the collision point.

These two conditions can be expressed in the form of inequality:

$$E = \frac{1}{2}mv_{tf}^2 - mgr(1 - \cos\theta) > 0 \tag{24}$$

In which $\cos\theta$ can be obtained from d and v_0 by equation 13 and 16.

To find the critical escaping velocity v_c at different d, we start from $d = 0$ and change the d little by little, e.g. 1mm per step. For each d, we increase v_0 from 0 and check if it satisfies escape condition. The value of v_0 satisfying the escaping condition is the critical velocity v_c. The relationship between v_c and d is shown in figure 7, with $\mu = 0.4$ and $e = 0.4$, taken as our experimental value.

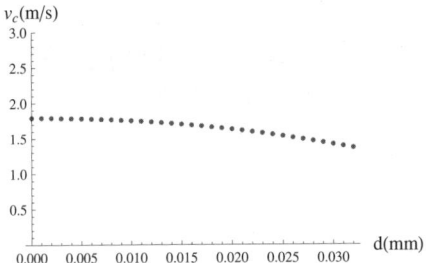

Fig. 7.   Simulation result of escape velocity, each deviation d corresponds to an initial velocity v_c.

2.4. Multi-Collision

Multi-collision frequently occurs when d becomes relatively large (e.g. bigger than $0.5R$). Consequently, the $v_c - d$ relation we previously get may fail in this region. In experiment, double and triple collisions were observed.

To examine a multi-collision condition, we have to trail the entire moving process of the ball. To simplify the simulation, we employ the expression of vector and metrics in Mathematica. After each collision, the dynamic equations about the ball's motion can also be expressed in the form of vector:

$$\vec{I}_N = -(1 + e)m\frac{\vec{v} \cdot \vec{r}_{OC}}{r^2}\vec{r}_{OC} \tag{25}$$

$$\vec{I}_f = \mu\vec{I}_N(\vec{r}_{OC} \times \frac{\vec{v} + \vec{\omega} \times \vec{r}_{OC}}{|\vec{v} + \vec{\omega} \times \vec{r}_{OC}|}) \times \vec{r}_{OC} \tag{26}$$

$$\vec{p}_{after} = \vec{p}_{before} + \vec{I}_N + \vec{I}_f \qquad (27)$$

$$\vec{L}_{after} = \vec{L}_{before} + \vec{r}_{OC} \times \vec{I}_f \qquad (28)$$

in which I is the impulse, p is the momentum and L is the angular momentum. $\vec{r}_{OC}$ is the position of the collision relative to center of the hole at ground plane, and r is the radius of the ball.

The collisions occur when the following conditions are satisfied (for above rim collision),

$$z > 0 \qquad (29)$$

$$(R - \sqrt{x^2 + y^2})^2 + z^2 < r^2. \qquad (30)$$

Or (for below rim collision)

$$z < 0 \qquad (31)$$

$$R - \sqrt{x^2 + y^2} < r. \qquad (32)$$

We also have equations of projectile motion to calculate ball's position and velocity between two collisions:

$$\vec{v}(t + dt) = \vec{v}(t) + \vec{g}dt \qquad (33)$$

$$\vec{r}_{OC}(t + dt) = \vec{r}_{OC}(t) + \vec{v}(t)dt. \qquad (34)$$

Thus we can track the motion of the ball. For example, Fig. 8 shows ball's motion at $d = 0.02m$ and $v_0 = 1.7m/s$. Fig. 9 demonstrates that "rolling on the rim" is an extremity of multi-collision situation, where $d = 0.02m$, v_0 is (from left to right) 1.7, 1.665 and 1.6 m/s, respectively.

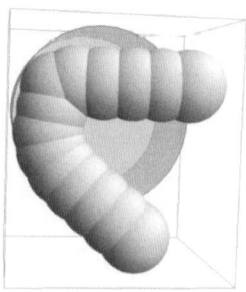

Fig. 8.   An example of multiple collision occurred in simulation, with $d = 0.02m$ and $v_0 = 1.7m/s$ a trail of the ball is shown.

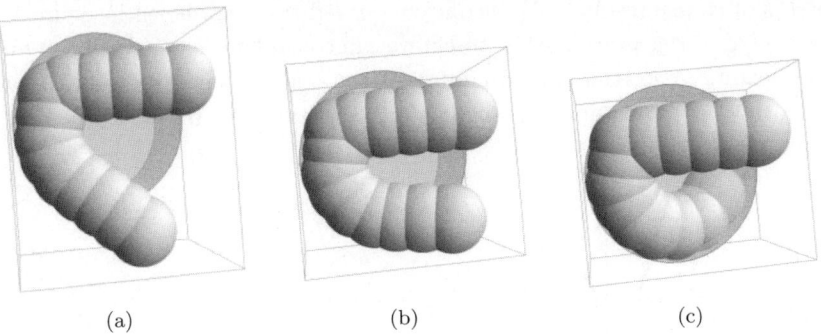

(a)   (b)   (c)

Fig. 9.   Examples of simulation result: the initial velocity of (a), (b) and (c) is respectively $1.7m/s$, $1.665m/s$ and $1.6m/s$, restitution coefficient $e = 0.4$, friction coefficient $\mu = 0.4$, deviation $d = 0.02m$; The result infers that the case of rolling on the rim could be a limit of multiple collision.

Based on the analysis above, we calculated the relation between the critical escaping velocity v_c and the deviation d considering multiple collision case, see Fig. 10.

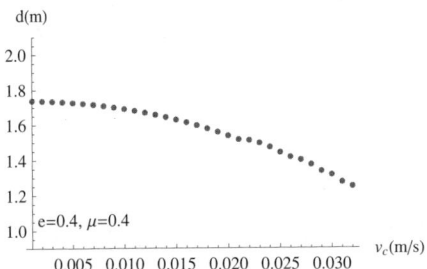

Fig. 10.   The relation between the critical escaping velocity v_c and the deviation d considering multiple collision case.

3.   Experiment

To verify our model, we measure the escaping velocity at different deviation.

3.1.   *Equipment*

According to USGA standard, a wooden hole with radius of 5.4 cm is made. Usually the hole is made of plastic, which differs from the wood in μ and e. The two parameters can be manipulated in simulation. A standard golf ball is of course needed. We also use a curved track as shown in figure 11. When ball is released from the track, we can control its velocity by changing the height. Also, the curved ending of the track ensures the ball to run smoothly. Other equipments include an iron stand, a camera, a ruler, and a photogate.

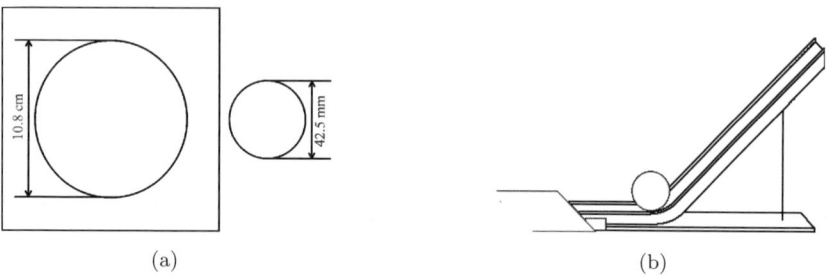

(a)   (b)

Fig. 11. Equipment of experiment: (a)The size of the golf ball and the hole, according to USGA standard. (b)A curved track is used to control the direction of ball's motion. Initial velocity is controlled by releasing the ball from different height.

3.2. Measuring Coefficient of Restitution and Friction

To measure the friction coefficient μ, we drag the ball on the board and measure the kinetic friction force.

$$f = \mu mg \tag{35}$$

For restitution coefficient, we release the golf ball from different height, and measure its bounce height, see figure 12.

$$e = \sqrt{h_0/h_r} \tag{36}$$

The friction coefficient is measured to be 0.4. The restitution coefficient has a lager uncertainty due to the unsmooth surface of the golf, but generally between 0.3 to 0.5.

3.3. Measuring Escaping Velocity under Different Deviation

Fig. 13 shows the method of measuring the relationship between escape velocity and deviation. For every given deviation d, we increase the height of releasing and observe the behavior of the ball. The velocity is measured and recorded by photogate. The result is shown in Fig. 14. In this diagram, diamonds correspond to escapable shot, and squares correspond to

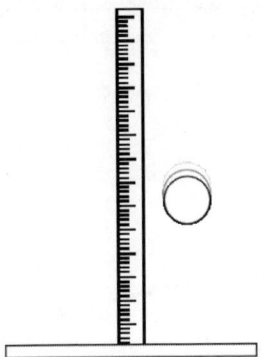

Fig. 12. Equipment to measure the restitution coefficient: a ruler and a high speed camera are used to catch the height of the ball's bounce.

unescapable shot. The solid line highlights the boundary between above-rim and below-rim collision.

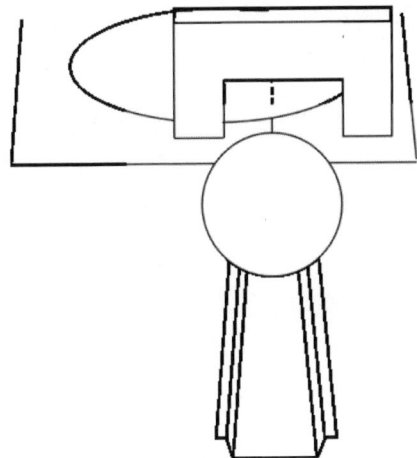

Fig. 13. Equipment of measuring the escaping velocity: a photogate is used to record ball's initial velocity at the edge of the hole. The deviation is controlled by changing the relative position of the hole and the track.

3.4. *Verification of the Model*

To verify our model, we compare the simulation result with the experiment data of $v_c - d$ relation. We use the simulation result from Section 2.2 and 2.3 to illustrate the case of single collision. The comparison is shown in

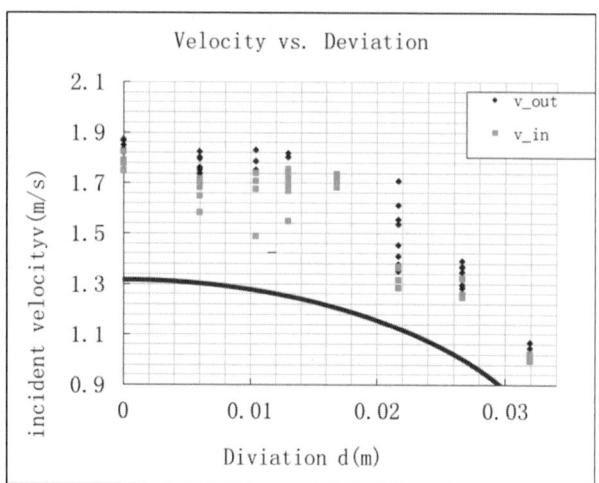

Fig. 14. Result of the experiment: The solid line below is the boundary of below-rim and above-rim collision. The diamonds correspond to escapable shot, and squares correspond to unescapable shot.

figure 15. In the figure, triangles are escapable shots, dots are unescapable shots. The solid curve stands for the theoretical prediction for the escape velocity. It is obvious that the calculation is consistent with experiment at the region where d is small, but differs from experimental data when d increases.

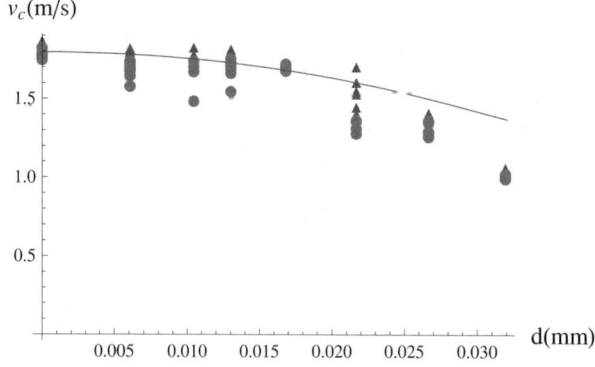

Fig. 15. Comparison of theoretical prediction and experiment: triangles are escapable shots, dots are unescapable shots. The solid curve stands for the theoretical prediction for single collision.

If we take multi-collision into account, the theoretical result fits the experimental data better even in region that d is large (See Figure 16). In the figure, the triangles are escapable shots, dots are unescapable shots. The dotted curve is the theoretical prediction of escape velocity when multiple-collision is considered.

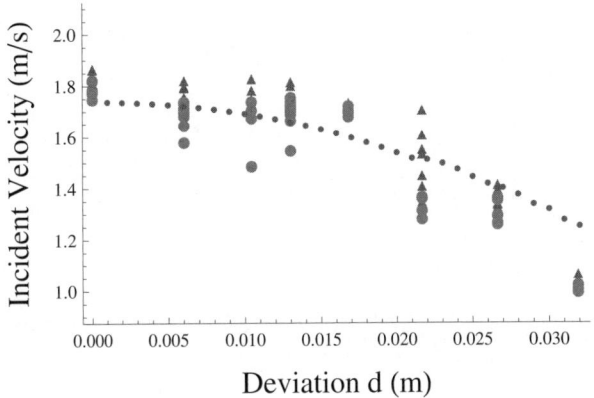

Fig. 16. Comparison of theoretical prediction and experiment: triangles are escapable shots, dots are unescapable shots. The dotted curve is the theoretical prediction of escape velocity when multiple-collision is considered.

4. Conclusion

With parameters of the ball and the hole given, initial velocity and deviation are the two dominant factors that determine whether the ball will escape from the hole. The collision between golf ball and the hole can be divided into two types: below-rim collision and above-rim collision. In below-rim collision, the ball is hard to escape. In above-rim collision, the critical escaping velocity v_c for each certain deviation d can be predicted for single collision and multiple collision by our model. The calculation results of our model are consistent with experimental results.

References

1. U. S. G. Association et al., *Rules of Golf.* US Golf Association (USGA) (1987).

Chapter 9

2013 Problem 2: Elastic Space

IIeng Ho*, Zhihang Qin, Sihui Wang and Huijun Zhou

School of Physics, Nanjing University

We define and measure the "gravitational constant" G geometrically by analogizing the world of membrane with the gravitational field in two dimensional space. Investigation on the dynamic behavior of the system also provides alternative measurement of the "gravitational constant". The G values we obtained from the geometric method and dynamic methods are consistent. The properties of the membrane's gravitational world and the dynamics of the system are investigated and compared to the real world gravity systematically.

1. Introduction

Problem Statement:

The dynamics and apparent interactions of massive balls rolling on a stretched horizontal membrane are often used to illustrate gravitation. Investigate the system further. Is it possible to define and measure the apparent "gravitational constant" in such a "world"?

According to the problem statement, our solution tries to investigate the dynamics and apparent interactions of massive balls rolling on a stretched horizontal membrane, and make comparison to the real world gravity. Theoretical and experimental methods are employed in the solution.

First of all, we will examine the properties of a membrane with a heavy ball M placed in the middle. The shape of the membrane which is associated with the mass of the ball (M) will determine the apparent interaction exerted on other massive balls rolling on the membrane. The potential energy corresponding to the deformation of membrane will be derived. The potential energy has a form analogous to that of the two-dimensional gravitational field. In this way, the apparent "gravitational

*E-mail: eliza1213me@gmail.com

constant" can be defined and measured geometrically by the stress on the membrane.

Then, we investigate the dynamics of massive balls in the "gravitational field" of M by simplifying the system into a one-body problem. The "gravity" will be inversely proportional to the distance from the center in the form $1/r$, rather than the inverse square law gravitation in the real world. Two typical orbits corresponding to "radial release" and "tangential release" are studied. The dynamic equations are given and verified by experimental results. In addition, the "gravitational constant" can also be found and compared to that measured geometrically.

Finally, the condition under which the "gravitational constant" exists will also be discussed. We compare the membrane's gravity world and gravitation in real world. We reexamine the conservation laws and Kepler's law of planetary motion. Some of the laws are no longer valid in the membrane's world, some others require modifications or supplementary conditions.

2. Properties of the Membrane

2.1. *The Shape of a Membrane with a Heavy Mass in the Middle*

A stretched horizontal membrane is required by the problem statement, see Fig. 1 . The membrane will be stretched uniformly before fixed to the circular frame, so that the stress is predetermined without the center mass M. We make further assumptions

- We assume that the membrane is homogeneous.
- A heavy ball with mass M placed in the middle of the membrane is stationary all the time.
- The angle θ between the tangent of the membrane and the horizontal plane is small.

With these assumptions, force analysis of a stretched membrane can be performed. See Fig. 1(a), we divide the membrane by a circle with radius r. By symmetry, we only need to consider the radial stress.

See Fig. 2, the equilibrium condition along the circular arc $d\alpha$ on vertical direction is:

$$r \cdot \sigma_r \cdot \sin \theta \cdot d\alpha = \frac{T}{2\pi} \cdot d\alpha, \qquad (1)$$

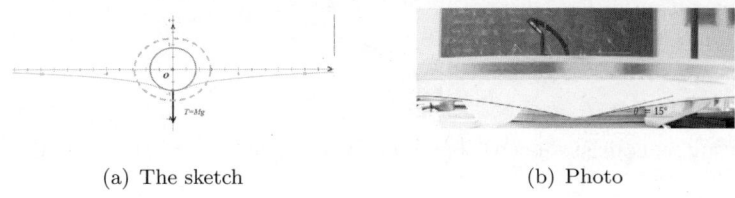

(a) The sketch (b) Photo

Fig. 1.   A stretched membrane with a heavy mass in the middle. The angle θ between the tangent of the membrane and the horizontal plane is small.

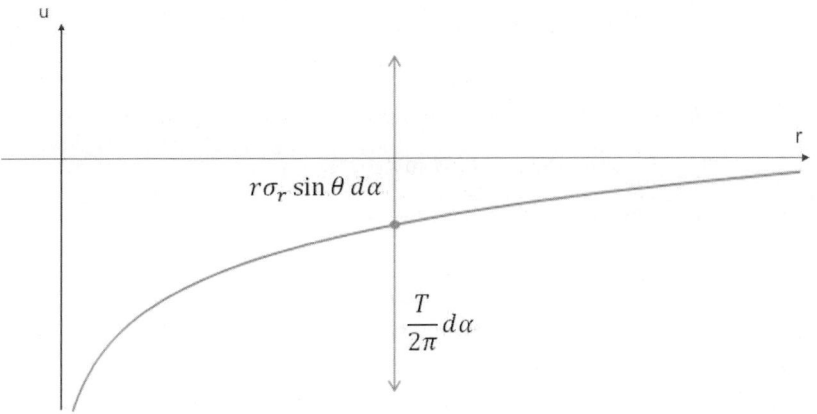

Fig. 2.   The force analysis of the membrane.

Table 1.   Variables used in the force analysis.

Variable	Meaning
M	The mass of the heavy massive ball.
g	The gravitational field.
r	The distance from the center point to a chosen point on the membrane.
σ_r	The radial stress per unit length on the homogeneous membrane.
σ	The stress per unit length on the homogeneous membrane.
θ	The angle between the tangent of the membrane and the horizontal plane.
$d\alpha$	Central Angle.
u	The displacement of the homogeneous membrane from its initial horizontal position.

where $T = Mg$ is the total tension on the circle surrounding the center mass M. Consider the small angle approximation, so that $\sin\theta = \tan\theta = \frac{du}{dr}$. The tension is homogeneous and uniform on the membrane, so that $\sigma_r = \sigma$.

An integration of Eq.1 yields:

$$u = \frac{g}{2\pi\sigma} M \ln r + u_0. \tag{2}$$

The result is consistent with Feynman's result for a stretched membrane.[1]

2.2. Defining "Gravitational Constant" G

Gravitational potential is $\phi = gu$, so the gravitational potential on the membrane at a distance r from the center M is

$$\phi = \frac{g^2}{2\pi\sigma} M \ln r + \phi_0. \tag{3}$$

Comparing Eq.3 to the gravitational potential in two dimensional spaces

$$\phi = GM \ln r + \phi_0. \tag{4}$$

The "gravitational constant" can be defined as

$$G = \frac{g^2}{2\pi\sigma}. \tag{5}$$

The geometrical definition of "gravitational constant" in Eq. 5 indicates that G is in inverse proportion to the stress of the stretched membrane. This means that "gravitational constant" not only depends on the elastic properties of the membrane but also depends on how much it is stretched. By measuring the stress of the stretched membrane, we can get the "gravitational constant" in experiments.

2.3. Experimental Verification on the Properties of the Membrane

2.3.1. *The Shape of the Membrane*

Fig. 3 shows the experimental set up. The membrane is a synthetic rubber sheet stretched uniformly before being fixed to a round frame horizontally. We placed a massive ball M in the middle of the membrane, with the angle between the surface of the membrane and the horizontal plane small. The inner diameter of the frame is $d = 29.2cm$, the mass $M = 27.9g$.

First of all, we measure the vertical displacement of the membrane from its horizontal position u and the distance r from the center mass M. According to Eq. 2, we plot u vs. lnr in Fig. 4 which shows a linear relationship as expected.

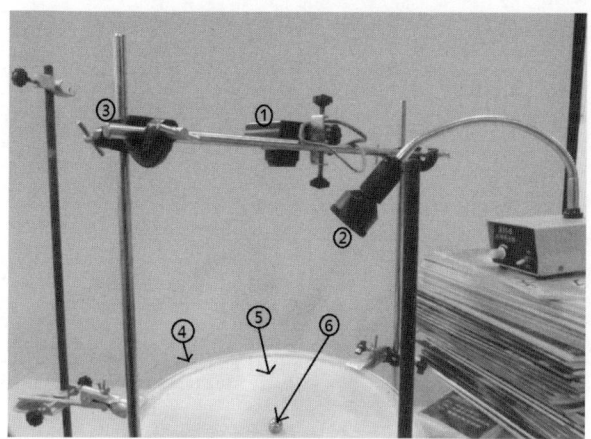

Fig. 3.   Experiment setup.

Table 2.   Experimental setup.

Number	Device
1	Casio EX-FC150 high speed camera
2	Lamp
3	Fixing devices
4	Round Frame; inner d = 29.2cm
5	A Synthetic rubber membrane
6	Steel balls M = 27.9g, m = 1.48g

2.3.2. *Geometric Measurement of "Gravitational Constant" G*

In our experiment, membrane's stress is measured by stretching it along both X and Y directions simultaneously. The corresponding stress is measured by force sensor, as shows in Fig. 5.

According to the experimental conditions, the stress of the stretched membrane is $\sigma = 54.6 \pm 2.0 N/m$. Consequently, we get the "gravitational constant" G

$$G = \frac{g^2}{2\pi\sigma} = 0.28 \pm 0.01 N \cdot kg^{-2} \cdot m \tag{6}$$

where we take $g = 9.8 m \cdot s^{-2}$.

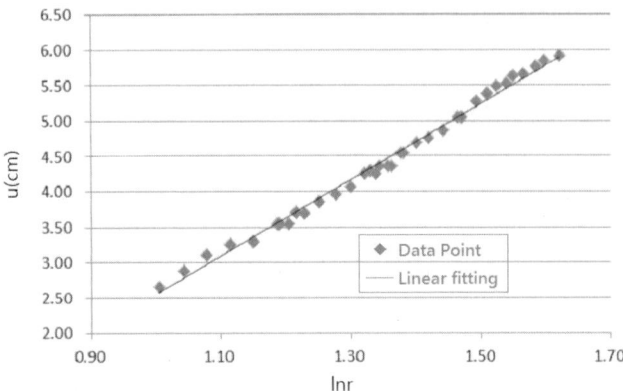

Fig. 4.　Linear relation between the displacement of the membrane from its horizontal position u and lnr.

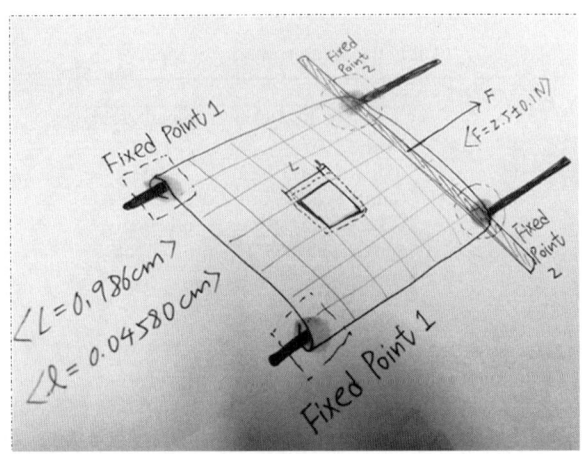

Fig. 5.　Measuring the stress of the membrane.

3.　The Dynamics

3.1.　*The Force in Membrane's Gravity World*

Suppose that a ball with mass m rolling in the "gravitational field" of M, $m << M$, so that the motion of m will not influence the state of M and the shape of the membrane. Thus the system can be simplified as a one-body problem. We can derive the apparent interaction force exerted on m from the membrane's gravitational potential Eq. 3

$$F = -m \frac{\partial \phi}{\partial r} = -\frac{Mmg^2}{2\pi\sigma r} = -\frac{GMm}{r} \propto \frac{1}{r}. \tag{7}$$

Here the "gravity" is inversely proportional to the distance rather than the square of it! It's a gravity different from the Newton's law.

3.2. Two Typical Orbits: "Radial Release" and "Tangential Release"

3.2.1. Motion of "Radial Release"

Now we consider the motion with two typical orbits. One is called "radial release" in which the ball rolling along the radial direction towards the center of the membrane, as shown in Fig. 1. The force analysis is shown in Fig. 2. Neglecting the sliding friction, we have

Fig. 6. The motion of radial release.

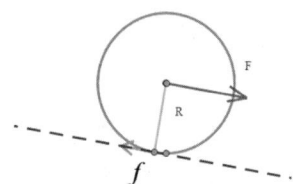

Fig. 7. The force analysis of the ball in radial release.

$$m\dot{v} = F - f$$
$$I\dot{\omega} = f \cdot R \tag{8}$$
$$v = \omega \cdot R$$

where $I = \frac{2}{5}mR^2$. Solving Eq. 8, we have

$$F = (m + \frac{I}{R^2}) \cdot \dot{v} = m'\dot{v} \tag{9}$$

where equivalent mass $m' = \frac{7}{2}m$.

Table 3.   Variables used in the equation.

Variable	Meaning
F	The force on the ball.
f	Rolling friction.
ω	Angular velocity.
v	Linear velocity.
v_0	Initial linear velocity.
R	The radius of the ball.
I	The rotational inertia of the ball.
m	The mass of the rolling ball.
m'	The equivalent mass of the rolling ball.

By integrating Eq. 9, we may derive the energy conservation equation as

$$\frac{1}{2}m'(v^2 - v_0^2) + m(\phi - \phi_0) = 0. \tag{10}$$

Set the initial velocity to be zero, $v_0 = 0$; and substitute the expression of potential in Eq. 4, the speed can be expressed as a function of radius

$$v(r) = \sqrt{2MG\frac{m}{m'}ln\frac{r_0}{r}}. \tag{11}$$

According to Eq. 11, the "gravitational constant" G can be measured by analyzing the motion of radial release.

In the experiment part, the motions of the rolling balls are recorded by using a Casion EX-FC150 high speed camera. The date are collected by using "Tracker", shown as the dots in Fig. 8. "Mathematica" is used to fit the data according to Eq. 11. The result of fitting is show as the curve in Fig. 8, from which the constant G is calculated as $G = 0.26N \cdot Kg^{-2} \cdot m$.

3.2.2.   Motion of "Tangentiul Release"

The other typical orbit is a circular motion with an ideal "tangential release", as shown in Fig. 1 and Fig. 2. So the equation of motion is

$$\frac{GMm}{r} - f = m(\frac{2\pi}{T})^2r. \tag{12}$$

According to the angular momentum theorem, friction f is needed to provide the torque for the procession of the rolling ball (or to produce the increase in the angular momentum), therefore

$$fR = \frac{Iv^2}{Rr}, \tag{13}$$

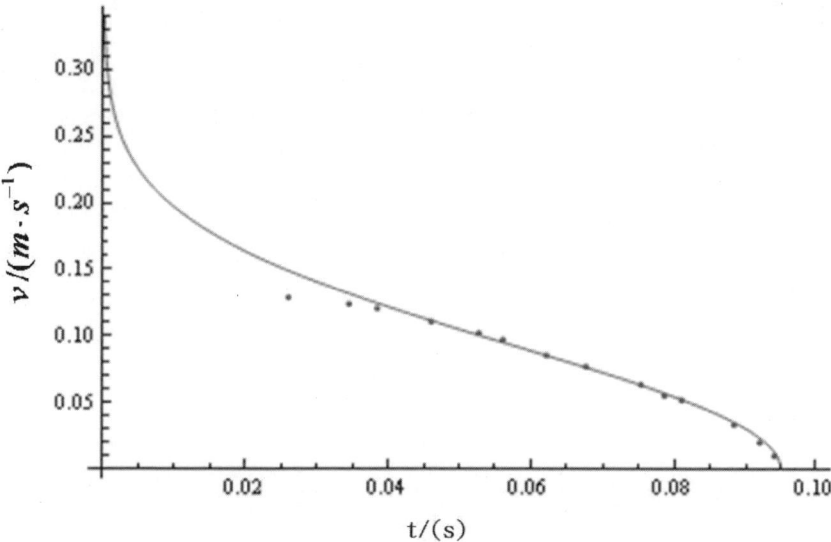

Fig. 8. The speed is measured as a function of radius. The dots are experimental data, and the curve is the fitting result according to theory.

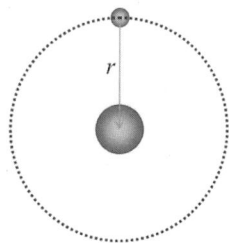

Fig. 9. The motion of tangential release.

Fig. 10. The circular orbit for tangential release.

where $I = \frac{2}{5}mR^2$. Combining Eq. 12 and 13, we have

$$G = \frac{7}{5}\frac{1}{M}(\frac{2\pi}{T})^2 r^2. \tag{14}$$

Here we find a new "Kepler's law": the orbital period of the circular motion is proportion to the radius of the orbit

$$T \propto r \tag{15}$$

From Eq. 14 the "gravitational constant" G can also be measured by analyzing the details of the motion. We use "Traker" to collect data of the orbital period and the radius of the circular motion. The data in Fig. 11 confirms the linear relation between period and the radius of the circular motion. Fitting the data according to Eq. 14, the "gravitational constant" G is found to be $G = 0.303N \cdot kg^{-2} \cdot m$

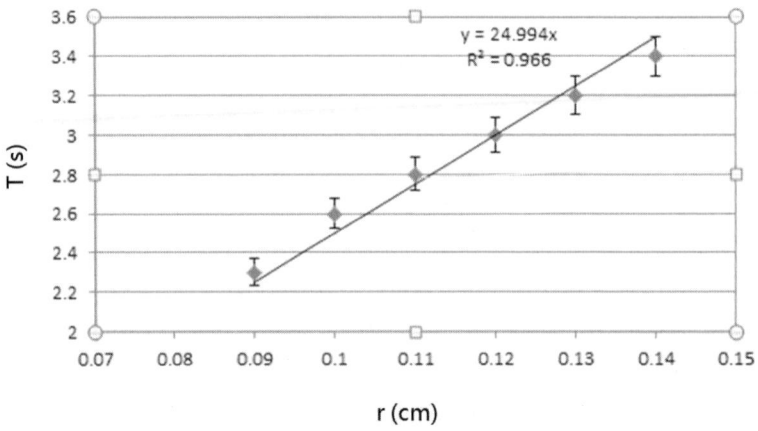

Fig. 11.   A new "Kepler's law": the linear relation between period and the radius of the circular motion. The diamonds are the experimental results, the solid line is the linear fitting result.

3.2.3. General Orbits

In theory, the energy and angular momentum along z direction are conserved for the dynamic system

$$E_0 = \frac{1}{2}m'[\dot{r}^2 + (r\dot{\theta})^2 + \dot{z}^2] + m \cdot \phi(r) = const \tag{16}$$

$$J_z = mr^2\dot{\theta} = const. \tag{17}$$

So we can find the orbital equation of the rolling ball

$$d\theta = \frac{J \cdot dr}{r^2 \cdot \sqrt{2m'(E_0 - m\phi(r) - \frac{J^2}{r^2})}}. \tag{18}$$

The solution of the orbit is calculated numerically using "Mathematica", see Fig. 12. One can see that unlike the closed elliptical orbit in the inverse square field of gravity, the orbit is open here!

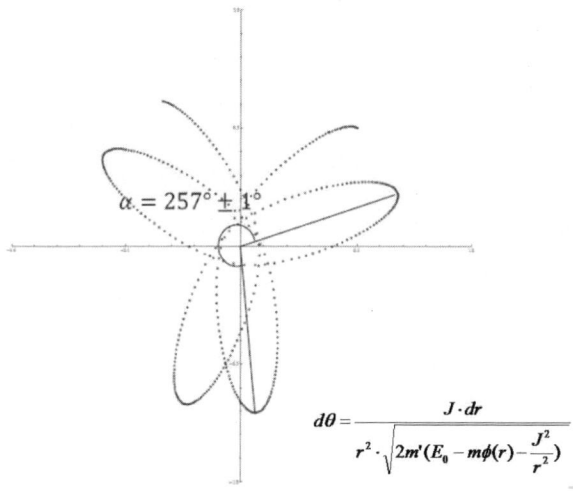

Fig. 12.   The orbit of a rolling ball on membrane is open. The numerical result is made by using "Mathematica".

We use "Tracker" to gain the motion trail of the rolling ball. Similar to the theoretical solution, the actual orbit is also open, see Fig. 13. Note that the experimental trial shrinks with time, indicating that mechanical energy is no longer conserved due to friction or other energy dissipation. Similarly, areal velocity also reduces due to energy losses, see Fig. 13.

In addition, even though in theory, the total angular momentum is not conserved for a rolling ball on the membrane. As a rigid object performing a motion of procession, friction is needed to provide the torque, in order to sustain the increase in angular momentum. So the angular momentum only conserves along z direction, but not conserved on x-y plane.

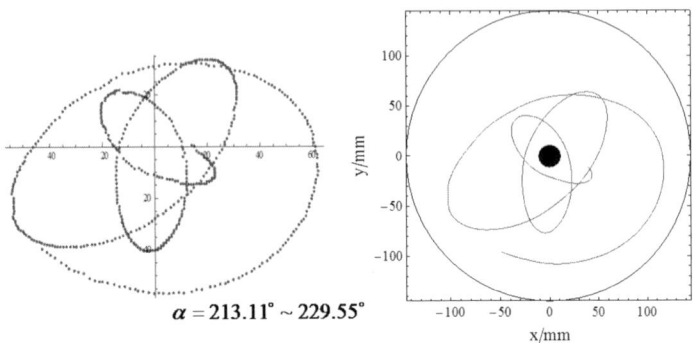

Fig. 13.   The experimental trial of a rolling ball, data collected by using "Tracker".

4. Discussion and Conclusion

The "gravitational constant" is determined by the shape and the gravitational potential of the membrane. The violation of any of the conditions (the small angle approximation, uniform stress assumption on the membrane, the stationary assumption of the center mass M, the pure rolling assumption of the masses) will make the definition of the "gravitational constant" meaningless and invalid. So practical measures are important to ensure an effective "membrane world of gravity", such as a suitable mass M so that the inclination angle will be small on most area of the membrane, meanwhile other masses will not cause deformation on the membrane. The membrane has to be stretched uniformly before fixed to the frame, so that the stress is predetermined without the center mass M.

We have defined and measured gravitational constant G geometrically by an analogy the world of membrane with the gravitational field in two dimensional space. Investigation on the dynamic behavior of the system also provides alternative measurement of the "gravitational constant". The G values we obtained from the geometric method and dynamic methods are consistent. The properties of the membrane gravitational world and the dynamics of the system have been investigated. We summarize and compare the membrane's gravity world with that of the real world gravity in the following table.

Table 4.   Comparison of different worlds.

Item	The real gravity world	The membrane's gravity world
The form of force	$F = -\frac{GMm}{r^2}$	$F = -\frac{GMm}{r}$
Kepler's 1st law	Elliptical orbit	Invalid : open orbit, rather than an elliptical orbit.
Kepler's 2nd law	Equal areal velocity	Valid with condition: equal areal velocity when sliding friction is negligible
Kepler's 3rd law	$T^2 \propto a^3$	Modified with condition: for circular orbit only. The orbital period of the circular motion is proportion to the radius of the circle. $T \propto r$
Energy	Mechanical energy is conserved	Valid with condition: mechanical energy is conserved when sliding friction is negligible.
Angular momentum	Angular momentum is conserved	Modified with condition: total angular momentum is not conserved for a rolling ball. Angular momentum conservation is valid along z directions only if sliding friction is negligible

References

1. R. Feynman, R. Leighton, M. Sands, and M. Gottlieb, *The Feynman lectures on physics, Volume II*. The Feynman Lectures on Physics, Pearson/Addison-Wesley (1963). ISBN 9780805390490. URL http://books.google.com.hk/books?id=_6XvAAAAMAAJ.

Chapter 10

2013 Problem 5: Levitation

Qiyuan Ruan*, Pei Zeng, Huijun Zhou and Sihui Wang

Nanjing University, School of Physics

In this work, we reproduce the phenomenon through a preliminary experiment. The main factors to optimize the system are identified as the mass of the ball, the flow velocity and distribution of the airstream. We propose a Gaussian velocity distribution model to describe the flow velocity field model quantitatively which is supported by COMSOL simulation and experimental data. Through force analysis, the supporting forces that balance the gravity of the ball are identified. Equation for the tilt angle has been found, from which the optimal tilt angle can be calculated and compared to experimental data. Our research also shows that levitation is more stable without rotation. So the method we used to adjust the mass of the ball by injecting water is also effective in preventing rotation and enhance stability. The theoretical result for the optimal tilt angle is consistent with experimental data.

1. Introduction

Problem Statement:

A light ball (e.g. a Ping-Pong ball) can be supported on an upward airstream. The airstream can be tilted yet still support the ball. Investigate the effect and optimize the system to produce the maximum angle of tilt that results in a stable ball position.

In this article, a preliminary experiment is done at first to reproduce the phenomenon and determine the main factors to optimize the system. The mass of the ball, the value and distribution of the airstream velocity are found to be important. In section 3.1, we build a cone-shaped flow velocity field model. Our quantitative equation by Gaussian velocity distribution is supported by experimental observation. In section 3.2, force analysis is made. The supporting forces that balance the gravity of the ball are

---

*E-mail: njaldehyde@gmail.com

identified as the radial lift force due to pressure difference and the drag force along the axial direction. Expression for the tilt angle can be figured out from force balance condition. The theoretical result is consistent with experimental data. Finally, the effect of rotation is discussed, and the consequent lift force is modified. It will be found that the levitation is more stable without rotation. So the method we proposed to adjust the mass of the ball by injecting water is also effective in preventing rotation and enhance stability.

2. Preliminary Experiment

The experiment setup is shown in Fig. 1. A tube is connected to the air pump with five levels of flow speed. A Ping-Pong ball is used to demonstrate the "levitation phenomenon". The mass of the ball can be adjusted by injecting water into the ball. In the experiment, the ball vibrates slightly without rotation. The equilibrium position deviated from the center of the air flow.

The experiment is recorded by high speed camera, and the video is analysis with software *Tracker*. The experiment result is shown in Fig. 2. We recorded the maximum tilt angle with respect to the extra mass of the ball. Each curve corresponds to one of flow speed levels. We find that the mass of the ball is important in optimizing the tilt angle. An extra mass of $10 - 15g$ gives rise to the maximum tilt angle around $60°$. The flow speed is also important, for example, at level three, a $60°$ supporting angle can be realized with extra mass ranging from $10g$ to $30g$. The mass of the ping-pong ball is 2.7g.

3. Solution

3.1. *Airflow Velocity Distribution*

Fluid projected from a nozzle into a infinite free space is called free jet.[1] The speed of flow is maximum at its center and decays rapidly in radial direction. This feature follows Gaussian Distribution model.[2] The concept figure is shown in Fig. 3.

According to Gaussian distribution, the velocity outside a certain domain approaches 0. Therefore, we may imagine a boundary of the flow field, which has the shape of a cone with radius (r_m) at different positions along z-axis. We can define relevent physical quantities in Fig. 4.

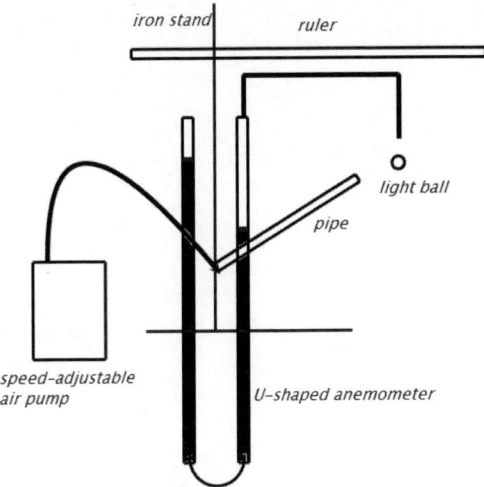

Fig. 1.   Sketch of experimental setup. The U-shaped anemometer uses connecting tubes to measure the pressure of the flow field.

Since the nozzle of the pipe is a circle, the general formula of Gaussian Distribution for a 2-D free jet is:

$$\frac{u_z(r)}{v(z)} = exp(-\frac{r^2}{r_e^2}). \tag{1}$$

where r_e is the boundary radius of cone-shaped flow field. So if $r > r_e$, the velocity can be taken as 0. A circular truncated cone satisfies:

$$r_e = z \tan \varphi + r_0. \tag{2}$$

Our Gaussian distribution within a cone-shaped flow field is justified by a COMSOL simulation shown Fig. 5. Notice that the flow forms a clear boundary, whose radius has been denoted by r_e in Gaussian distribution.

The flow field is stable, so the momentum flux will be conserved in every cross-section,[2] which means:

$$\rho \int_0^\infty u_z^2(r)2\pi r dr = \rho v^2(0)\pi r_0^2. \tag{3}$$

From Eq. 1 and Eq. 3 we have:

$$\frac{v(z)}{v(0)} = \frac{r_0}{r_e}. \tag{4}$$

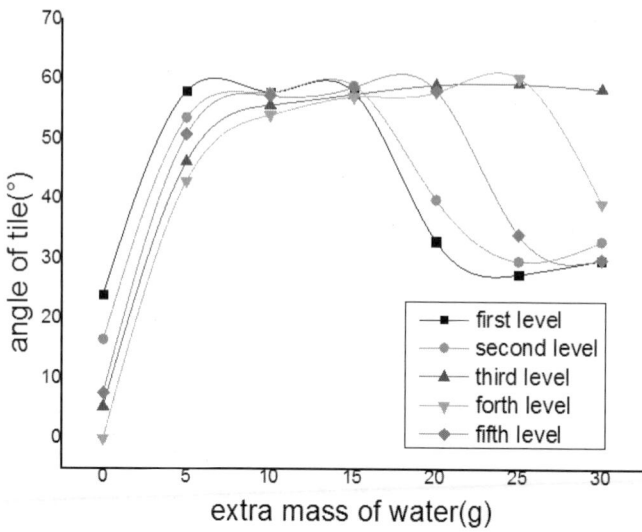

Fig. 2.   Preliminary experiment. Five curves correspond to five speed levels of the air pump.

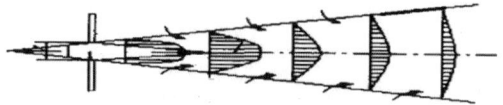

Fig. 3.   Concept figure of Gaussian velocity distribution in every cross-section.

Substitute Eq 2 into Eq. 4 yields velocity distribution along axial direction:

$$v(z) = v(0)\frac{r_0}{z\tan\varphi + r_0}. \tag{5}$$

Combining of Eq. 1 and Eq. 5 we get velocity distribution along radial direction:

$$u_z(r) = v(0)\frac{r_0}{z\tan\varphi + r_0}exp\left(-\frac{r^2}{(z\tan\varphi + r_0)^2}\right). \tag{6}$$

We use an anemometer to measure the velocity distribution along axial direction at different air flow levels, which is shown in Fig. 6 by the discrete

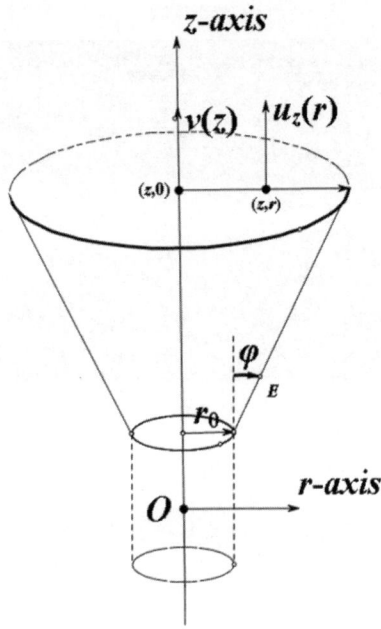

Fig. 4. Physical quantities of the airstream field. $v(z)$ is the velocity along z-axis and $u_z(r)$ is the velocity on radial direction. r_0 is the radius of the nozzle. φ is the apical angle of the cone.

dots. The solid curves are the theoretical results fitted by Eq. 6 where $\tan \varphi$ is a fitting parameter corresponding to the flow field of each level.

The radial velocity distribution is also measured and compared to the Gaussian distribution by Eq. 6, indicated by dots and curve in Fig. 7 respectively. The result shows that Eq. 6 provides an appropriate description of the flow field. We will estimate the force provided by the air jet on the basis of the Gaussian distribution.

3.2. Forces Analysis

With large Reynolds Number ($Re > 10^4$), viscosity of the air can be neglected. Consider a ball without rotation, the sphere is subject to three forces as shown in Fig. 8: gravity G which is downward in direction; lift force F_L (caused by non-homogeneous flowing velocity) that is perpendicular to flow axial direction; drag force F_D which is paralleled to axial direction.

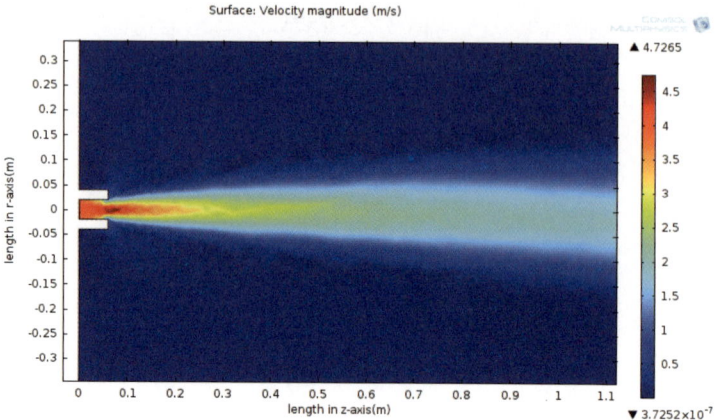

Fig. 5. Velocity distribution for a free jet. The magnitude of speed is indicated by the colors.

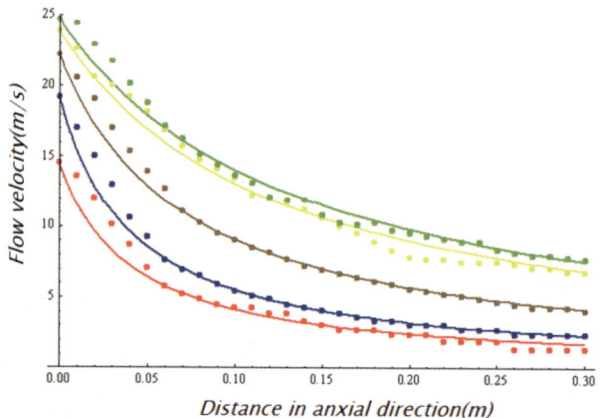

Fig. 6. Theoretical and experimental results of flow velocity along axial direction. When the ball is close to the nozzle, the size of the ball cannot be neglect, the ball may affect the fluid field a lot. So the first few dots don't fit well with theoretical curve.

The pressure distribution surrounding a sphere subject to an air flow is shown in Fig. 9 by COMSOL simulation.

Fig. 9 shows that how lift force and drag force are created by the pressure difference in the flow field.

Now we may calculate the magnitude of the forces. The gravity is given by:

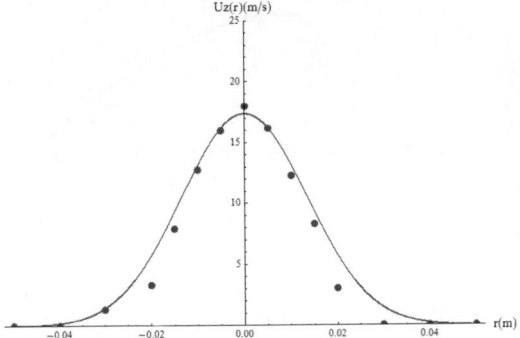

Fig. 7. Velocity distribution along radial direction. The dots shows the experimental velocity, and the solid curve represents the theoretical result.

$$G = mg. \tag{7}$$

The lift force can be estimated by the Bernoulli's equation. The sphere we have used is much smaller than the scale of flow field, so the change of flow field caused by the existence of sphere can be neglected. For high speed incompressible airstream we have:

$$p_z(r) + \frac{\rho u_z^2(r)}{2} = p_z(r+d) + \frac{\rho u_z^2(r+d)}{2}.$$

Therefore,

$$F_L = \pi r^2 [p_z(r+d) - p_z(r)] = \frac{\pi \rho d^2}{8}(u_z^2(r) - u_z^2(r+d)) \tag{8}$$

The high speed drag force[3] has the form

$$F_D = 1/2 C_D \rho \pi U^2 r_{ball}^2 = 1/8 C_D \rho \pi d^2 u_z^2(r+d/2) \tag{9}$$

where C_D is drag coefficient.

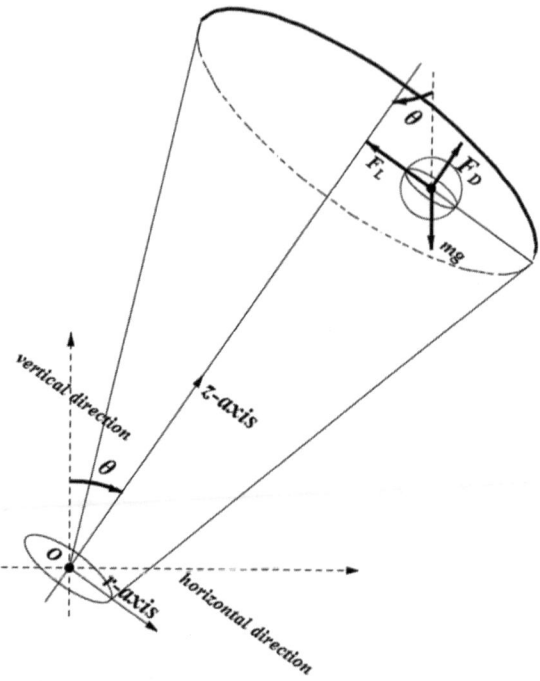

Fig. 8. Forces analysis. In which θ is the angle of tilt, F_L is the lift force and F_D is high speed drag force.

Owing to equilibrium of three forces, we can derive:

$$\tan\theta = F_L/F_D = \frac{exp\left(\dfrac{(r+d/2)^2 - r^2}{(z\tan\varphi + r_0)^2}\right) - exp\left(\dfrac{(r+d/2)^2 - (r+d)^2}{(z\tan\varphi + r_0)^2}\right)}{C_D}$$

(10)

$$\cos\theta = \frac{F_D}{mg}$$

$$r = (z\tan\varphi + r_0)\sqrt{\ln\frac{v^2(z)\,C_D\rho\pi d^2}{8mg\cos\theta} - \frac{d}{2}}.$$

(11)

Therefore, we can figure out the expression for the tilt angle:

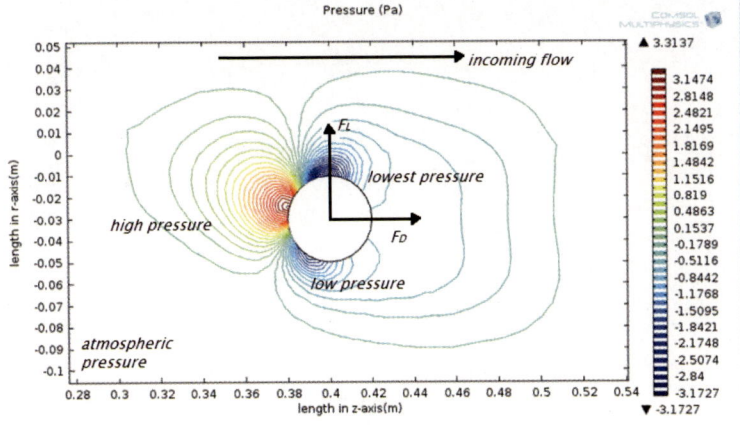

Fig. 9.   Pressure distribution near the ball. The magnitude of pressure is indicated by colors.

$$\tan\theta = \frac{1}{C_D}\left[exp\left(\frac{d(z\tan\varphi + r_0)\sqrt{\ln\frac{\left(v(0)\frac{r_0}{z\tan\varphi+r_0}\right)^2 C_D\rho\pi d^2}{8mg\cos\theta} - \frac{d^2}{4}}}{(z\tan\varphi+r_0)^2}\right)\right.$$

$$\left. - exp\left(\frac{-d(z\tan\varphi+r_0)\sqrt{\ln\frac{\left(v(0)\frac{r_0}{z\tan\varphi+r_0}\right)^2 C_D\rho\pi d^2}{8mg\cos\theta} - \frac{d^2}{4}}}{(z\tan\varphi+r_0)^2}\right)\right]. \tag{12}$$

Since the value of drag coefficient[4] C_D and air density ρ are known, $((C_D = 0.44), (\rho = 1.71 kg/m^3, 26°C))$, Eq. 12 has only four variables: θ, $v(0), m, z$. So the adjustable variables $v(0)$ and m can determine the angle of tilt under a certain speed level by Eq. 12, in which measured parameter z is related to $v(0)$ and m. So the adjustable variables $v(0)$ and m can decide the angle of tilt under a certain speed level by Eq. 12.

The theoretical results match well with the experimental values, as shown in Fig. 10.

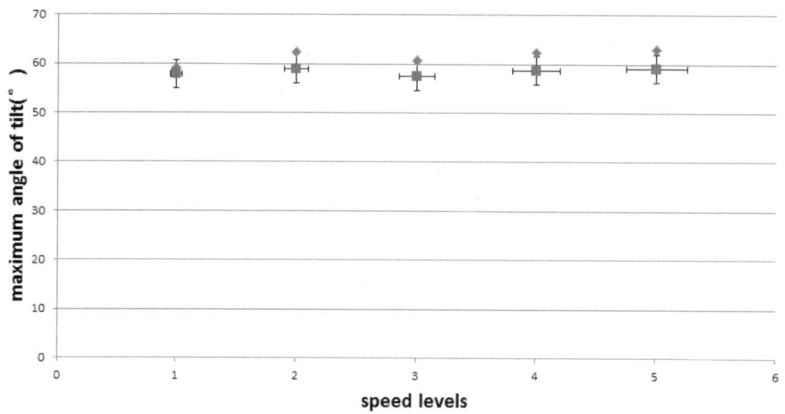

Fig. 10.   Maximum tilt angle corresponding to the five speed levels, the diamonds are theoretical results, the squares are experimental data.

3.3. *Further Discussion: Rotation*

A rotating ball in a flow field is subject to an extra force, the magnus force[5]

$$\overrightarrow{F_M} = s\left(\overrightarrow{\omega} \times \overrightarrow{v}\right).$$

The ball may rotate with angular velocity paralleled or perpendicular to the plane determined by three forces F_L, F_D and G. When the angular velocity is paralleled to the plane, the magnus force is perpendicular to the plane, it may cause the ball to deviate from equilibrium.

When angular velocity is perpendicular to the plane, which is shown in Fig. 11 (a), the magnus effect modifies the lift force in magnitude. Usually, the magnus lift force in a uniform field is given by:

$$F_{ML} = \frac{1}{2}\rho v^2 AC_L$$

which is no longer valid for our nonuniform flow field. However we can estimate the magus effect by a modification to Eq. 8:

$$F_L = \frac{\pi \rho d^2}{8}\left((u_z(r) - \omega r)^2 - (u_z(r + d) + \omega r)^2\right) \tag{13}$$

where the rotation angular velocity ω is measurable. The sign is taken according to experiment observation. According to Eq. 13, the life force is reduced, which is unfavorable to levitation at large tilt angles.

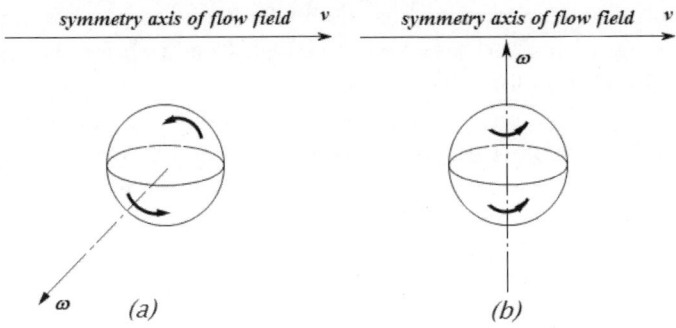

Fig. 11. The ball may rotate with angular velocity perpendicular or paralleled to the plane.

When the angular velocity is paralleled to the plane, which is shown in Fig. 11 (*b*), magnus Force is perpendicular to the plane of the three forces which are mentioned before, it can only bring side effect to the system.

So we may conclude that levitation is more stable without rotation. In our experiment, adjusting the mass by injecting water has another effect in optimizing the system: the water's viscosity prevents the ball from rotation and can help enhance the stability of the levitation.

4. Conclusion

The task of the problem to "investigate the effect and optimize the system to produce the maximum angle of tilt that results in a stable ball position." In this work, we have investigated the effect and optimize the system by experimental and theoretical approach.

Through preliminary experiment, we reproduce the phenomenon and find the main factors to optimize the system: the mass of the ball; the flow velocity and distribution of the airstream. The rotational state of the ball also affects the stability of levitation.

We examined the airstream velocity field with care. We proposed a cone-shaped flow velocity field model quantitatively described by Gaussian velocity distribution. The model is supported by COMSOL simulation and experimental data.

Through force analysis, the supporting forces that balance the gravity of the ball have been identified as the lift force due to pressure difference and the drag force along the axial direction. Condition for the tilt angle has been found from force balance equation. Finally, the effect of rotation is discussed. Levitation is found to be more stable without rotation. So the method we proposed to adjust the mass of the ball by injecting water is also effective in preventing rotation and enhance stability.

The theoretical result for the optimal tilt angle is consistent with experimental data.

Acknowledgments

I'd like to thank my friends Zheyuan Zhu, Yaohua Li for their help in the simulation part and the preparation of this report.

References

1. DB Spalding. Concentration fluctuations in a round turbulent free jet. *Chemical Engineering Science*, 26(1):95–107, 1971.
2. Xiaodong Ruan Jianzhong Lin. *Fluid Dynamics*. Tsinghua University Press, Beijing, 2005.
3. Yunhao Qin. *Thermaldynamics*. Higher Education Press, 3 edition, 2011.
4. Jianfeng Zou Anlu Ren, Guangwang Li. Numerical study of uniform flow over sphere at intermediate reynolds numbers. *Journal of Zhejiang University (Engineering Science)*, 38(5):644–648, 2004.
5. Frank Gaitan. Microscopic analysis of the nondissipative force on a line vortex in a superconductor: Berrys phase, momentum flow, and the magnus force. *Physical Review B*, 51(14):9061, 1995.

Chapter 11

2013 Problem 6: Colored Plastic

Lan Chen*, Youtian Zhang, Sihui Wang and Huijun Zhou

School of Physics, Nanjing University

The phenomenon concerning "colored plastics" are investigated extensively. It is ascribed to the "stress birefringence" of the polymer transparent materials. First of all, we obtained colored fringe patterns with three different plastic samples. The isochromatic and isoclinic fringes can be identified by rotating the polaroids and by their colors. Five different light sources are used including monochromatic ones and white light. Colorful patterns can be observed with white light sources. The isochromatic and isoclinic fringes can be identified by rotating the polaroids and by their colors. For monochromatic light, the isochromatic patterns become denser for shorter wavelengths.

We also explain the "colored plastics" phenomenon in everyday life without polaroids. By theoretical simulation on uniaxial crystal 2D plate, we prove that polarization in front or behind the birefringence sample is unnecessary in producing colors.

1. Introduction

Problem Statement:

In bright light, a transparent plastic object (e.g. a blank CD case) can sometimes shine in various colors. Study and explain the phenomenon. Ascertain if one also sees the colors when various light sources are used.

A transparent plastic object can shine in various colors under different conditions. The origin of phase difference and polarization is the key to understand the phenomenon. In general, the birefringence of anisotropic materials accounts for colored plastics in most cases. Among them, "stress birefringence", the base of photoelasticity, is the common cause of polymer transparent materials' anisotropy.

---

*E-mail: chenlanphy@gmail.com

A typical system to show colorful patterns in plastic samples consists of three parallel parts: two polaroids with mutually perpendicular axes (polarizer and analyzer) and the specimen (plastic plate) between them. When white light propagates through them successively, one can observe colored fringes which are relevant to the property of the specimen.[1]

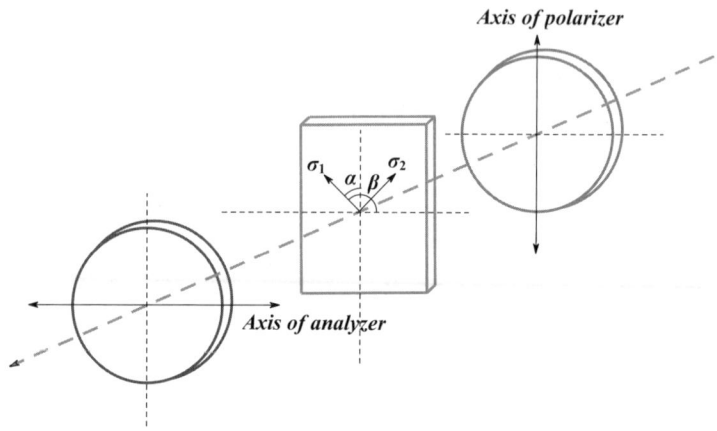

Fig. 1. Experiment system: polariscope. Light passes through the polarizer, specimen and analyzer successively. The orientations of two polaroids are perpendicular. σ_1 and σ_2 are two principle stresses in specimen.

However, colored patterns appear in everyday life more easily. For example, when the light is unnormally incident (incident angel is greater than 0), patterns can be observed even without polaroid (see Fig. 2). Similar phenomena can also be observed more frequently under reflection condition.

We will investigate the origin of phase difference and polarization separately. In Sec. 2 we will investigate the colored patterns of anisotropic birefringence materials through polariscope. In Sec. 3, we will discuss the formation mechanism of colored patterns without polaroids.

2. Colored Patterns through Polariscope

2.1. Theory

Polymer transparent materials acquire birefringence under mechanical stress. The stress may be applied externally or "frozen" inside the modeling

Fig. 2.   Patterns in reflection without polaroids under sunlight.

product without external loads, namely, the residual stresses.[2] Stress-induced birefringence provides a practical alternative to analyze the stress distribution inside a sample optically. This method called photoelasticity relates the material's mechanical properties to optical ones.

When the light passes through a photoelastic material, its electromagnetic wave components decompose along the two principal stress directions and two components correspond to different refractive indices due to birefringence. The difference in the refractive indices leads to a relative phase retardation between the two components. The magnitude of the relative retardation is given by the *stress-optic law*:

$$n_1 - n_0 = A\sigma_1 + B\sigma_2 \tag{1}$$

$$n_2 - n_0 = A\sigma_2 + B\sigma_1, \tag{2}$$

where n_0 is the refractive index of the specimen without stress and σ_1, σ_2 are two principal stresses, respectively. n_1 and n_2 are the refractive indices for the light vector parallel to σ_1, σ_2. A and B are material constants named *absolute stress-optical coefficient*.[2] Then we have:

$$n_1 - n_2 = C(\sigma_1 - \sigma_2), \tag{3}$$

which shows the relationship of principal stresses and the principle refractive indices. "$C = A - B$" is *stress-optical coefficient*.[2] Besides, the light vector is just decomposed to the directions of principle stresses.

As the polariscope (Fig. 1) has two cross polaroids, no light can transmit through the system without anisotropy; nevertheless, with stresses on the specimen, the light vectors decomposed along principle stresses will carry an phase difference and will not cancel each other after passing the second polaroid, forming the interference patterns, which is colorful with white light source.

Light vector after passing through the polarizer is:

$$L = A \cdot e^{i\omega t}. \tag{4}$$

Two components corresponding to two principle stress directions are:

$$L_1 = A \cos \alpha \cdot e^{i\omega t} \tag{5}$$

$$L_2 = A \sin \alpha \cdot e^{i\omega t}. \tag{6}$$

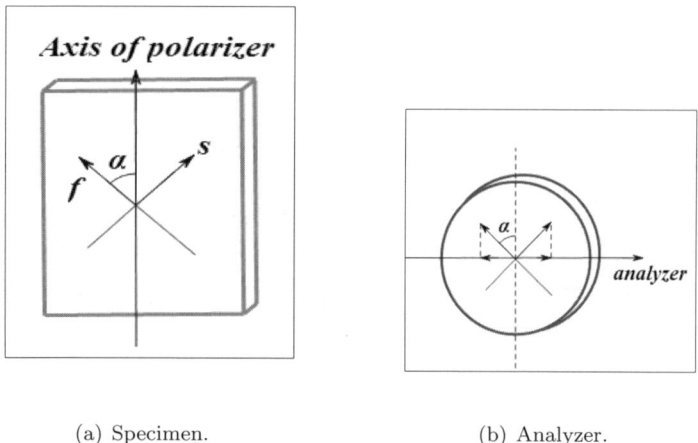

(a) Specimen.   (b) Analyzer.

Fig. 3.   Light vector decomposes when passing through the specimen and analyzer.

Two component waves will get different phases after travelling through the specimen:

$$L_1 = A \cos \alpha \cdot e^{i(\omega t - \delta_f)} \tag{7}$$

$$L_2 = A \sin \alpha \cdot e^{i(\omega t - \delta_s)} \tag{8}$$

with:

$$\delta_1 = \frac{2\pi d}{\lambda}(n_f - 1) \tag{9}$$

$$\delta_2 = \frac{2\pi d}{\lambda}(n_s - 1), \tag{10}$$

where d is the thickness of specimen, λ is the wavelength of the light source in the air and n_f is the smaller one in n_1 and n_2. Then two components project to the analyzer's orientation and the final light wave is:

$$L' = A \sin \alpha \cos \alpha \cdot e^{i\omega t} \cdot (e^{-i\delta_f} - e^{-i\delta_s}). \tag{11}$$

The light intensity[1] of the transmission light is:

$$I = A^2 \sin^2 2\alpha \sin^2(\frac{\delta_f - \delta_s}{2}). \tag{12}$$

These deduction is for single wave length with perpendicular polaroids, while the intensity of polychromatic[3] light is:

$$I = A^2 \cos^2(\alpha - \beta) - A^2 \sin 2\alpha \sin(2\beta) \sin^2(\frac{\delta_f - \delta_s}{2}). \tag{13}$$

β is the angle between analyzer orientation and fast axis of specimen. $A^2 \cos^2(\alpha - \beta)$ is usually called background item. Compare Eq. (13) with Eq (12) and it is obvious that we can observe the brightest fringes on black background with perpendicular polaroids for the contrast is the highest.

2.2. Experiment

All samples we used satisfy the condition of plane-stress system in that the thickness is much smaller than dimensions in the plane. Sample A is a plastic plate (a piece of candy box) with nearly symmetry stress distribution. Sample B is a plastic triangular rule that exhibits irregular stresses due to modeling process. Sample C is a photoelastic ring which will be tested under applied load in comparison to our theoretical calculation.

The problem requires to "ascertain if one also sees the colors when various light sources are used." In this part, we will take sample A as the mian example and various light sources are listed in Tab. 1.

2.2.1. Colored Patterns with Various Light Sources

First of all, we choose white light source (light source 5) to show colored patterns. Samples are set between two polaroids as shown in Fig. 1. A bright colorful pattern can be observed behind the second polaroid. In

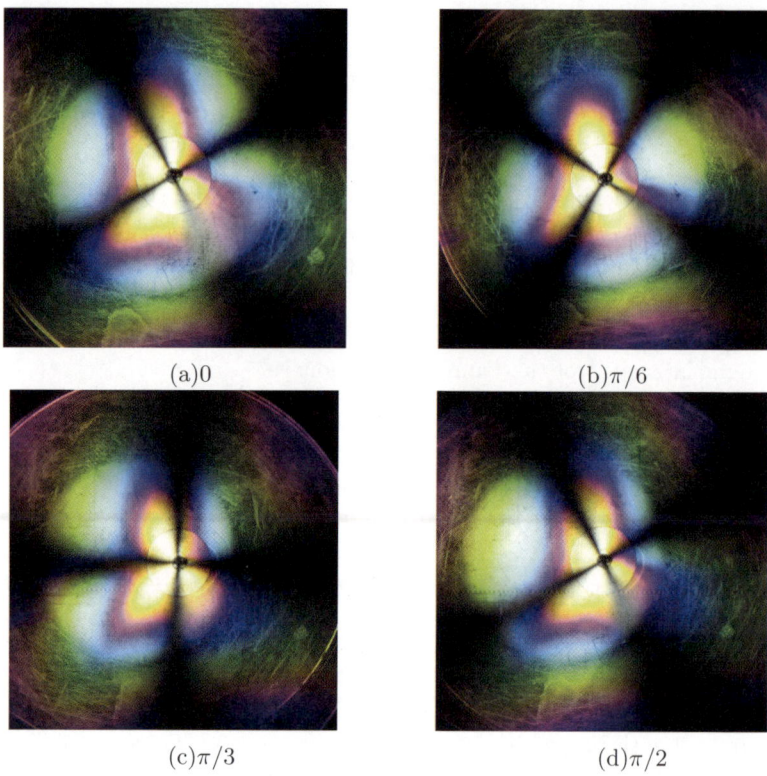

(a)0   (b)$\pi/6$

(c)$\pi/3$ (d)$\pi/2$

Fig. 4. Sample A: Rotating polaroids. Because stress distribution in sample A is nearly centrosymmetric, the isoclinic patterns can also rotate when we rotate two polaroids together, while isochromatic does not.

Fig. 5 we can observe colored fringe patterns and dark strips which are called isochromatic and isoclinic respectively.[1]

According to Eq. (12), if $\alpha = 0, \frac{\pi}{2}$, when either σ_1 or σ_2 is aligned with the polarizer's axis, extinction will occur irrespective of light wavelength. This fringe pattern is known as *isoclinic*, which can be used to determine the directions of the principal stress at any point in the plane of the specimen. If we rotate the polarizer and the analyzer together with their axes of polarization vertical until an isoclinic fringe is coincident with a given point on the plane. The directions of the principal stress are determined by inclination of the axes of the polarization of the polarizer and the analyzer.[1]

The isoclinic fringes in Fig. 4 of sample A are always two cross dark bands, which is different to other samples (e.g. isoclinic fringes in Fig. 8(a)).

Table 1.   Various light sources.

No.	color	wavelength/nm	light source
1	red	633	He-Ne laser
2	yellow	589	sodium lamp
3	green	532	laser pointer
4	blue	460~474	LED
5	white	NULL	sun light

LED light bulbs are not ideal monochromatic.

This implies that the internal stresses of sample A are centrosymmetric in another aspect.

Extinction also occurs if $\sin^2(\frac{\delta_1-\delta_2}{2}) = 0$ which requires

$$\delta_1 - \delta_2 = \frac{2\pi d}{\lambda} \cdot C(\sigma_1 - \sigma_2) = k\pi \qquad (14)$$

and particular wavelengths satisfies Eq. (14) locally.

These patterns in sequence of light wavelength are called *isochromatic*. For monochromatic light source, the isochromatic fringe pattern appears as a series of dark and bright bands (see Fig. 5). For white light source, the isochromatic fringes are colored, since the difference of principal stresses generally produces extinction only for a particular wave length.

Then we replace the white light with monochromatic light sources (light source 1 ~ 4). The patterns with shorter wavelengths are denser to the same sample in Fig. 5.

This is reasonable because the light with shorter wavelength can experience more periods within certain optic path.

Besides, when we rotate either polarizer or analyzer, only light intensity changes with monochromatic light source. The patterns dim and bright twice in a round. If we use white light source, complementary colors appear when two polaroids are perpendicular or parallel. However, the patterns' shapes are independent of rotation or color of the light sources.

In Fig. 6 sample A shows complementary colors when two polaroids are vertical and parallel. This two conditions correspond to $\alpha - \beta = \frac{\pi}{2}$ or $\alpha - \beta = 0$ in Eq. (13):

$$I = \begin{cases} A^2 \sin^2 \alpha \, \sin^2(\frac{\delta_1-\delta_2}{2}) & \alpha - \beta = \frac{\pi}{2} \\ A^2 - A^2 \sin^2 \alpha \, \sin^2(\frac{\delta_1-\delta_2}{2}) & \alpha - \beta = 0. \end{cases} \qquad (15)$$

The summary of the intensities is just A^2.

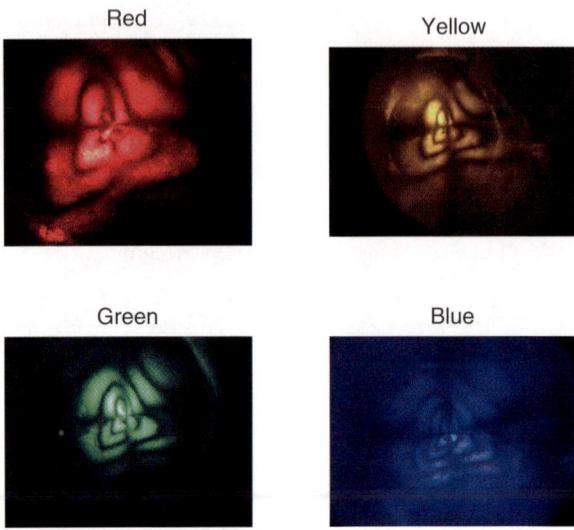

Fig. 5.　Sample A: Patterns become denser as wavelength becomes shorter. The cross black bands are isoclinic patterns while the irregular black bands belong to isochromatic patterns.

From the roughly symmetric pattern, we can deduce that the internal stress in sample A is nearly centrosymmetric which may be caused by the centrosymmetric molding process. But for sample B (see Fig. 7) the pattern is nonsymmetrical and much more complicated, revealing a much more complicated stress distribution.

2.2.2. *Photoelastic Pattern under Applied Load*

From Eq. (14), we see $(\delta_1 - \delta_2) \propto (\sigma_1 - \sigma_2)$. The lines of the same principle stresses' difference coincide with the isochromatic fringes. Therefore, the interference patterns can be derived from the internal stress distribution. However, the stress distribution, in most cases, needs numerical calculations except for some simple symmetrical problems.

ϕ *Ariy stress function*[4] is usually used to calculate the stress distribution. It is a special case of the Maxwell stress functions and is widely applied in stress calculation for two-dimensional problems. According to

(a) Two vertical polaroids.   (b) Two parallel polaroids.

Fig. 6. Sample A: Complementary colors appear when two polaroids are perpendicular or parallel. Two cross black bands in Fig. 6(a) are isoclinic patterns and they always exist when two polaroids rotate together (see Fig. 4).

materials mechanics we can get:

$$\sigma_y = \frac{\partial^2 \phi}{\partial x^2}$$

$$\sigma_x = \frac{\partial^2 \phi}{\partial y^2} \tag{16}$$

$$\tau_{xy} = -\frac{\partial^2 \phi}{\partial x \partial y}.$$

The above equations can be reduced to a fourth degree, biharmonic, partial differential equation:

$$\nabla^4 \phi = 0. \tag{17}$$

Thus, the solution of a 2D problem reduces to finding a solution for Eq. (17). Any stress function which satisfies this equation is a valid stress function from which stresses, strains, and displacements can be calculated. The stresses must also satisfy the boundary conditions for any particular problem. The concentrated load on the ring can be expressed by the *Gaussian function* with the form of $e^{-\lambda x^2}$ and the λ describes the distribution of the load.

The black lines in Fig. 8(a) are isoclinic fringes and the colored bands are isochromatic fringes. Fig. 8(b) shows the distribution of isochromatic fringes from calculation which is the same as Fig. 8(a).

Fig. 7. Sample B: Patterns under sunlight with perpendicular polaroids. The bands with the same color are isochromatic fringes. There is no evident isoclinic pattern here which indicates that the principle stresses' directions in Sample B is different with polarizer and analyzer's orientations.

3. Hidden Polaroids

3.1. *Isotropic Fresnel Equation*

In Sec. 2 we have studied and explained how colored patterns are related to the properties of different light sources and specimens. With standard experiment system (polariscope), the interference condition is well satisfied. The first polaroid (polarizer) makes the incident light linearly polarized. The second polaroid (analyzer) recombines the two light components which are separated by the phase retardation.

However standard experiment system is artificial and laboratory based. The marvelous fact is that *plastic with colors* can be observed in everyday life without polaroids under both transmission and reflection conditions, and the phenomenon is more apparent under reflection condition. Why? We have comprehended the significance of polaroids so far. It is nature to consider some mechanism acting as hidden polaroids.[5]

The natural light can be partial polarized after scattering by particles in atmosphere. For isotropic particles, the forced vibration of induced dipole moment radiates subwaves. Scattered light is polarized in the direction perpendicular to the original light, partially polarized in other directions. For anisotropic particles, the direction of induced dipole moment and electric

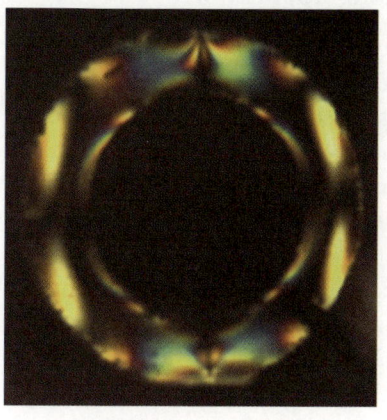

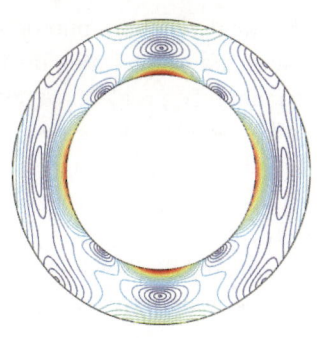

(a) Experiment result.   (b) Theory of $\sigma_1 - \sigma_2$.

Fig. 8. Loaded ring. Fig. 8(b) is the distribution of $\sigma_1 - \sigma_2$, which is the same to the distribution of isochromatic fringes. The extra black bands in experiment result (Fig. 8(a)) is isoclinic patterns which is related with the orientation of polaroids.

vector of incident light can be different and the subwaves are usually partial polarized.[3] A linear polaroid can be used to check the partial polarization state of the nature light for a clear blue sky.[6]

Fresnel equations[7] can help understand why the phenomenon is more apparent under reflection. For non-magnetic isotropic media:

$$R_s = \left| \frac{n_1 \cos\theta_i - n_2 \cos\theta_t}{n_1 \cos\theta_i + n_2 \cos\theta_t} \right|^2$$

$$R_p = \left| \frac{n_1 \cos\theta_t - n_2 \cos\theta_i}{n_1 \cos\theta_t + n_2 \cos\theta_i} \right|^2 \qquad (18)$$

$$T_s = 1 - R_s$$

$$T_p = 1 - R_p.$$

Polarization state can be enhanced by reflection, for example, at Brewster's angle the reflected light is totally polarized and the transmitted light partially polarized. In application to birefringence material, Fresnel equations need to be revised, but it helps us understand that light will be "analyzed" when penetrating through the interface every time.

Actually the Fresnel equation can be directly utilized when incident light is S or P polarized. The reflection coefficient of nature light is:

$$R = \frac{R_s + R_p}{2}. \qquad (19)$$

3.2. *Anisotropic Fresnel Equations and Application*

So far, the analysis about photoelastic material can be analogous to uniaxial crystal with its optical axis on the surface. Refractive index ellipsoid is useful to simplify problem. Arbitrary section containing center point is a ellipse. The lengths of its two principle axes can represent the refractive indices of light propagating to the normal direction of this section. The orientations of the principle axes are parallel to the dielectric displacements separately.

Take a uniaxial crystal with its optic axis on the surface as an example. The angle between the optics axis and the incidence plane is γ. The incident angle is θ_i and the refractive angles are θ_1 and θ_2, which are determined by ϵ_o, ϵ_e and the ϵ_0 of air. The relation[7] of electric displacement field $\mathbf{D}$ and electric field $\mathbf{E}$ in the crystal is:

$$\mathbf{D} = \vec{\vec{\epsilon}} \cdot \mathbf{E}. \tag{20}$$

$\vec{\vec{\epsilon}}$ is a tensor here. Then we can calculate the polarization and intensity of refraction light. Optical path difference of two transmission lights is:

$$\Delta = (n_1 \cos\theta_1 - n_2 \cos\theta_2) \cdot d, \tag{21}$$

where d is the thickness of specimen.

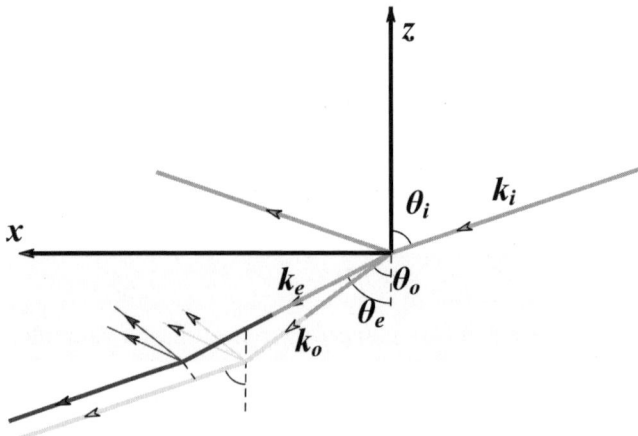

Fig. 9.   Propagation of light in anisotropic materials. Birefraction and bireflection must be both taken into consideration.

Due to the Maxwell equations at boundary conditions, we can finally get the relation of the colors and incident angles with Matlab p-polarization and s-polarization. And the condition of nature light is the average of them analogous to Eq. 19.

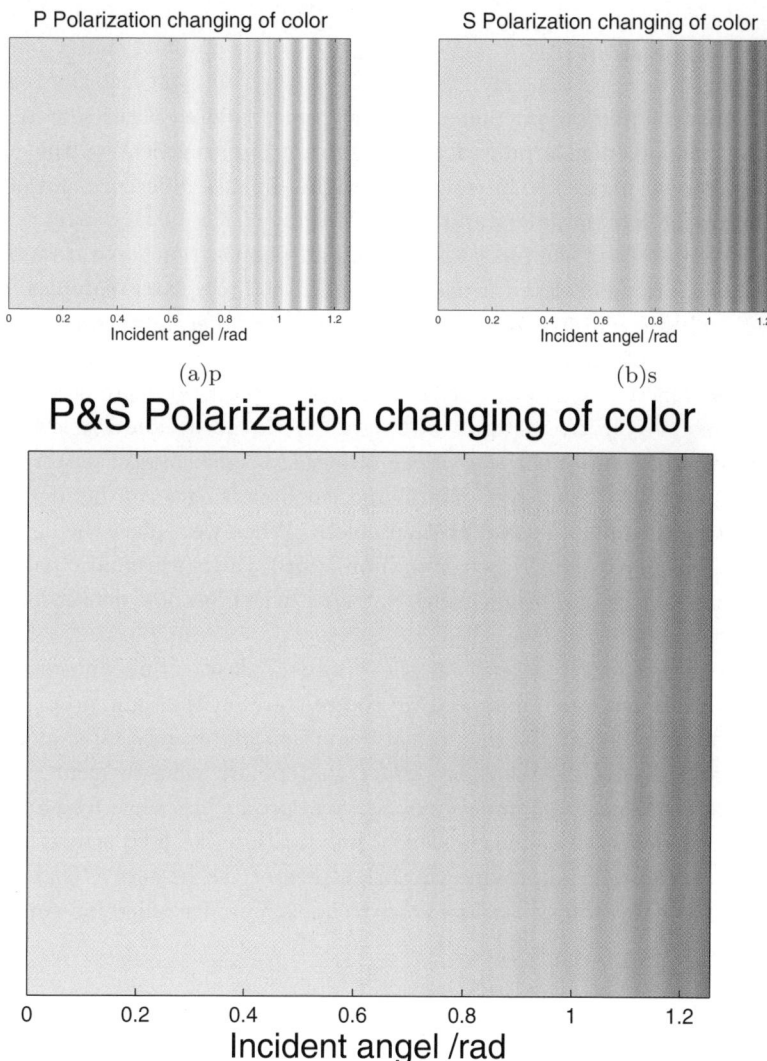

Fig. 10. Color of uniaxial crystal in different incident angle under sun light. $\gamma = \frac{\pi}{4}, d = 2mm, n_o = 1.5, n_e = 1.51$ and the light source intensity is unit here.

Fig. 10 indicates that polarization of light source is unnecessary. There is a defect that only three colors (red green blue) are utilized to simulate the nature light's continuous spectrum. The result can simulate the experiment better with more colors.

4. Conclusion

We investigated the phenomenon concerning "colored plastics" extensively. The phenomenon is related to the "stress birefringence" of the polymer transparent materials, in which the origin of phase difference and polarization is the key to understand the problem.

We discussed the phase difference and polarization separately. First of all, we obtained colored fringe patterns with three plastic samples: sample A, a plastic plate (a piece of candy box) with nearly symmetry residual stress; sample B, a plastic triangular rule with irregular residual stresses; and sample C, a photoelastic ring. Various light sources are used including 4 monochromatic ones and white light. With the standard polariscope set up and the white light source, we observed bright colorful patterns for all three samples. The isochromatic and isoclinic fringes can be identified by rotating the polaroids and by their colors. When we replace the light with a monochromatic light (laser or sodium lamp), the isochromatic patterns all appear as dark and bright fringe patterns, which become denser for shorter wavelengths.

Then, in order to explain the "colored plastics" phenomenon in everyday life without polaroids, we concentrate on the light propagation in anisotropic media. By theoretical deduction and numerical simulation on uniaxial crystal 2D plate, we prove that polarization in front or behind the birefringence sample is unnecessary in producing colors for non-vertical light incidence. Any mechanism of polarization like light scattering or reflection can help improving the brightness of the pattern. That explains why CD cases shine in colors more frequently under reflection condition.

References

1. A. S. Khan and X. Wang, *Strain measurement and stress analysis*. VCH Publisherss,New York (2001).
2. J. W. Phillips, *Experimental Stress Analysis*. Univ. of Illinois at Urbana-Champaign (1998).
3. K. Gåsvik, *Optical Metrology*. Wiley (2002). ISBN 9780470843000.
4. F. P.Beer, J. E.Russell Johnston, and J. T.DeWolf, *Mechanics of Materials*, 3th edn. Tsinghua University Press (2003).

5. D. W. H. M. M. Bond, Photoelasticity without polaroids, *Phys. Educ.* **9**(411) (1974).

6. G. S. Smith, The polarization of skylight: An example from nature, *Am. J. Phys.* (2007).

7. E. W. Max Born, *Principles of Optics*. Cambridge University Press (1999).

Chapter 12

2013 Problem 8: Jet and Film

Pei Zeng*, Lan Chen, Kejing Zhu

School of Physics, Nanjing University

In this article, we investigate the interaction between the water jet and soap film under different jet speeds and incident angles. We consider two different phenomena– penetrating and non-penetrating, and their corresponding conditions. In the case of penetration, we seek for the relationship between the parameters of incident jet and emergent jet, calculate the shape of the film under specific occasions. In the case of non-penetration the jet may adhere to the surface of the film or bounce off the film several times. Depending on the incident angle and velocity of the jet, the film will be found in stable and unstable patterns. We calculate the shape of the jet and the film under different conditions and found the patterns in experimental observations. Finally we portrait a 'phase diagram' illustrating the conditions for different forms of jet and film interaction.

1. Introduction

Problem Statement:

"A thin liquid jet impacts on a soap film (see Fig. 1). Depending on relevant parameters, the jet can either penetrate through the film or merge with it, producing interesting shapes. Explain and investigate this interaction and the resulting shapes."

The interaction between water jet and film is common in daily life. To produce the phenomena "penetration and mergence" mentioned by the problem, we will try different parameters like the incident velocity and angle of the jet.(See Fig. 2 and Fig. 3)

In the case of penetration, the shape of a soap film and the relationship between incident angle and emergent angle are investigated. By momentum

---

*Peizeng@live.com

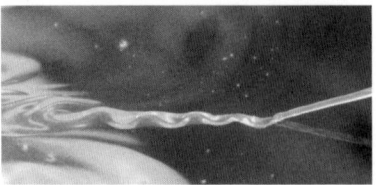

Fig. 1.   Jet and film interaction.

and flux conservation and force equations, we will figure out the emergent angle and discuss the condition at which the "refractive index" exist.

For a non-penetration interaction, the jet may adhere to the surface of the film (merge with the film as mentioned by the problem); or bounce off the film several times, exhibiting another interesting form of interaction. Depending on the incident angle and velocity of the jet, the film will be found in stable and unstable patterns. We will calculate the shape of the jet and the film under different conditions and found the patterns in experimental observations.

The critical condition is found to classify the cases of penetration and non-penetration in terms of incident angle and velocity of the jet.

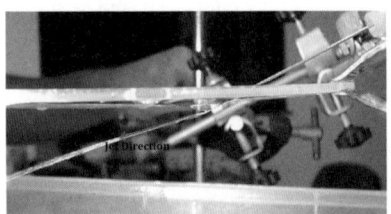

Fig. 2. Penetration.

Fig. 3. Mergence.

2. Experiment

In our student laboratory, a 'Jet-and-Film' device is constructed as shown in Fig. 4. To produce stable and elastic film, the liquid sample is prepared by mixing 10g of detergent with a liter of pure water. The liquid surface tension coefficient is measured as $0.29 \times 10^{-2} N/m$.

The jet velocity can be taken as stable over long time. The flux of the jet is measured and verified to be uniform by a measuring cup and timer.

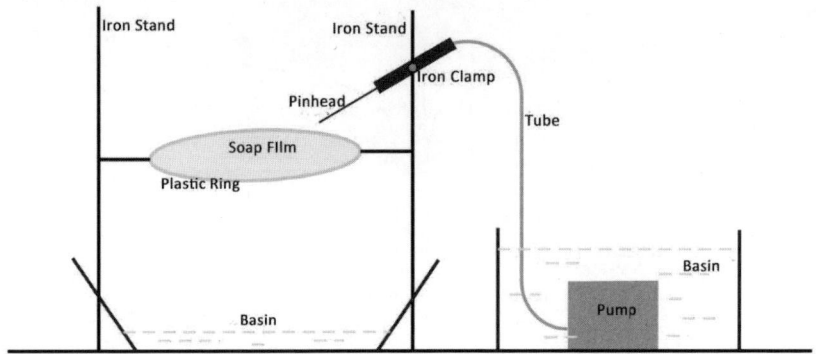

Fig. 4.   Experiment Setup.

As the diameter of the pin is given, the velocity of the jet is easily controlled and calculated.

2.1. *Penetration*

Penetration will happen when the incident angle is adequately small and the velocity is large enough. We'll first analyze the specific case of vertical injection, then extend the situation to the general condition. Finally, we'll discuss the validity of the so-called 'refraction index'.

2.1.1. *Vertical Injection*

When vertically injected, the jet will reduce its kinetic energy during penetration, thus become thicker according to the conservation of flux as illustrated in Fig. 5.

Analyze the force on a small segment of the jet through the film, see Fig. 5. Denote F_R as the force applied by the film, F_i the force applied by the incident part of the jet (above the film), F_r the force exerted by the emergent part (below the film); σ as the coefficient of surface tension, R_i and R_r as the incident and emergent radius of the jet, v_i and v_r as the

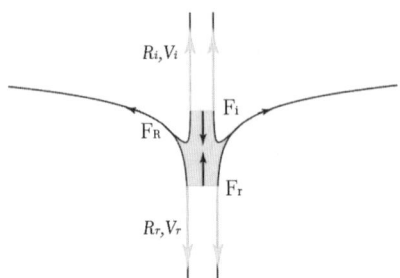

Fig. 5.   Vertical Injection.

incident and emergent velocity of the jet, p_i and p_r the inner pressure of
the jet, ρ the density of the liquid. Accordingly we can write:

$$F_R = 4\pi R_r \sigma; F_r = 2\pi R_r \sigma - \pi R_r^2 p_r; F_i = 2\pi R_i \sigma - \pi R_i^2 p_i$$

$$\pi R_r^2 p_r = \pi R_r \sigma, \pi R_i^2 p_i = \pi R_i \sigma.$$

According to the momentum theorem:

$$F_R - F_r + F_i = \pi R_i^2 \rho v_i^2 - \pi R_r^2 \rho v_r^2.$$

Thus we have:

$$3\pi R_r \sigma + \pi R_i \sigma = \pi R_i^2 \rho v_i^2 - \pi R_r^2 \rho v_r^2. \tag{1}$$

Moreover, by the continuity equation of fluid:

$$\pi R_i^2 v_i = \pi R_r^2 v_r. \tag{2}$$

There are four variables in Equation 1 and 2 -- R_i, R_r, v_i, and v_r. Given
the incident velocity and radius, one can figure out the emergent values.

Now we try to solve the shape of the film during a jet penetration.
Establish a cylindrical coordinate frame (r, ϕ, z) to describe the shape of
the film. Ignoring the mass of the film, and divide the film into circular
areas, using the force balance supplied by the surface tension on the top
and lower sides of the film segment, we have

$$2\pi \sigma r_0 = 2\pi \sigma r \sin \theta.$$

Obviously,

$$\sin \theta = \frac{z'}{\sqrt{1 + z'^2}}.$$

Thus we can obtain:

$$z(r) = r_0 \ln\left(\frac{r}{r_0} + \sqrt{\frac{r^2}{r_0^2} - 1}\right). \tag{3}$$

The result is shown in Fig. 6. By recording the side view of a film, and recording the data along the curve by analyzing the video using *Tracker*, as shown in Fig. 7, the shape of the film can be measured in experiment, which matches well with our theoretical result.

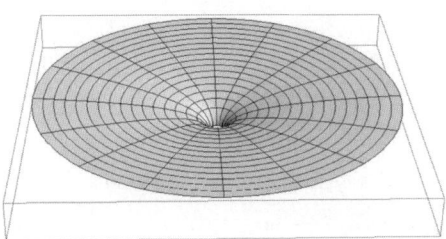

Fig. 6. Theoretical result of film shape in vertical injection.

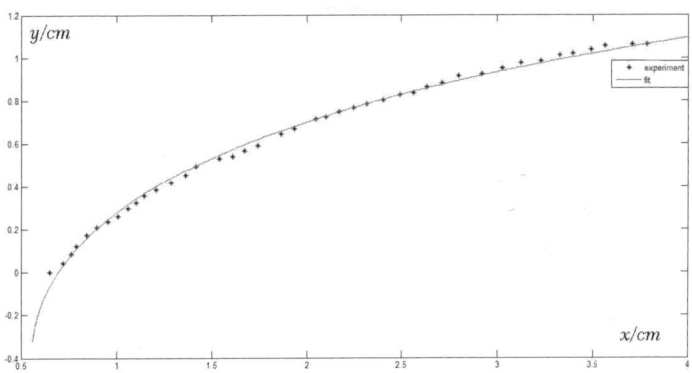

Fig. 7. Experimental result of film shape in vertical injection.

2.1.2. *Inclined Injection*

As Fig. 8 shows, it is more complicated when the jet is injected with an angle θ_i. Like the analysis above, we can write the momentum theorem and the fluid continuous equation:

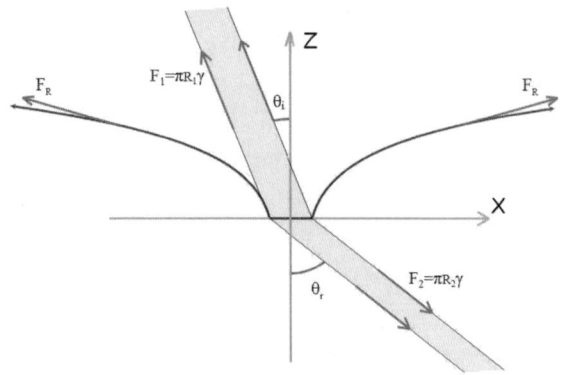

Fig. 8.   Inclined injection.

$$\pi R_i^2 v_i = \pi R_r^2 v_r$$

$$\rho \pi R_r^2 v_r^2 \sin \theta_r - \rho \pi R_i^2 v_i^2 \sin \theta_i = F_r \sin \theta_r - F_i \sin \theta_i \qquad (4)$$

$$\rho \pi R_r^2 v_r^2 \cos \theta_r - \rho \pi R_i^2 v_i^2 \cos \theta_i = F_r \cos \theta_r - F_i \cos \theta_i - F_R. \qquad (5)$$

A dimensionless number in fluid dynamics – the Weber number is defined as:

$$We = (\rho v_i^2 R_i)/\sigma. \qquad (6)$$

From Eqn. (4), (5) and (6), we can obtain:

$$(We - 1)\sin(\theta_r - \theta_i) = F_R \frac{\sin \theta_r}{\pi \sigma R_i} \qquad (7)$$

where F_R is the force applied by the film. An approximate expression for F_R is: [2]

$$F_R = \frac{4\sigma \pi R_i}{\cos \theta_i}. \qquad (8)$$

Eqn. (7) and Eqn. (8) describe the relationship between the incident angle and emergent angle.

To solve the shape of the film, two conditions must be satisfied: the contact position of film and jet ought to have the same tangential as the velocity of jet;and the surface area should be minimized.(Fig. 9)

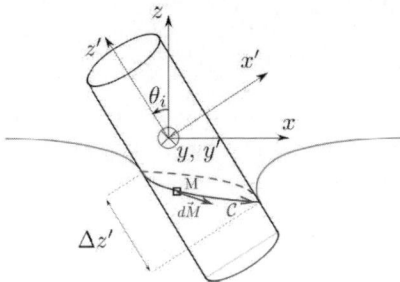

Fig. 9. A simplified model to calculate penetrate force applied by the film(from reference(2))

Ignoring the mass of film and the substance exchange between jet and film, simulation can be made as shown in Fig. 10 and compared to experimental result (Fig. 11). The simulation in this article is done using finite element method, programmed independently in C++ and *Mathematica*. Jet in the simulation is divided into tiny droplets with radius $5 * 10^{-4}m$. The flux is set as $1 * 10^{-5}s$ per droplet. The droplets' motion without the film is simplified to parabolic motion.

Fig. 10. The simulation of inclined injection.

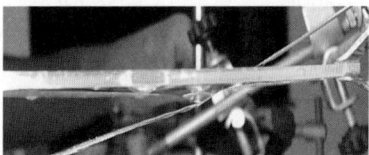

Fig. 11. The experiment of inclined injection.

2.1.3. *The Validity of 'Refraction Index'*

The 'refraction index' means that the incident and emergent angle obey the law of refraction. From Eqn.7 and 8, we cannot find the refraction law in terms of 'refraction index'.

Nevertheless, provided that the angles are sufficiently small, we may obtain:

$$\sin\theta_i \approx \theta_i; \sin\theta_r = n\sin\theta_i; \cos\theta_i \approx \cos\theta_r \approx 1$$

$$F_R \approx 4\pi\sigma R_i.$$

Then we can figure out:

$$n = \frac{We - 1}{We - 5}.$$ (9)

which is valid when the angle is small enough. While the angle is relatively large, this formula has lost its accuracy. In that case, the concept of 're-fraction index' will be meaningless.

2.2. Non-Penetration

For a non-penetration interaction, the jet may jump off the film or experience a wavelike motion. The type of motion depends on the condition of the incident jet. We will consider some special conditions according to incident angle and speed. In the end, an interesting phenomenon "snake walk" will be discussed.

2.2.1. Wavelike Motion-Low Speed Large Angle Injection

At large incident angle with low injection speed, the jet experience a wave-like motion. It's not easy for us to measure the wavelength and amplitude of the wave accurately, thus we merely explain it qualitatively.

The wavelike motion is formed as the force applied by the film is positively related to the vertical displacement. The deeper the jet sinks into the film, the more deformation occurs on the film. As the contact area becomes larger, a larger surface tension is applied to the jet. A restoring force is formed in this way. Meanwhile there are also other factors contribute to the amplitude of wavelike motion. The viscous force both in tangential and normal direction reduces the amplitude of the wave. On the contrary, as the film sinks most the center, the gravitational potential energy compensates the energy lose and tends to increase the wave amplitude.

In our experimental observations in Fig. 12, the amplitude of vibration is getting larger, which illustrates that gravity here contributes more than the viscous effects.

2.2.2. Arc-like Motion-High Speed, Large Angle Injection

In this case, the jet will jump off the film. Firstly, a simulation has been carried out. Comparing the simulation result (Fig. 13) with the experiment (Fig. 14), we find that the contact area of jet and film takes an arc-like

Fig. 12.   Wavelike motion-low speed large angle injection.

shape. Thus we suppose that the shape is an arc and the velocity remains unchanged. The jet segment moves in circular motion, with the centripetal force provided by the film and the segments beside.

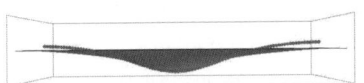

Fig. 13. The simulation result of arc-like motion.

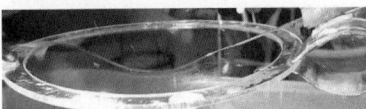

Fig. 14. The experiment result of arc-like motion.

We can make the force analysis, (see Fig. 15)

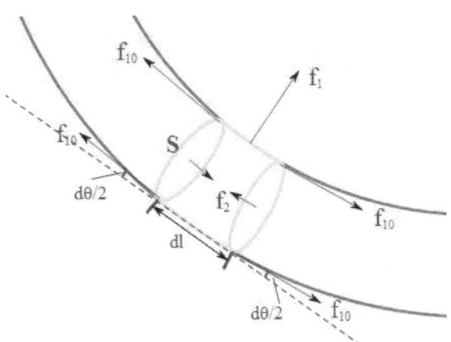

Fig. 15.   Force analysis of high speed jet with large incident angle.

The contact force by adjacent liquid segment is:

$$f_1 = \pi R_i \sigma d\theta.$$

And the surface tension applied by the film is:[2]

$$f_2 = k\sigma \cdot dl (k = 0 \sim 4).$$

The combined centripetal acceleration can be expressed as:

$$f_n = \rho \pi R_i^2 dl \cdot \frac{v_i^2}{R}. \tag{10}$$

Therefore,

$$f_n = f_1 + f_2 \Leftrightarrow \rho \pi R_i^2 dl \cdot \frac{v_i^2}{R} = \pi R_i \sigma d\theta + k\sigma \cdot dl$$

With $R = dl/d\theta$, we obtain:

$$R = \frac{\pi R_i \cdot (We - 1)}{k} (k = 0 \sim 4)$$

$$\lambda \approx 4R \cos \theta_i = \frac{4\pi R_i (We - 1)}{k} \cos \theta_i \tag{11}$$

where λ is the equivalent wavelength of the arc. In Fig. 16, the measured the value of λ in experiment at different incident angles is shown as squares. Our theoretical result according to Eqn.11 shown in solid line is in good agreement with the experimental data.

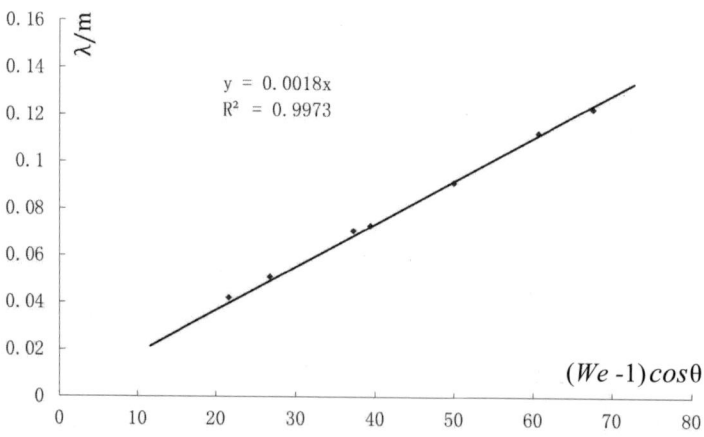

Fig. 16. Equivalent wavelength λ at different incident angles for high speed jet with large incident angle.

2.2.3. *Multiple Bouncing-Low Speed and Smaller Incident Angle*

When we reduce the speed and incident angle gradually, the jet may bounces many times as shown in Fig. 18. Quantitative experiment is hard under this occasion, so we made a simulation (Fig. 17) which resembles the experiment phenomenon (Fig. 18).

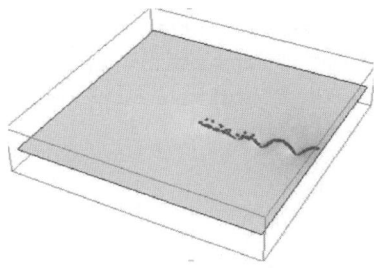

Fig. 17. The simulation result of jet with low speed and smaller incident angle.

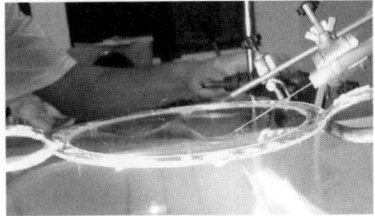

Fig. 18. The experiment result of jet with low speed and smaller incident angle.

2.2.4. *Snake Walk*

An interesting phenomenon called "snake walk" is a'transverse vibration' as Fig. 19 shows, which is rarely paid attention to. A qualitative explanation is that fluid of the film sinks and rotates under the interaction of jet when the jet doesn't go through the center of film, which in return impacts the motion, especially the vibration of the jet.(See Fig. 20)

Fig. 19. An interesting phenomenon— "snake walk".

Fig. 20. The rotation and sink of film contribute to "snake walk".

2.3. *Critical condition*

The critical condition that determines whether penetration occurs or not is found by setting the emergent angle $\theta_r = 90°$. From Eqs. 7 and 8, and set the emergent angle $\theta_r = 90°$, the critical incident angle satisfies

$$We = \frac{4}{\cos^2 \theta_i} + 1. \tag{12}$$

According to Eq. 12, the relation between $(Wb - 1)$ and $Cos^{-2}\theta$ is linear, as shown in Fig. 21. Again the experiment data (diamonds in Fig. 21) are found in good agreement in comparison to theoretical value(the straight line).

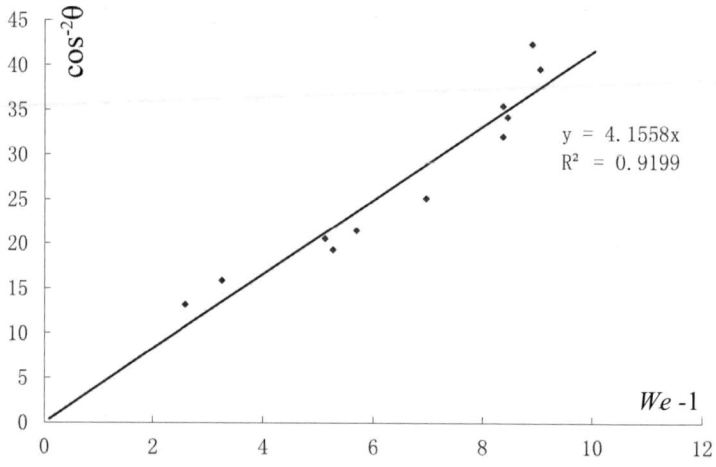

Fig. 21. Experiment and theoretical results of critical condition.

From Equation 12, we obtain:

$$v = \sqrt{\frac{\sigma(\frac{4}{\cos^2 \theta} + 1)}{\rho R_i}}. \tag{13}$$

Thus we can plot a 'Phase Diagram' of the jet and film interaction with respect to injection angle and velocity, as shown in Fig. 22. Although a whole grasp of jet and film interaction are rather complicated, investigations on the special cases help us understand the essential and construct a relatively simplified picture.

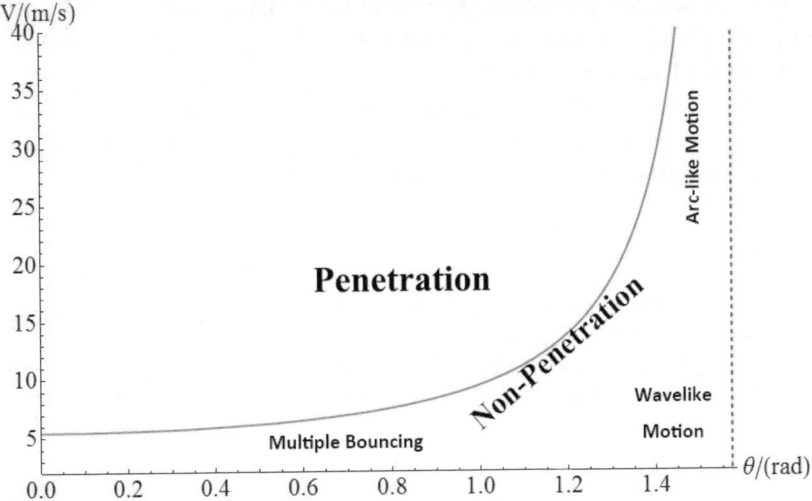

Fig. 22. 'Phase Diagram' that summarizes the conditions and corresponding jet and film interactions.

3. Conclusion

We have investigated the interaction between the water jet and soap film under different jet speeds and incident angles. In our investigation, there are two main different phenomena – penetration and non- penetration.

In the case of penetration, we firstly analyze the vertical injection and obtain Eqn.1 and Eqn.2, which reflects the relation of jet's radius R and the velocity v. Then we theoretically derive the shape of film (Eqn. 3) and verify it experimentally. Similarly, we deduce the formula in the inclined case, and obtain the relationship between incident angle and emergent angle of the jet. As for the shape of film in inclined injection, we made a simulation which is in accord with the experimental result.(Fig. 10 and Fig. 11). The validity of 'refraction index' is discussed and found effective at small incident angles.

For a non-penetration interaction, we discuss three extreme occasions: wave-like motion when incident angle θ is large and jet velocity v is small enough; arc-like motion (when both θ and v are large enough); and multiple bouncing when both θ and v are smaller. The wave-like motion and multiple bouncing are qualitatively explained and a simulation result was given for the latter occasion. As for the arc-like motion, we proposed a simplified model and a equation (Eqn. 11) to describe the relationship between the wavelength and the properties of jet and film, which is supported by experimental data. Then we describe an interesting 'snake walk' phenomenon with a tentative explanation.

Finally, we derive the critical condition to classify the penetration and and non-penetration interaction. A 'Phase Diagram' is plotted to summarize the conditions for different jet and film interactions.

References

1. Kirstetter G, Raufaste C, Celestini F. Jet impact on a soap film[J]. Physical Review E, 2012, 86(3): 036303.
2. Raufaste C, Kirstetter G, Celestini F, et al. Deformation of a free interface pierced by a tilted cylinder[J]. EPL (Europhysics Letters), 2012, 99(2): 24001.

Chapter 13

2013 Problem 12: Helmholtz Carousel

Jia'an Qi, Zhihang Qin, Sihui Wang and Huijun Zhou

Nanjing University, School of Physics

In our solution, we try to investigate the origin of the force that propels the Helmholtz Carousel. We build a theoretical model from the fundamental hydrodynamics. As the propelling force is in fact a *nonlinear effect*, we examine the nonlinear terms with care. The nonlinear terms in our equation corresponding to two origins of the force: one origins from the nonlinearity of the impedance, the other origins from nonlinear "restoring force". We will prove that the true origin of the force effect is the nonlinear impedance, which resembles the recoil force of a balloon that breathes air in and out unsymmetrically in a period of motion. Modification to linear impedance is also made which is important in giving a correct resonant frequency. Finally, we optimize the Helmholtz carousel according to the predictions of our model. Among the parameters we investigated, specific frequency, neck length, neck radius values exist to optimize the Helmholtz carousel, while larger resonator volume produces larger driving force. Predictions based on our model are supported by experimental results.

1. Introduction

Attach Christmas tree balls on a low friction mounting (carousel) such that the hole in each ball points in a tangential direction. If you expose this arrangement to sound of a suitable frequency and intensity, the carousel starts to rotate. Explain this phenomenon and investigate the parameters that result in the maximum rotation speed of the carousel.

The problem introduces a rather fascinating phenomenon: the rotation of Christmas tree balls under the impact of external sound. Intrigued by the promise of moving things around only by playing suitable "sonata", we design a bearing bridge as the carousel. We place two identical Christmas tree balls symmetrically on each arm of the bridge and place it near two opposing loudspeakers (see Fig. 1). By adjusting the frequency and intensity of the sound, we find that rotation is possible only under a certain range of frequencies, at around 200Hz for our Christmas balls. The Christmas ball or similar structures with a cavity and a "neck" combined together is called *Helmholtz resonator*.

Fig. 1.   The "Helmholtz Carousel" we built.

A common model of Helmholtz resonator is to treat the air in the neck as a block, the air in the cavity as a spring that provides restoring force, and that the air friction as the damping. Thus the device can be reduced to a spring-mass-damp system. When the frequency of driving force, i.e. the sound frequency, coincides the nature frequency of the system, resonance occurs and the amplitude of the oscillation reaches its maxima.

However, this model is oversimplified when we try to explain the rotation of the Helmholtz Carousel. First of all, the net force over an oscillation period is zero for a harmonic oscillation. Secondly, the amplitude response over frequencies cannot be predicted since the damping is unknown.

In our solution, we try to investigate the origin of the force that propels the Helmholtz Carousel. We build a theoretical model and the equation of motion from the very fundamental hydrodynamics. The equation contains a term that describes the effect of nonlinear restoring force that originates from the adiabatic compression(expansion) of air inside the cavity. This term gives rise to asymmetric solution to the equation of motion and a resultant propelling force,[1] however fails to give a correct magnitude of the force effect. In our equation, we examine the effect of nonlinear impedance with care, and modify the corresponding term. We will prove that the true origin of the force effect is the nonlinear impedance, which resembles the recoil force of a balloon that breathes air in and out unsymmetrically in a period of motion.

With a clear understanding based on our theoretical model, we can optimize the rotation velocity of the "carousel". Relevant parameters are classified as:

- the frequency of external sound f;
- the sound intensity, or the sound pressure level SPL;
- the geometry of the Christmas tree ball, including the neck length L, volume of the cavity V_0, and the radius of the neck a;
- the friction of the mounting device and air.

Systematic experiments will be done to explore the optimal conditions. The results are found consist with the predictions of our theoretical model.

2. Pre-Experiments

Instead of measuring the rotation velocity of the Christmas balls, which is apparently imprecise and clumsy, we use an electric scale to measure the propulsion force exerted on the Helmholtz resonator under different sound frequencies by measuring the apparent additional mass. The accuracy of the scale is $5 \times 10^{-3} g$.

We measure the force-frequency response of a spherical resonator, as shown in Fig. 2. The geometry of a spherical cavity (diameter $8\,cm$ and neck length $4.3\,cm$) is similar to that of a Christmas tree ball. The frequency force response is tested over a large range of frequencies. The force is apparent near the natural frequency around $200\,Hz$. The same effect is also observed in the rotation of Helmholtz Carousel, in which the rotating speed achieves maximum at the nature frequency. The rotation speed soon drops down to zero as the sound frequency shifts away by a few Hertz.

3. Model

In the following investigation, we consider the structure of a Helmholtz resonator as shown in Fig. 3. The model contains a spherical cavity with a cylindrical neck as required by the problem. Denote the cross-section

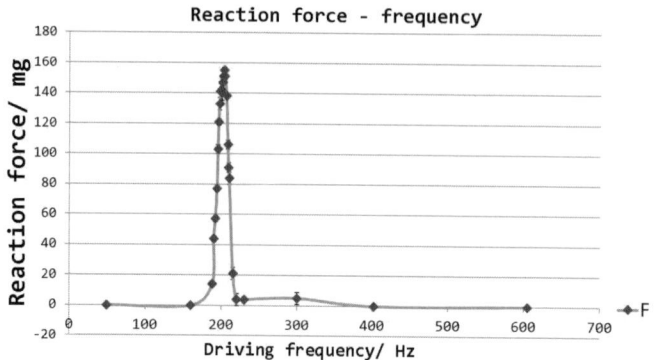

Fig. 2.   The measured propulsion force exerted on the Helmholtz resonator as a function of sound frequencies.

area as A, length of the tube as L, volume as V, the internal and external pressure as p_{in} and p_{out} respectively.

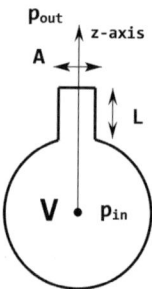

Fig. 3.   The structure of a typical Helmholtz resonator.

3.1. *Theoretical Model and First Order Approximation*

Consider the momentum equation for an infinitesimal volume element of air, we have:

$$\rho\frac{d\mathbf{u}}{dt} = \rho\mathbf{f} + \nabla \cdot \mathbf{P} \tag{1}$$

where ρ is the density of air, $\mathbf{u}$ is the venosity vector, $\mathbf{f}$ is body force density and $\mathbf{P}$ is the stress tensor.[2] We can further expand the equation under certain conditions. Given the fact that sound field is irrotational[3]

$\nabla \times \mathbf{u} = 0$, the body force is negligible so that $\mathbf{f} = 0$, and air is Newtonian fluid with dynamic viscosity is denoted here as μ, we have

$$\rho \left(\frac{\partial \mathbf{u}}{\partial t} + \nabla \frac{\mathbf{u}^2}{2} \right) = -\nabla p + \mu \nabla^2 \mathbf{u} + \frac{1}{3} \mu \nabla (\nabla \cdot \mathbf{u}). \qquad (2)$$

Integrate Eq. 2 from the cavity to the outside along the the pipe, the length of integration path is $\tilde{L}$, which is a larger than the length of the pipe due to radiation effect.[4] As will be proved in section 3.2.1, $L \approx L + 1.7a$. Following Melling,[5] we denote $R\mathbf{u} = - \int_{in}^{out} \mu \nabla^2 \mathbf{u} \, ds$ as the linear impedance term, which will be discussed in detail in the appendices, we have

$$\frac{d\mathbf{u}}{dt} \cdot \tilde{L} + \frac{1}{2} \rho (u_{out}^2 - u_{in}^2) = p_{in} - p_{out} - R\mathbf{u} + \frac{1}{3} \mu \left(\nabla \cdot \mathbf{u}_{out} - \nabla \cdot \mathbf{u}_{in} \right). \qquad (3)$$

Define x as the average displacement over cross-section, so that $\dot{x} = u$. Use the adiabatic equation for the inner pressure, $p_{in} V^\gamma = p_0 V_0^\gamma$. The outer pressure under a harmonic sound signal is given by $p_{out} = p_0 + p_s \exp(-i\Omega t)$. Neglect the air compressibility outside the pipe $(\nabla \cdot \mathbf{u})_{out} = 0$. For the inner side, the quasi-static process ensures:

$$(\nabla \cdot \mathbf{u})_{in} \approx \frac{1}{V_0} \frac{dV}{dt} = \frac{Au}{V_0} \qquad (4)$$

where $A = \pi a^2$ is the cross-section area of the neck. With all these conditions, the vibration equation of a Helmholtz resonator can be reduced as:

$$\rho \tilde{L} \ddot{x} + \frac{1}{2} \rho \left(-u_{in}^2 + u_{out}^2 \right) + \left(R + \frac{A\mu}{3V} \right) \dot{x}$$

$$+ \frac{\gamma p_0 A}{V_0} \left(x - \frac{A(1+\gamma)}{2V_0} x^2 \right) = p_s e^{-i\Omega t}. \qquad (5)$$

Eq. 5 is the equation of motion describing the behavior of a Helmholtz resonator. The two undetermined terms in the equation R and $-u_{in}^2 + u_{out}^2$ are the linear and nonlinear impedance respectively. When x and $\dot{x}$ are both small, first order vibration equation can be obtained by neglecting the high order terms:

$$\rho \tilde{L} \ddot{x} + R' \dot{x} + \frac{\gamma p_0 A}{V_0} x = p_s e^{-i\Omega t}. \qquad (6)$$

Hence, the nature frequency of the system is:

$$f = \frac{c_0}{2\pi} \sqrt{\frac{A}{V_0 \tilde{L}}} \qquad (7)$$

where c_0 is the speed of sound.

To verify the relation of Eq. 7, we make our own resonators with different geometric parameters. The resonant frequencies versus volume are measured, see Fig. 4. The linear relation shows the validity of the first order approximation Eq. 7, with a relative uncertainty of less than 5%.

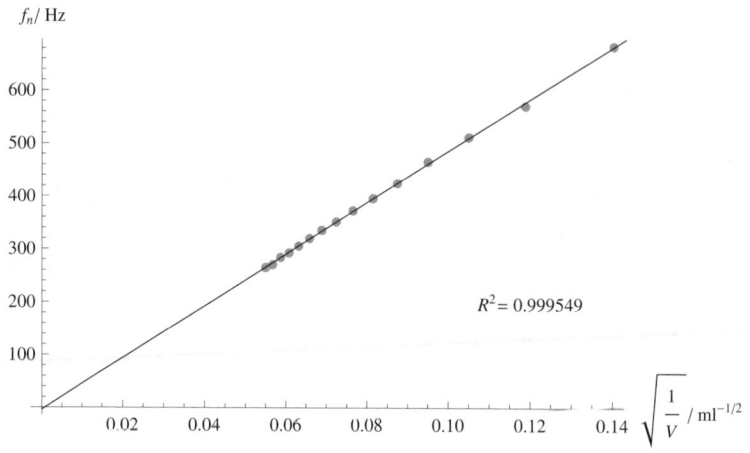

Fig. 4.   The resonant frequency versus $\sqrt{1/V_0}$.

3.2. Effect of the Impedances

In Eq. 5, the linear and nonlinear impedances R and $-u_{in}^2 + u_{out}^2$ have significant impact on the behavior of the resonator. In this section, we will examine these terms with care and later substitute them back into Eq. 5 to derive the full form of vibration equation.

3.2.1. *Linear Impedance: Modification to the Neck Length*

The impedance of a mechanical system is defined by dividing the driving force by the velocity, which in general takes a complex form:

$$Z = \frac{F}{u} = z_r + iz_i. \tag{8}$$

The linear impedance of a Helmholtz resonator origins from three parts: the air friction inside the neck, the inward radiation and the outward radiation at both ends of the neck. By analogy of the sound system to the electric circuit, the linear impedance of the resonator can be taken as three parts is "in series", the external part, the tube and internal part. We may calculate the three terms individually and get the total impedance by adding them up.

$$Z = z_{in} + z_{tube} + z_{out}. \tag{9}$$

From the deduction in appendices A and B, we have:

$$z_{in} \approx z_{out} = \rho_0 c_0 \pi a^2 \left(\frac{1}{2}(ka)^2 + i\frac{8ka}{3\pi} \right) \tag{10}$$

$$z_{tube} = i\omega\rho_0\pi a^2 L + \pi a L \sqrt{2\omega\rho_0\mu}(1 + i).$$

The radiation at both sides is approximately treated as the radiation of an acoustic piston embedded in an infinite large wall.

All the terms of linear impedance are acquired by assuming a harmonic oscillation, hence the time differentiation to a variable is to multiply it by an $i\omega$, time integration is to divide it by $i\omega$. So a linear vibration equation takes the form:

$$F_{ext} = (z_r + iz_i)u = z_r u + i\omega\frac{z_i}{\omega}u. \tag{11}$$

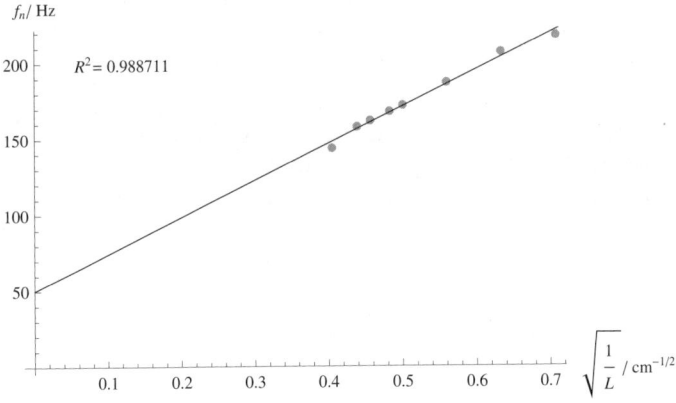

Fig. 5. The natural frequencies of a Helmholtz resonator with varying neck lengths. The non-zero interception in the vertical axis reveals the effect of additional mass.

One can see that the real part of the impedance z_r serves as a damping term, the imaginary part of the impedance, known as *reluctance*, contributes an additional mass z_i/ω to the vibration system.[3] This means an

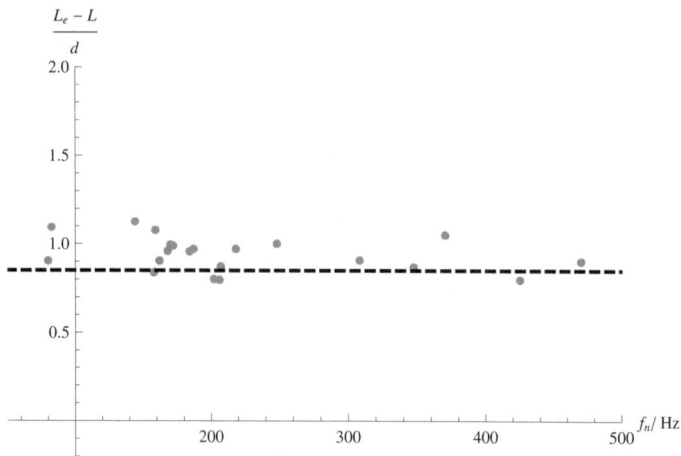

Fig. 6. The additional length is measured at different frequencies. The dashed line is the theoretical value, dots are experimental results.

additional mass is added to the mass inside the neck ρAL. Equivalently, we can modify the neck length by an additional value, as we have done in Eqs. 5 and 7, effective neck length $\widetilde{L} = L + L^*$ is used instead of L.

The effect of additional mass or additional neck length is supported by our experiment. We measure the natural frequencies of a group of Helmholtz resonators with varying neck lengths, see Fig. 5. The frequency doesn't approaches zero when $1/L \to 0$. The non-zero intercept verifies the correction of neck length $\widetilde{L} = L + L^*$.

By collecting the reactance calculated above, we can estimate neck length correction. The reactance due to air friction in the neck is much smaller than that of radiation effect. So the final result is:

$$L^* = \frac{8d}{3\pi} \approx 0.85d. \tag{12}$$

The validity of this result is also proved by our experiment. In figure 6, dots indicate the additional length measured at different frequencies, the dashed line is the theoretical value from Eq. 12.

3.2.2. *Nonlinear Impedance: Asymmetric Air Flow*

The propulsion force in a rocket motion is produced by the recoil force of the fuel flow, $\frac{1}{2}\rho u^2$. Similarly, the nonlinear impedance term $\frac{1}{2}\rho \left(-u_{in}^2 + u_{out}^2\right)$ in Eq. 5 will produce a net force during one period of the "breathing" for the Helmholtz resonator if the outward air flow is stronger or weaker than the inward airflow. The actual air flow distribution need to be described by a velocity field. But our goal here is to build a simplified model to express u_{out} and u_{in} in terms of the average speed $\dot{x}$.

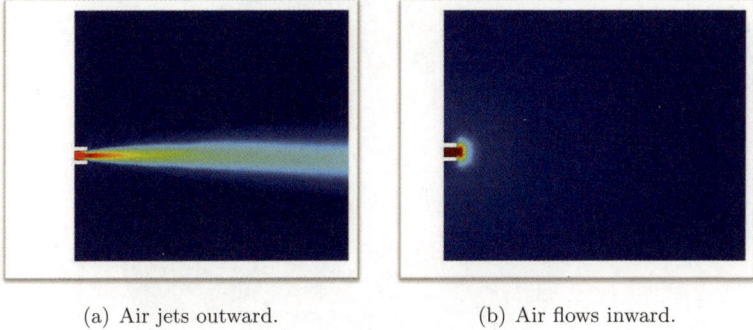

(a) Air jets outward.   (b) Air flows inward.

Fig. 7.   The velocity field outside the cavity when air flows outward and inward the resonator. In the picture, brightness represents the magnitude of air velocity.

To get an idea of the air flow distribution, we employ COMSOL stimulation at first. Figure 7(a) and Figure 7(b) demonstrate the velocity distribution when air is flowing outward and inward of the resonator respectively.

When inner pressure is larger than external pressure, air rushes out. As it goes through the straight pipe, a compact air jet is formed. It is appropriate to suppose approximately:

$$u_{out} = \dot{x}, \qquad \text{when} \quad \dot{x} > 0. \tag{13}$$

When external pressure takes advantages, air flow gathers from all directions. It is an inverse radiation because the speed distribution origins from the negative pressure inside the cavity. Here the cross-section of the neck behaves like an acoustic piston. So we assume the velocity distribution similar to that of a piston.[3] According to equation B.7, we have:

$$v = \frac{A}{i\omega\rho_0 r^2} \exp(i\omega t). \tag{14}$$

Take u_{out} at a point at the semi-sphere with radius $r = a$, we'll get:

$$u_{out} = \frac{\dot{x}}{2} \qquad \text{when} \quad \dot{x} < 0. \tag{15}$$

To find the inner velocity u_{in}, COMSOL stimulation is also done to demonstrate the velocity distribution, as shown is Fig. 8. The gray level represents magnitude of the velocity, and the arrows represent the local

Time=0.002   Surface: Velocity magnitude (m/s)
Arrow Surface: Velocity field

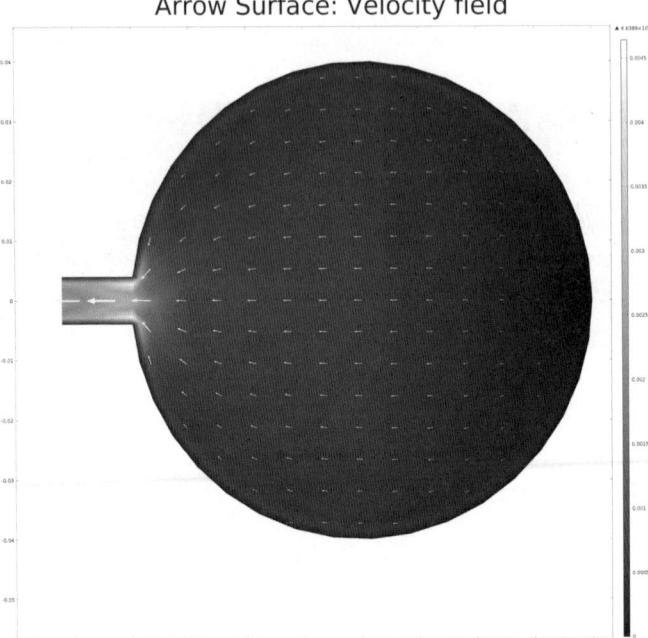

Fig. 8.   The velocity field inside a cavity. The gray level represents magnitude of the velocity, and the arrows represent the local velocity (direction and magnitude).

velocity (direction and magnitude). Notice that arrows vanish near the wall. Figure 8 shows that air jet is not formed inside the resonator due to the confine of cavity. Of course the speed distribution will depend on the geometry of the cavity. But when the magnitude of a is not as large as that of R, it is still reasonable to regard it as a piston radiation. Again we may assume u_{in} for both outward and inward airflow

$$u_{in} \approx \frac{\dot{x}}{2} \qquad \text{when} \quad \dot{x} < 0, \text{ or } \dot{x} > 0 \qquad (16)$$

which can also be demonstrated by computer simulation.

4. Propelling Force

With equations 13, 15 and 16, the Helmholtz resonator equation 5 is finally expressed in an explicit form. So we are able to evaluate the net force acting upon the Helmholtz resonator. Let's simplify the vibration equation 5 in the form:

$$\ddot{x} + f_1(\dot{x}) + f_2(x) = Ae^{-i\omega t}. \tag{17}$$

The first order approximation is linear, so we assume a harmonic solution:

$$x = x_M e^{-i(\omega t + \varphi)}. \tag{18}$$

The net force can be estimated by calculating the periodic residue in the acceleration of the oscillator.

$$< F > \propto < \ddot{x} > = -\frac{1}{T} \int_0^T (f1 + f2)dt. \tag{19}$$

After integration, all linear terms vanish and only nonlinear terms remain. So that the propelling force is in fact a *nonlinear effect*.

Equation 5 has two nonlinear terms with x^2 and $\dot{x}^2$. The corresponding forces are of two origins: one origins from the nonlinearity of the impedance, the other origins from nonlinear "restoring force".

To estimate the net force, we firstly apply the approximate solution by solving equation 6. The amplitude of a forced oscillation is given by

$$x_M = \frac{p_s/(\tilde{L})\rho_0}{\sqrt{(\omega_0^2 - \omega^2)^2 + (R'\omega/(\tilde{L})\rho_0)^2}} \tag{20}$$

where

$$R' = \rho_0 c_0 \cdot (ka)^2 + L\sqrt{2\omega\rho_0\mu/a} + \frac{A\mu}{3V} \tag{21}$$

$$\tilde{L} = L + \frac{16a}{3\pi}.$$

So the force effect owing to the nonlinear restoring force of the cavity is:

$$\begin{aligned} F_1 &= -\frac{1}{\tau} \int_0^\tau \frac{\gamma p_0 A^2}{V_0} \cdot \frac{A(1+\gamma)}{2V_0} \cdot (x_M \exp(-i\Omega t))^2 \, dt \\ &= -\frac{\gamma(1+\gamma)p_0 A^3}{4V_0^2} x_M^2. \end{aligned} \tag{22}$$

The force effect owing to the nonlinearity of impedance is:

$$F_2 = -\frac{1}{\tau}\int_0^\tau -\frac{1}{2}\rho_0 A(u_{out}^2 - u_{in}^2)dt$$

$$= -\frac{1}{\tau}\int_0^{\frac{\tau}{2}} -\frac{1}{2}\rho_0 A\frac{3}{4}\left(-i\Omega \cdot x_M \exp(-i\Omega t)\right)^2 \qquad (23)$$

$$= \frac{3}{32}\rho_0 A\Omega^2 x_M^2.$$

The total force is:

$$F = F_1 + F_2. \qquad (24)$$

Apparently the ratio of the two forces at the natural frequency F_1/F_2 is the ratio of the volume of the neck to the cavity volume, normally $F_1/F_2 <<$ 1. So we suggest that the nonlinear impedance is the dominant origin of the propelling force.

With parameters according to experimental conditions, the total net force can be approximately estimated according to Eqs. 22 and 23. See Fig. 9, the solid curve is the theoretical result of the force frequency response, the dots indicate the experimental data.

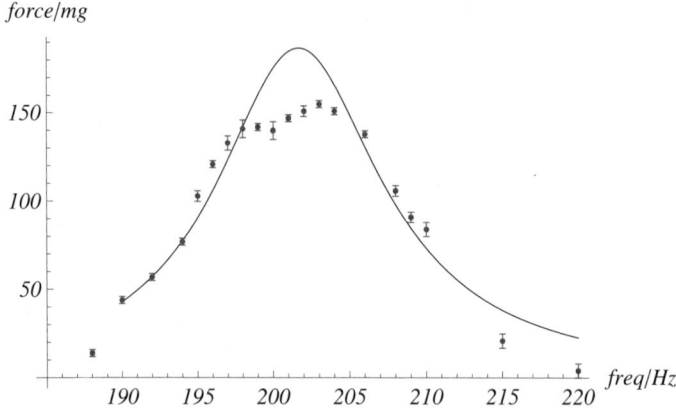

Fig. 9. The force frequency response of a Helmholtz resonator, the solid curve is the theoretical result of the force frequency response, the dots indicate the experimental data.

On can see the our theoretical result gives a good description to the force effect only aside from the nature frequency. A large deviation exists near nature frequency. This is inevitable. Our solution employs a linear first order solution, where nonlinear terms become dominant. Even so, our solution provides a clear understanding to the cause of the force effect, and a semi-quantitative estimation to the force magnitude.

5. Optimize the Helmholtz Carousel

Finally, we can optimize the Helmholtz carousel according to the predictions of our model, which can be compared to our experimental results.

Relevant parameters include the frequency of external sound f, the sound intensity, or the sound pressure level SPL; the geometry of the Christmas tree ball, including the neck length L, volume of the cavity V_0, and the radius of the neck a; the friction of the mounting device and air. Among these parameters, the effect of the sound frequency is already discussed. The effects of sound pressure level and the friction of the device are evident and trivial. We will concentrate on the geometrical parameters of the resonator, including the neck length L, volume of the cavity V_0, and the radius of the neck a.

Figure 10 shows the result with fix volume, $V = 268\,mL$, fixed neck radius, with cross sectional area $A = 1.49\,cm^2$. We still test the driving force instead of the rotation speed. The neck length is a varying parameter. Both experimental and theoretical curves are similar in shape, and suggest a $2cm$ optimal neck length. Owing to the practical shape of the resonator and the simplification of our model, the result is satisfactory.

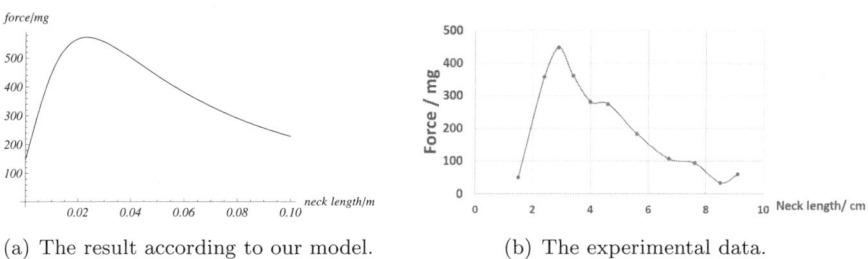

(a) The result according to our model.   (b) The experimental data.

Fig. 10.   Force vs. neck length.

Figure 11 shows how the driving force change with the volume of resonator. Here the neck radius and length are constant. A cylindrical neck with radius $a = 0.83\,cm$ and length $L = 2.0\,cm$ is adopted. The solid curve is the theoretical result, and the dots indicate the experimental result. They coincide well with each other.

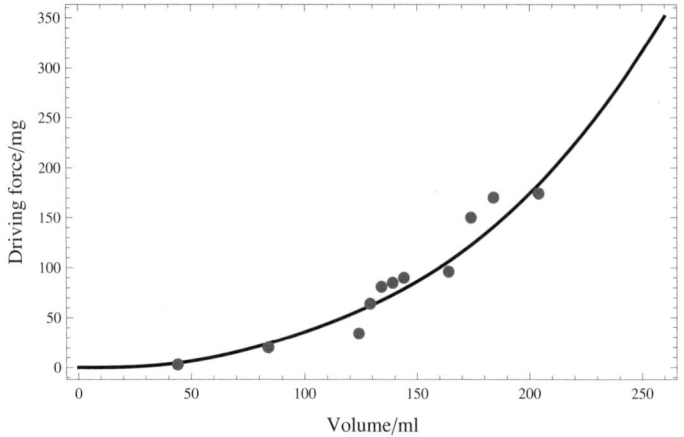

Fig. 11. Driving force vs. volume of the cavity. The solid curve is the theoretical result, and the dots indicate the experimental result.

Being successful in describing the parameters of the neck length and the cavity volume, we can rely on our model to make further prediction to optimize other parameters. Here we keep the neck length $L = 2cm$ and cavity volume $V = 268mL$ fixed, and change the radius of the neck. Our model suggests that an optimal neck radius exist at around $1cm$, see Fig. 12.

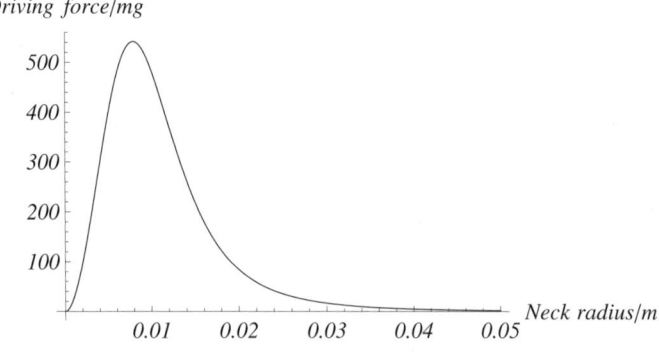

Fig. 12. Driving force vs. neck radius.

Among the parameters we have studied, specific frequency, neck length, neck radius values exist to optimize the Helmholtz carousel. While larger resonator volume produces larger driving force, it also reduces the resonant frequency.

Appendix A. Impedance due to air friction in the neck

The maximum air velocity is around $10m/s$, corresponding to a maximum Reynolds number of 2000. So we'll consider the motion as laminar flow. Air is a Newtonian liquid, we have $\tau = \mu\sigma$.

Consider the friction acts between adjacent layers:

$$f = \mu \cdot 2\pi r \frac{\partial u}{\partial r} dz. \tag{A.1}$$

With the driven force being the pressure gradient, we'll have the force equilibrium:

$$i\omega\rho_0 u - \frac{\mu}{r}\frac{\partial}{\partial r}\left(r\frac{\partial u}{\partial r}\right) + \frac{\partial p}{\partial z} = 0 \tag{A.2}$$

the non-slip boundary condition:

$$u(r,t)|_{r=a} = 0. \tag{A.3}$$

Solution of the differential equation gives the velocity distribution:

$$u = -\frac{1}{i\omega\rho_0}\frac{\partial p}{\partial z}\left[1 - \frac{J_0(\sqrt{-i}k_0 r)}{J_0(\sqrt{-i}k_0 a)}\right] \tag{A.4}$$

where $k_0 \equiv \sqrt{\omega\rho_0/\mu}$ and J_0 is the zeroth order Bessel function. Correspondingly, the cross-sectionally averaged velocity can be found by:

$$\begin{aligned}
\bar{u} &= \frac{1}{\pi r^2}\int_0^a u 2\pi r dr \\
&= -\frac{1}{i\omega\rho_0}\frac{\partial p}{\partial z}\left[1 - \frac{2}{\sqrt{-i}k_0 a}\frac{J_1(\sqrt{-i}k_0 a)}{J_0(\sqrt{-i}k_0 a)}\right].
\end{aligned} \tag{A.5}$$

Consider the wavelength being much larger than the length of neck, we can substitute $\partial p/\partial z$ by $\Delta p/\Delta z$, to get the impedance:

$$Z = \frac{\Delta p S}{\bar{u}} = i\omega\rho_0 SL\left[1 - \frac{2}{\sqrt{-i}k_0 a}\frac{J_1(\sqrt{-i}k_0 a)}{J_0(\sqrt{-i}k_0 a)}\right]^{-1}. \tag{A.6}$$

$k_0 a \sim 10^1$ in our case, after a series expansion, we can get:

$$Z = i\omega\rho_0\pi a^2 L + \sqrt{2}\pi\mu(k_0 a)(1+i)L. \tag{A.7}$$

Hence the impedance due to air friction is:

$$Z_f = \pi a L\sqrt{2\omega\rho_0\mu}(1+i). \tag{A.8}$$

Appendix B.   Radiation impedance

Acoustic radiation is a redistribution of energy over space. This redistribution process, generate impedance as pressure drops while velocity remains unchanged. Outward radiation is relatively easy to calculate because it is not dependent on the geometry of the cavity. One can just take it as radiation towards an infinite space. The piston radiation on an infinite wall to infinite space is first calculated by H. Levine and J. Schwinger in 1948. Here we will use the impedance of piston radiation on an infinite wall as an approximation to our case. More rigorous derivation involves solving the boundary value problem which is determined by the shape of the cavity can be found in the work of Ingard in 1953.[6]

For linear sound wave travelling at speed c_0, the pressure satisfies following wave equation:

$$\frac{\partial^2 p}{\partial t^2} - c_0^2 \nabla^2 p = 0. \tag{B.1}$$

Assuming the signal is harmonic, we'll have Helmholtz equation.

$$\nabla^2 p + k^2 p = 0. \tag{B.2}$$

Consider a spherical wave generated from the origin, due to symmetry, only radial part is kept:

$$\nabla^2 \rightarrow \frac{\partial^2}{\partial r^2} + \frac{2}{r}\frac{\partial}{\partial r}. \tag{B.3}$$

Substitute (B.3) into equation B.2, we get:

$$\frac{\partial^2 (rp)}{\partial r^2} + k^2 (rp) = 0 \tag{B.4}$$

the solution is:

$$p = \frac{A}{r}\exp(-ikr + i\omega t). \tag{B.5}$$

Here A is an integration constant. Notice that the time related term is harmonic in our case. Substitute it into the momentum conservation equation:

$$\rho_0 \frac{d\mathbf{u}}{dt} + \nabla p = 0 \tag{B.6}$$

we'll get the speed distribution in the vicinity of the point sound source:

$$
\begin{aligned}
\mathbf{u} &= \frac{A\mathbf{e_r}}{i\omega\rho_0 r^2}(1 + ikr)\exp(-ikr + i\omega t) \\
&\approx \frac{A\mathbf{e_r}}{i\omega\rho_0 r^2}\exp(i\omega t) \qquad \text{for} \quad kr << 1
\end{aligned}
\tag{B.7}
$$

where $\mathbf{e_r}$ is the radial normal vector pointing outwards. The sound wave generated by the point source thus form a pulsating sphere. The volume of air pumped out per second

$$q = 4\pi r^2 u = \frac{4\pi A}{i\omega\rho_0} \exp(i\omega t)$$

$$= Q\exp(i\omega t).$$

(B.8)

Substitute A with the maximum volume speed Q to get the final pressure distribution for a point source:

$$p = \frac{i\omega\rho_0 Q}{4\pi r} \exp(-ikr + i\omega t).$$

(B.9)

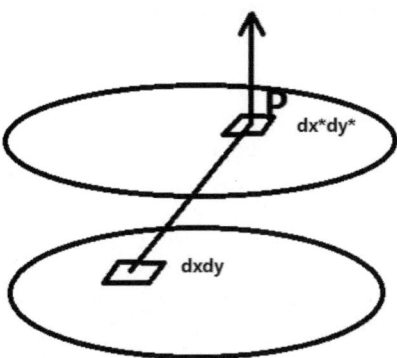

Fig. B.1. The pressure at a random point P is calculated by integrate the distribution from all the point source over the area. The force is acquired by integrate that pressure on the same surface.

The radiation generated by the Helmholtz resonator can be compared with that of a piston embedded on an infinite wall. Consider every small surface area on the piston as a point source whose maximum volume velocity $Q = u_0 dxdy$. The pressure at point P is the integral of all the contributors $dxdy$ over the neck area. $u_0 = (dV/dt)_M$ is the maximum velocity of air oscillating inside the tube. Effect of the wall is to double the pressure.

The impedance due to radiation can be gained by integrating pressure in every $dxdy$ and divide with average velocity (see Figure B.1). In this way we can find radiation impedance:

$$Z_{rad} = \frac{f}{u}$$

$$= i\omega\rho_0 \iint_S \iint_S 2\frac{\exp(-ikr)}{4\pi r} dSdS \tag{B.10}$$

$$= i\omega\rho_0 \frac{\pi}{2} \int_0^a r(1 - J_0(2kr) + K_0(2kr))dr$$

where J_0 is the zeroth order Bessel function and K_0 is the zeroth order Struve function . Because the wave length is much larger than the radius of the neck, ka is a small quantity in our experiment, we can thus get a series expansion of Z_{rad} :

$$Z_{rad} = z_r + iz_i$$

$$\approx \rho_0 c_0 \pi a^2 \left(\frac{1}{2}(ka)^2 + i\frac{8ka}{3\pi}\right). \tag{B.11}$$

Hence the end correction for extra neck length at one side of the tube is $8a/3\pi \approx 0.85a$. The case is slightly different for the end of tube, which takes on the value of $0.61a$.[3]

References

1. R. R. Boullosa and F. O. Bustamante, The reaction force on a helmholtz resonator driven at high sound pressure amplitudes, *American Journal of Physics*. **60**(8), 722–726 (1992).
2. W. Wu, *An Introduction to Fluid Dynamics*. Peking University Press (2011).
3. D. Ma, *Basics of Modern Acoustics*. Science Press, China (2004).
4. D. K. Singh and S. W. Rienstra. A systematic impedance model for non-linear helmholtz resonator liner (2013).
5. T. H. Melling, The acoustic impendance of perforates at medium and high sound pressure levels, *Journalof Sound and Vibration*. **29**(1), 1–65 (1973).
6. U. Ingard, On the theory and design of acoustic resonators, *The Journal of the Acoustical Society of America*. **25**(6), 1037–1061 (1953).

Chapter 14

2013 Problem 13: Honey Coils

Youtian Zhang[*] and Jiaan Qi[†]

Nanjing University, School of Physics

The objective of this article is to "study and explain" the phenomenon of honey coils. The process of *stable coiling* is studied at first. The corresponding equation is derived, from which the solution of stable coiling has been obtained numerically. The viscous, gravitational and inertial regimes of coiling are studied. We provide our own derivation of the coiling frequencies according to the three regimes. We also carry out experiment, from which the behavior of three different regimes and a complicated "multiple coexisting state" can be identified. Finally, the *onset* of the coiling is also discussed.

1. Introduction

Problem Statement:

A thin, downward flow of viscous liquid, such as honey, often turns itself into circular coils. Study and explain this phenomenon.

When a stream of vicious liquid is poured down from some distance onto a solid surface, it will turn itself into a specific pattern. The phenomenon of liquid rope coiling can be easily produced in daily life. For example, when you pour honey onto your bread for breakfast, the honey may start coiling. In most of the cases, the liquid will stably turn into a helical coil. Such phenomenon resembles the act of an elastic rope falling onto a surface. So it was named "liquid rope coiling" after it was first discovered and reported in 1957.[1] However, the coiling of vicious liquid differs from that of a rope, which is driven only by the elastic force and the gravity.[2] The radius of the liquid stream changes as it moves downwards, also, the motion of liquid

---

[*]E-mail: ricardo.ytchang@gmail.com
[†]E-mail: jiaanq@163.com

stream involves bending, twisting, and stretching, which makes it harder to analyze and describe.[3]

Fig. 1. "Liquid rope coiling" is common in our daily life.

In the early days, the coiling frequency was considered proportional to the effective falling height of the liquid through experimental observation.[1] But such conclusion was declined later by more experiments. In 1968, an important theoretical advance was made by Taylor who realized the importance of fluid buckling instability.[4] Subsequently, the critical fall height and frequency at the onset of coiling were determined using linear stability analysis.[5,6] In 1998, Mahadevan et al proposed scaling law to determine the relationship between coiling frequency, flow rate and filament radius.[7] It was not until quite recently that a relative complete solution was proposed by Ribe, who reduced it into a 17th-order nonlinear two-point boundary-value problem with two free parameters and 19 boundary conditions, under the case that slenderness of the stream is negligibly small.[8] According to balance conditions among the three principal forces acting on the liquid rope, the coiling process can be classified into 3 different regimes, viscous, gravitational and inertial. In addition, a complicated stage between gravitational and inertial regime where multiple states coexist was found.[9] Together they form four distinct coiling regimes, with remarkably diverse behavior.[10]

As a solution to the 2013 IYPT problem, our objective is to "study and explain" this phenomenon. According to the results of current literatures and our own understanding of the problem, we will study the process of *stable coiling* at first. The corresponding equation will be derived, from which the solution of stable coiling can be obtained numerically. Then, we study the 3 different regimes proposed by Ribe, viscous, gravitational and inertial. We will derive the coiling frequency with respect to other parameters in an alternative method according to balance conditions among the three principal forces. We also carry out experiment to verify our theoretical results. Finally, the *onset* of the coiling, i.e. how the coiling starts is discussed.

2. Equation and Solution for Stable Coiling

We start with the stage of stable coiling with a constant angular frequency Ω. Compared with Ref. 8, we adopt a simplified model in that the filament is taken as a one dimensional string. Such approximation holds when the filament is very slender.

We adopt a co-rotating reference frame. The shape of the filament is described by the coordinates at the jet's central axis $\boldsymbol{x}(s,t)$. Where s is the arc length from the injection point. We also define a set of material coordinates at point $\boldsymbol{x}(s,t)$ by

$$\mathbf{d}_3 \equiv \partial\mathbf{x}/\partial s \tag{1}$$

and $\mathbf{d}_1$, $\mathbf{d}_2$ as two orthogonal material unit vectors in the normal plane of $\mathbf{d}_3$ (See Fig. 2).

On each point of the filament, the speed of fluid particles are defined as $\mathbf{U} = d\mathbf{x}/dt$. In the co-rotating frame, we have the speed on each point parallel to the tangential vector of the string.

$$\mathbf{U} = \frac{\partial\mathbf{x}}{\partial s} \cdot \frac{ds}{dt} + 0 = \frac{ds}{dt}\mathbf{d}_3 = U\mathbf{d}_3. \tag{2}$$

The internal force $\mathbf{N}$ acting on the cross section of the string can be obtained by integrating the stress tensor over the cross section area A. It is composed of three components N_1, N_2 and N_3 in the material reference frame $\{\mathbf{d}_1, \mathbf{d}_2, \mathbf{d}_3\}$.

$$\mathbf{N} = \sum_{i=1}^{3} N_i \mathbf{d}_i \tag{3}$$

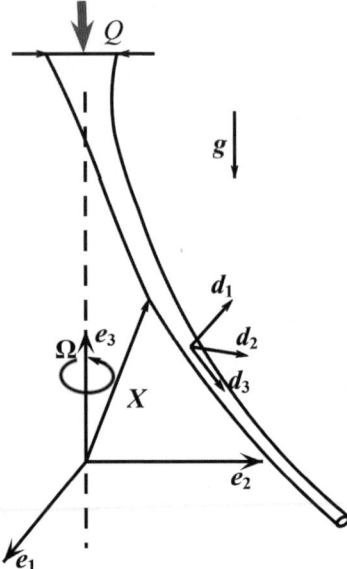

Fig. 2.   The coordinates to describe the motion of one dimensional filament.

N_1, N_2 are responsible for the shear stress arisen by twisting and bending of the liquid column. In the case that one dimension is taken into account, these components can be omitted. This is supported by a more rigorous model in which the magnitude of N_1, N_2 is very small compared to N_3,[8] which originates from the stretching of liquid column. Hence,

$$\mathbf{N} \approx N_3 \mathbf{d}_3. \tag{4}$$

The case that the liquid column elongates from the both ends is a typical uniaxial extension problem (or simple extension). The strain rate tensor, in material reference frame $\{\mathbf{d}_1, \mathbf{d}_2, \mathbf{d}_3\}$, can be written as:

$$\Delta = \begin{pmatrix} 2\dot{\varepsilon} & 0 & 0 \\ 0 & -\dot{\varepsilon} & 0 \\ 0 & 0 & -\dot{\varepsilon} \end{pmatrix} \tag{5}$$

where $\dot{\varepsilon} \equiv |\partial U/\partial s|$ is the principal rate of extension.

It can be rigorously deduced from Ref. 11 that:

$$\tau = 3\mu\dot{\varepsilon} \tag{6}$$

where μ is viscosity coefficient and τ is the viscous force on the cross-section per unit area. Now we can write down the force equation for a small segment of liquid column.

$$\rho A \left(\mathbf{\Omega} \times \mathbf{\Omega} \times \mathbf{x}(s) + 2\mathbf{\Omega} \times \mathbf{U} + U\frac{\partial \mathbf{U}}{\partial s} \right) = \frac{\partial}{\partial s} \left(3\mu A \frac{\partial U}{\partial s} \right) + \rho A \mathbf{g} \quad (7)$$

where ρ is the density of liquid and $\mathbf{g}$ is the gravitational acceleration.

$A(s)$ and U satisfy the flux conservation condition along the string:

$$Q \equiv U(s)A(s) \quad (8)$$

where Q is the rate of the flow.

Eqn. 7 describe the motion of the liquid. A stable solution of the motion can be found through numerical calculation by using *Mathematica*.

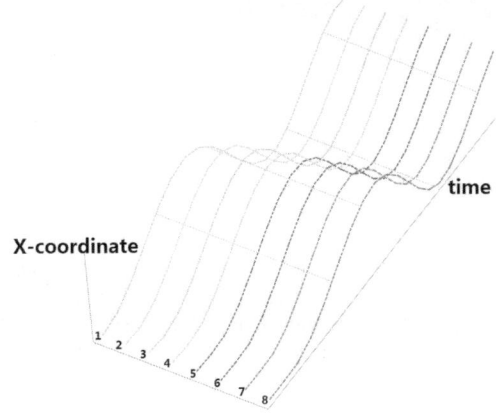

Fig. 3. The oscillation solution in x direction along a liquid column. Each line stands for a motion along the successive positions denoted by 1 to 8 upwards on the column.

Fig. 3 shows that the stable solution of x coordinate is oscillation. In the figure, each line stands for a motion along the successive positions denoted by 1 to 8 upwards on the column. The upper position has phase lag over lower positions.

Fig. 4(a) shows the motions on x-y plane of different points along the column (also represented by different lines) evolve with time. Both x and y ordinates oscillate with time. Fig. 4(b) shows the top view for a chosen point. Fig. 4 shows the coiling behavior of liquid filament.

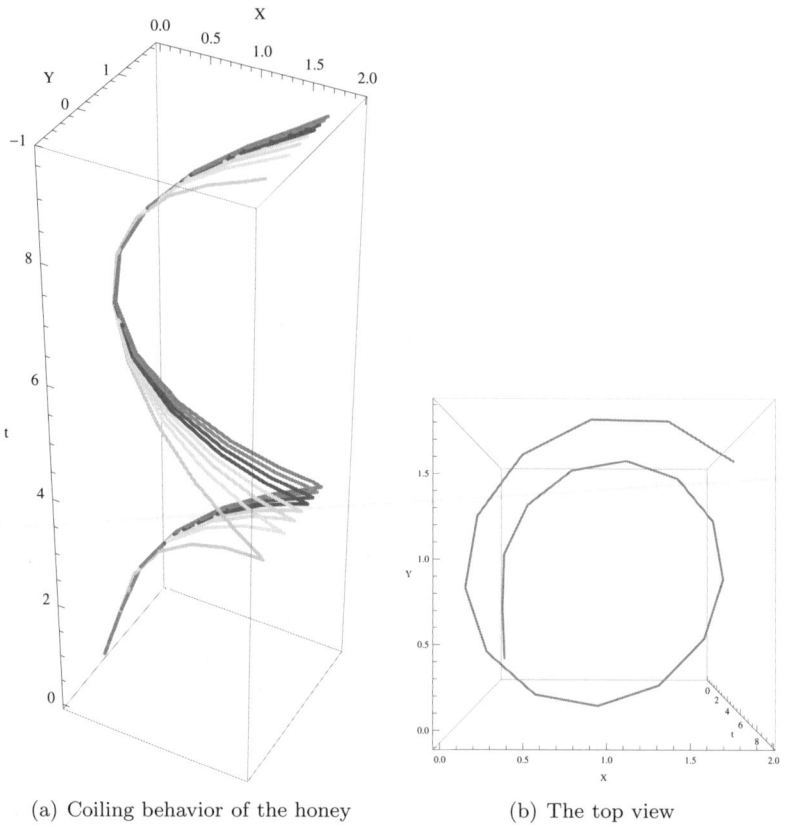

(a) Coiling behavior of the honey   (b) The top view

Fig. 4.   Solution for position versus time on x-y plane, showing a stable coiling behavior.

3. Three Regimes of Coiling

The classification of viscous, gravitational, and inertial regimes by Ribe is significant in revealing the mechanism of coiling. However, the derivation contains 17th-order nonlinear equations with free parameters, which makes it too difficult to understand. We propose an alternative method to derive the same relations.

The regimes are defined by the relative magnitudes of the viscous F_v, gravitational G, and inertial F_i forces per unit length in the helical portion of the rope near the plate. Viscous coiling occurs when both gravity and inertia are negligible, $G, F_i \ll F_v$. Coiling in this regime is driven entirely by the injection of the fluid, much like toothpaste squeezed from a tube. Gravitational coiling occurs when inertia is negligible and the viscous force is balanced by gravity $G \approx F_v \gg F_i$. Finally, inertial coiling occurs when gravity is negligible and viscous forces are balanced by inertia, $F_i \approx F_v \gg G$.

Now, we derive the coiling frequencies in these three regimes. First of all, we consider the *"viscous" coiling regime.*

At equilibrium, the moment created by external forces (and external moments) must be balanced by the couple induced by the bending moment. The bending moment can be calculated as $M = EI/R$, where E is Young's modulus and R denotes the curvature radius of neutral surface. I is the area moment of inertia which describes the capacity to resisting bending. For a filled circular area of radius a, $I = \pi a^4/4$. Young's modulus, E, can be calculated by dividing the tensile stress by the extensional strain in the elastic region of the stress-strain curve. Without velocity gradient along the honey coil, according to Eq.(6), the tensile stress does not exist. The coiling angular velocity is $\Omega = \frac{U}{R}$. In reality, velocity gradient causes tensile stress and the length of honey coil becomes longer due to the stretch. Assume the radius of coil is R, and we take the arc Δs corresponds to unit central angle for convenience. Obviously, $\Delta s = 1\text{rad} \times R = R$. The change of the ark length Δs is denoted by dl, so that $dl = dR$. Since $U = \Omega R$, the difference of velocities dU at two ends of Δs is $dU = \Omega dR$. According to definition, the Young's modulus is

$$
\begin{aligned}
E &\equiv \frac{3\mu A(\partial U/\partial s)/A}{dl/ds} \\
&= \frac{3\mu(dU/\Delta s)}{dl/\Delta s} \\
&= \frac{3\mu dU}{dU/\Omega} = \frac{3\mu U}{R}.
\end{aligned}
\tag{9}
$$

Then

$$
M = E\frac{I}{R} = \frac{3\pi \mu a^4 U}{4R^2}.
\tag{10}
$$

Neil M. Ribe proposed this formula in a different way of derivation.[8,10]

Euler's Formula gives the critical load for slender compression bars when the bending moment is balanced with the flexion moment.

$$
\frac{\pi^2 EI}{(\nu H)^2}R = \frac{EI}{R}
\tag{11}
$$

where ν is a constant coefficient for a specific constraint of the rod ends. From this condition, $R \propto H$ and

$$\Omega = \frac{Q}{\pi a^2 R} \propto \frac{Q}{a^2 H}. \tag{12}$$

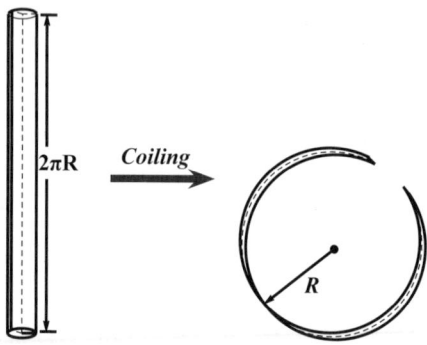

Fig. 5.   The filament with length $2\pi R$ is bend into a circle of R.

Consider the "gravitational" coiling regime. In this regime, the inertial force can be neglected, the gravity and the viscous force balance each other. Instead of applying the force equation directly, we evaluate the energy balance during a coiling process. As shown in Fig. 5, the filament with length $2\pi R$ is bend into a circle of R. The work done by flexion moment equals the change in gravitational potential energy. For one single turn of coil, we have

$$\rho(2\pi R \cdot \pi a^2)g \cdot (\frac{1}{2} \times 2\pi R) = \frac{3\pi\mu a^4 U}{4R^2} \cdot 2\pi. \tag{13}$$

Recalling $Q \equiv UA$, then,

$$\Omega = \frac{Q}{\pi a^2 R} = Q(\frac{3a^8 \mu Q}{8\pi^2 g \rho})^{-\frac{1}{4}} = (\frac{8\pi^2 Q^3 \rho g}{3\mu a^8})^{\frac{1}{4}}. \tag{14}$$

We consider the "inertial" regime similarly. Neglect the gravity, instead of setting the inertial force and the viscous force equal to each other, we calculate the work done by flexion moment which is provided by the inertial centrifugal potential energy. For one single turn of coil with radius R we have

$$\frac{1}{2}\rho(2\pi R \cdot \pi a^2)\Omega^2 R^2 = \frac{3\pi\mu a^4 U}{4R^2} \cdot 2\pi. \tag{15}$$

Then, $R = \left(\frac{3\mu a^2}{2\rho U}\right)^{\frac{1}{3}}$, so that

$$\Omega = \frac{Q}{\pi a^2 R} = Q(\frac{3\mu a^8}{2\rho U})^{-\frac{1}{3}} = (\frac{2\rho Q^4}{3\pi \mu a^{10}})^{\frac{1}{3}}. \qquad (16)$$

Eqs.(12)(14)(16) give the relation between rotating frequency and the radius of the end of the coil and the flux rate in all three regimes, which are consistent with Ribe's expressions.[8]

To get the relation between falling height and rotating frequency, we still need the relation between falling height and the radius a of the coil. Consider a circular orifice of radius a_0 ejecting a flux Q of fluid of density ρ. The resulting jet is shot downwards, and accelerates under gravity $-g\mathbf{e}_3$. We proceed by deducing the shape $a(z)$ and speed $U(z)$ of the evolving jet.

Point A denotes the outflow position and B is an arbitrary point along the falling liquid column. Applying Bernoulli's Theorem at points A and B:

$$\frac{1}{2}\rho U_0^2 + \rho g z + P_A = \frac{1}{2}\rho U^2(z) + P_B. \qquad (17)$$

The local curvature of slender column can be expressed in terms of the two principal radii of curvature, R_1 and R_2 :

$$\nabla \cdot \vec{n} = \frac{1}{R_1} + \frac{1}{R_2}. \qquad (18)$$

Thus, the fluid pressures within the jet at points A and B may be simply related to that of the ambient, P_0:

$$
\begin{aligned}
P_A &= P_0 + \frac{\sigma}{a_0} \\
P_B &= P_0 + \frac{\sigma}{a}
\end{aligned}
\qquad (19)
$$

where σ is the coefficient of surface tension.

Substituting (19) into equation (17) yields

$$\frac{1}{2}\rho U_0^2 + \rho g z + P_0 + \frac{\sigma}{a_0} = \frac{1}{2}\rho U^2(z) + P_0 + \frac{\sigma}{a} \qquad (20)$$

from which one finds

$$\frac{U(z)}{U_0} = [1 + \frac{2}{Fr}\frac{z}{a_0} + \frac{2}{We}(1 - \frac{a_0}{a})]^{\frac{1}{2}} \qquad (21)$$

where we define two dimensionless parameters as:

$$Fr = \frac{U_0^2}{g a_0} \quad \text{and} \quad We = \frac{\rho U_0^2 a_0}{\sigma}. \qquad (22)$$

With the flux conservation, we obtain

$$\frac{a(z)}{a_0} = \left(\frac{U_0}{U_z}\right)^{\frac{1}{2}}$$

$$= \left[1 + \frac{2}{Fr}\frac{z}{a_0} + \frac{2}{We}\left(1 - \frac{a_0}{a}\right)\right]^{-\frac{1}{4}}. \tag{23}$$

If B is the end of the coil, $a(z)$ then is the radius at the end of the coil. Now we can find the relation that determines the radius of the filament as a function of falling height H.

$$\frac{Fr}{2}\left[\frac{a_0^4}{a^4} - \frac{2}{We}\left(1 - \frac{a_0}{a}\right) - 1\right] = H. \tag{24}$$

However, when viscosity is taken into account, we need replace (17) by the equation blow,[13]

$$P + \frac{\alpha}{2}\rho U^2 + \rho g z + \sigma\left(\frac{\partial A}{\partial V}\right) + w_{los} = const. \tag{25}$$

where P is the pressure and U is the velocity averaged over the jet cross section and the coefficient α accounts for the velocity profile distribution. $\sigma\left(\frac{\partial A}{\partial V}\right)$ represents the jet interfacial energy density, w_{los} denotes the dissipation energy density due to the viscous resistance.

Eq. (12), eqs. (14) and (16) combining with eq. (25), we have the relation between falling height and rotating frequency in all three regimes theoretically. Figure 6 is the result that we calculated.

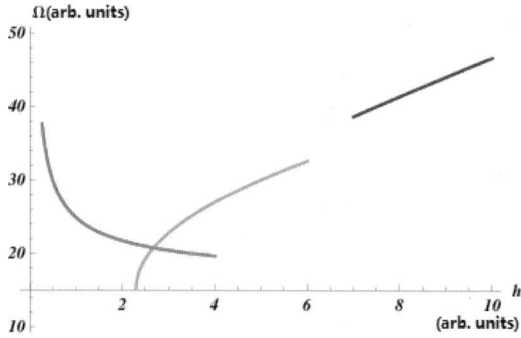

Fig. 6. The relation between the releasing height and coiling frequencies for three regimes

We also explore it through experiment straightway. The liquid we chose is honey. We made a device to control the flux of the honey.(See Fig. 7) An electro-motor is used to propel the piston of injection syringe in a uniform speed, so that the honey can be extruded uniformly.

During our experiment, we change falling height and use high speed camera to record the motion of the honey coils. The angular velocities are obtained by analyzing the videos by Tracker. In our experiments, both clockwise and counterclockwise coils can be observed. It hints that the Coriolis forces caused by earth rotation can be neglected. The result is shown in Fig. 8. Each data is obtained by an average of repeated observations.

It can be compared with the theoretical result in figure 6. Due to the limitation of the experiment condition, the initial release height can't be too small. But we can still find out the "viscous" regime, where the frequency decrease with the release height. Then comes the "gravitational" regime and "inertial" regime. Between these two regimes, there is a complicated regime which is called "multiple coexisting state".[9] The relation between frequency and release height in this stage is unstable and complicated. Under our experimental conditions, we are unable to describe the details in this stage. However, is it already observed and recorded by us in figure 6, notice those data points with extremely large error bars.

4. Onset Problem

We wish to discuss how the honey column starts coiling. Imagine the process of a stream of highly vicious liquid falling down on to a hard surface. Coiling will be formed after three stages of motion:

(I) At first, the liquid forms a column as it touches the surface. The diffusion rate is smaller than the vertical flowing rate, so that the column of liquid experiences a vertical compressive stress.

(II) As the stress inside the column accumulates, instability increases, driving the column to buckle towards a random direction. The straining energy inside the column is released and the column acquires an initial velocity. With the uniaxial dragging force acting as the restoring force, oscillation in one direction gets started.

(III) As the oscillation in one direction continues, the honey piles up rapidly. Meanwhile, gravitational energy and the energy required to bend the honey column also accumulates rapidly. As instability increases the motion on the other direction is soon started. An even distribution of energy along the two directions is favorable in consuming the gravitational

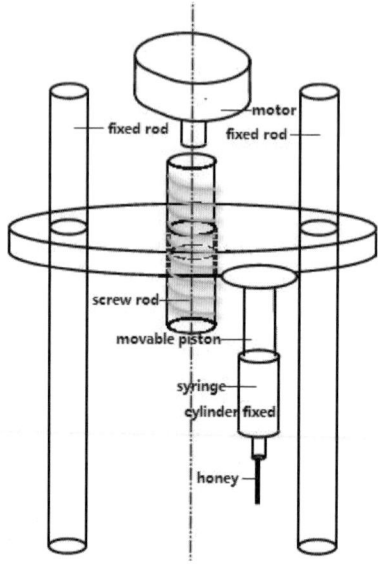

Fig. 7.   The device to control the flux of the honey. An electro-motor is used to propel the piston of injection syringe in a uniform speed, so that the honey can be extruded uniformly

energy. This corresponds to circular motion in plane, or the coiling behavior we observed.

The system would soon find its stable point by involving into the second dimension and satisfy the energy conservation.

5. Conclusion

Our objective of the article is to "study and explain" the phenomenon of honey coil. According to the results of current literatures and our own understanding of the problem, we study the process of *stable coiling* at first. By taking the viscous force acting on the cross section of the liquid as uniaxial stretching, the corresponding equation has been derived, from which the solution of stable coiling has been obtained numerically.

We study the three different regimes proposed by Ribe, viscous, gravitational and inertial regimes respectively. We derive the coiling frequencies with respect to the falling height in an alternative method according to

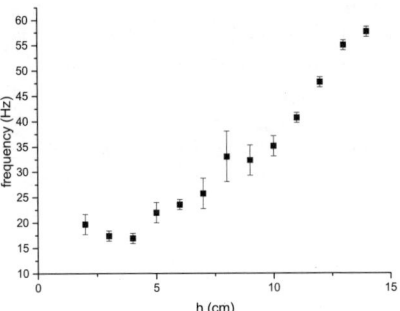

Fig. 8. The experimental results between the releasing height and coiling frequency.

balance conditions among the three principal forces. Our simplified derivation can help understand the problem better. We also carry out experiments, from which the behavior of three different regimes can be identified. Moreover, the complicated "multiple coexisting state" is also verified under our crude experimental conditions.

Finally, the *onset* of the coiling, i.e. how the coiling starts has been discussed.

References

1. G. Barnes and R.Woodcock, Liquid Rope-Coil Effect, *J. Phys. A.* **26**, 205 (1958).
2. M. Habibi, N. M. Ribe, and Daniel Bonn, Coiling of Elastic Ropes, *PhysRevLett* **99**, 154302 (2007)
3. M. Maleki *et al*, Liquid Rope Coiling on a Solid Surface, *PhysRevLett* **93**, 214502 (2004)
4. G. I. Taylor, Instability of jets, threads and sheets of viscous fluid, *Proc. Intl. Congr. Appl. Mech.* Springer.
5. J. O. Cruickshank, Low-Reynolds-number instabilities in stagnating jet flows, *J Fluid Mech.* (1988), vol 193, pp 111-127
6. B. Tchavdarov, A. L. Yarin and S. Radev, Buckling of thin liquid jets, *J. Fluid Mech.* (1993), vol. 253, pp. 593-615
7. L. Mahadevan, W. S. Ryu, & A.D.T. Samuel, Fluid "rope trick" investigated. *Nature* **392**, 140
8. N. M. Ribe, Coiling of viscous jets, *Proc. R. Soc. Lond. A* 2004 **460**.
9. N. M. Ribe et al, Multiple coexisting states of liquid rope coiling, *J. Fluid Mech.* (2006), vol. 555, pp. 275−297.

10. N. M. Ribe, M. Habibi and D. Bonn, Liquid Rope Coiling, *Annu. Rev. Fluid Mech.* 2012. **44**: 249–66
11. J. M. Dealy, Extensional Flow of Non-New tonian Fluids- A Review, *Polymer Engineering And Science*, November, 1971, Vol. 11, No. 6
12. F. T. Trouton, On the Coefficient of Viscous Traction and Its Relation to that of Viscosity, *Proc. R. Soc. Lond. A* 1906 **77**.
13. T. Massalha, R. M. Digilov, The shape function of a free-falling laminar jet: Making use of Bernoulli's equation *Am. J. Phys.* **81** (10), October 2013
14. http://en.wikipedia.org/wiki/List_of_area_moments_of_inertia

Chapter 15

2013 Problem 14: Flying Chimney

Youtian Zhang*, Zheyuan Zhu and Lan Chen

Nanjing University, School of Physics

This paper studies the motion of a top-burning, cylindrical tea bag and explains the origin of the lift force. After the top of cylindrical tea bag is lit, it will suddenly accelerate and lift off, gradually reaching a uniform motion until the flame extinguishes. We analyze the temperature distribution around a point heat source considering heat convection, and demonstrate how this temperature field contributes to the lift force on the tea bag and affects its motion.

1. Introduction

Problem Statement:

Make a hollow cylindrical tube from light paper (e.g. from an empty tea bag). When the top end of the cylinder is lit, it takes off. Explain the phenomenon and investigate the parameters that influence the lift-off and dynamics of the cylinder.

Make a hollow cylindrical tube with light paper (e.g. an empty tea bag). It takes off when the top end of the cylinder is lit. Finding the origin of the lift force is the key to explaining this phenomenon. Traditional flight model like balloon or sky lantern dating back to centuries ago all employs buoyancy to float. They consists of gasbags or canopies, filled with a relatively low-density gas, such as Helium or hot air, to provide buoyancy that overcomes their own weights. In contrast, the taking off of tea bags is mysterious because they are virtually open at both ends, which cannot enclose low-density gas. We will try to understand how combustion, the only source of power, results in the rising of tea bag.

In the following sections we will first present the phenomenon by analyzing a record of the rising process of tea bag. The motion can be divided

---

*E-mail: bide.cheung@gmail.com

acceleration stage, the uniform stage and a sudden stop followed by free
falling. Then a number of experiments are designated to determine the primary source of the lift force. Therefore, our theoretical analysis is devoted
to the temperature filed of a point heat source, with the air flow taken into
consideration.

2. Preliminary Experiments and Phenomena

2.1. *Phenomena*

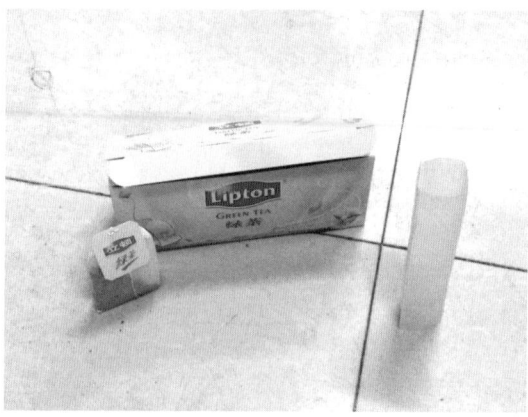

Fig. 1.   Tea bags used in experiments.

We use tea bags as shown in Fig. 1. Cut the tea bag into a cylindrical tube
standing on the ground and light its top edge. After a few seconds, the
tea bag with flame suddenly rises vertically and enters a uniform motion.
The tea bag stops rising as soon as the flame extinguishes and begins to
fall down. The motion of rising process is recorded in Fig. 2.

The accelerating stage, the uniform stage can both been observed in
sequence from the curve shown in Fig. 2. During the flying process last for
almost 5 seconds, uniform motion takes the longest time which starts after
1s or so. During the uniform process the lift force must cancel the gravity
and air resistance impeding its ascending motion. Evidently the lift force
is more predominant during the accelerating process. In order to reveal

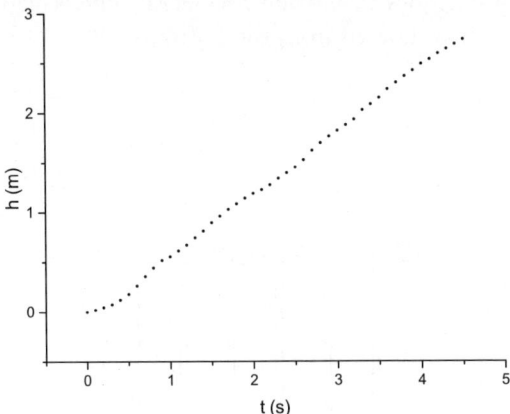

Fig. 2.   Motion of chimney. The vertical axis represents height and the horizontal axis represents time.

the origin of lift force and factors affecting its motion, we will introduce the following experiments in Section 2.2 which might offer some beneficial hints.

2.2.   *Further Experiments*

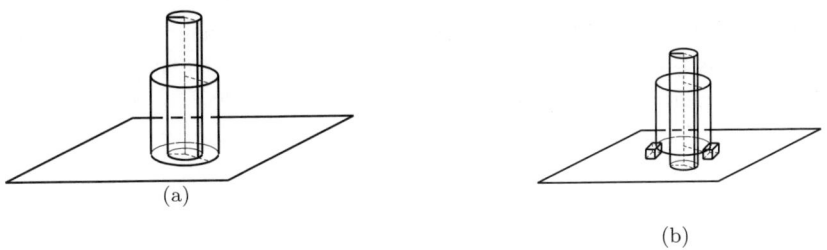

Fig. 3.   Cover the tea bag in two ways: (a) A larger cylinder surrounding the bottom of tea bag. (b) The peripheral cylinder is slightly lifted over ground, allowing air to flow in from the bottom.

Fig. 3 displays an experiment that suggests the rising of tea bags might be associated with the surrounding air flow. If we cover the bottom of cylindrical tea bag with another larger hollow cylindrical tube with air-tight wall, the tea bag cannot rise even when it has been burnt to ash completely. However, the tea bag does rise, if the peripheral tube is slightly lifted over ground, allowing air to flow in from the bottom.

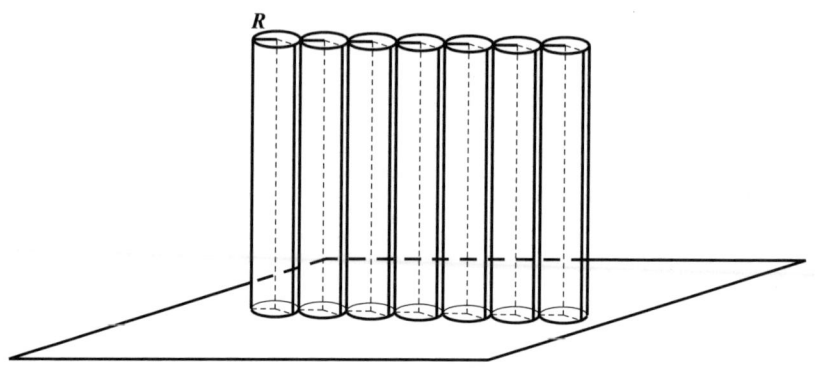

Fig. 4.   Several tea bags made of same material aligned in a line with different radius, which is the crucial factor determining whether it can fly or not.

If we arrange tea bags with the same radius in a line Fig. 4. We can observe that the bags at two ends rise before the bags at the center. The temperature gradient is larger at two sides while it is nearly zero at the center through it may be hotter. What influences the magnitude of buoyancy directly is the difference of temperature with the surrounding air instead of temperature itself.

Moreover, we can change the height of tea bags, or the cross sectional area and of tea bag by adopting different configuration, as shown in Fig. 5. Ascending always happens and the phenomena dont differ too much from each other, when the height of tea bags changed from 3cm to 15cm. The establishment of temperature field is fast the extra burning time doesnt change the lift process evidently. Another experiment on the cross sectional area of tea bags is implemented by using tea bags with different shapes (see Fig. 5). The rising of the tea bag with smaller sectional area is tardy. For its compact shape hinders the burning, its temperature rises slower too.

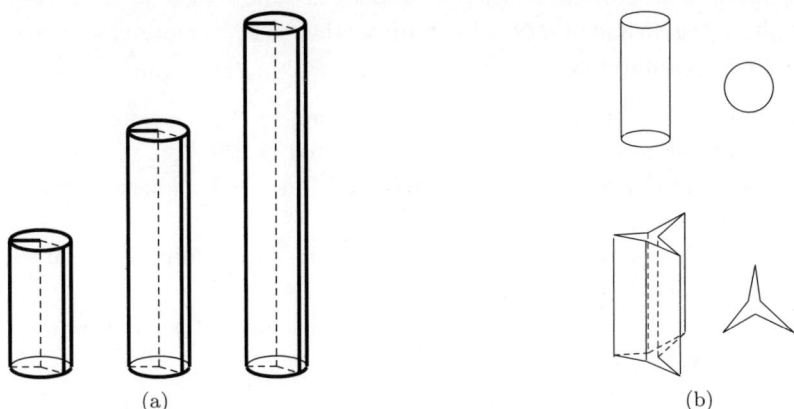

Fig. 5. Changing other tea bag parameters. (a) Tea bags made with different height. (b) Tea bags shaped into different configurations.

3. Theory

3.1. *Qualitative Analysis*

A heated bulk of air around the flame expands and thus has lower density than its surroundings. According to the rule of buoyancy, air with lower density tends to move up. Meanwhile as hot air around the flame moves up, convection channels cool air to the flame. As cool air heated and flows up again, more cool air comes to compensate for the loss of hot air around the flame. This circulation of air, whose velocity always points upward near the flame, drives the tea bag up, and thus creates the lift force we expect.

To explain this process quantitatively, we construct a model in Section 3.2.

3.2. *Quantitative Analysis*

3.2.1. *Model and Assumptions*

Compared to the cool, motionless, ambient air, we refer to an area around the flame as "plume region", which consists of hot and flowing air, called "plume", as a result of the heating from the flame. Ambient air may enter the plume region, and the velocity distribution inside plume region might not be uniform. In order to study the temperature and velocity distribution inside plume region and observe how it affects the tea bag's motion, we make the following assumptions:

(1) During the combustion, the flame of burning tea bag is considered as a point heat source and all energy generated by the flame is used to heat up the plume around the flame, which is axial-symmetric about the vertical axis of the cylindrical tea bag;

(2) The density difference between plume and its surrounding air is small and only contributes to the buoyancy;

(3) The air enters the plume at a rate u is proportional to the axial velocity v of plume, i.e. $u = kv$;

We establish a coordinate moving together with the tea bag. The center of flame is considered as a point source on the origin of our coordinate, as shown in Fig. 6. Since the plume region around the point heat source is assumed axial-symmetric, we denote the radius of plume region versus height z according to $r(z)$.

3.2.2. *Temperature Distributions of a Point Heat Source*

The mass flow rate through a cross-section at height z is:

$$\dot{m} = \frac{dm}{dt} = \pi r(z)^2 \rho v. \tag{1}$$

According to the continuity of mass, the mass flow includes the air entered into plume region from its side

$$d\dot{m} = 2\pi r dz \rho' kv \tag{2}$$

where ρ' is the density of ambient air, and $d\dot{m}$ has considered the flow rate difference between the top and bottom cross-sections of layer dz. Taking $\rho' = \rho$ approximately, this constrain implies the following equality:

$$\frac{d(r^2 v)}{dz} = 2rkv. \tag{3}$$

The buoyancy contributes to the external force on the infinitesimal air element:

$$dF = (\rho' - \rho)\pi r^2 g\, dz = \pi^2 r^2 g \Delta\rho\, dz.$$

According to assumption (1), the heat emitted by flame per unit time is used to heat up additional air that flow into the plume region:

$$\dot{Q} = c\dot{m}\Delta T = \pi r^2 \rho v c \Delta T \tag{4}$$

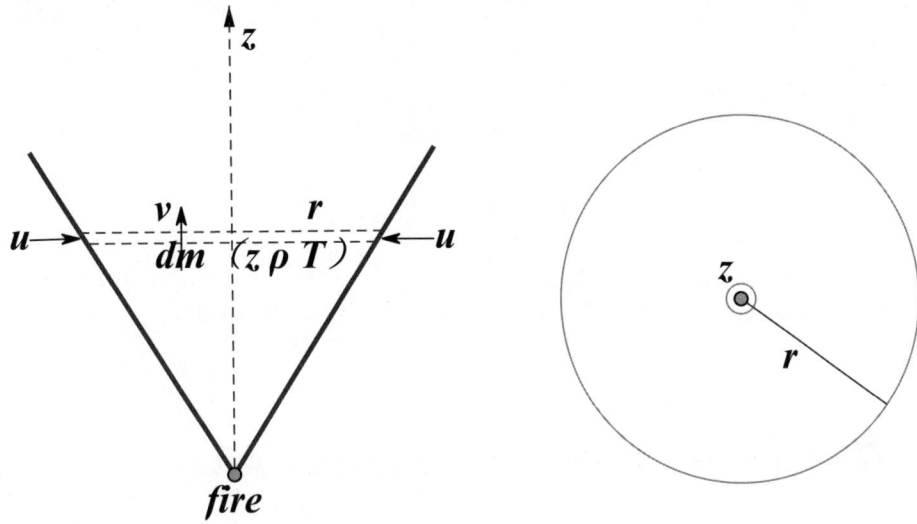

Fig. 6. Plum region. (a) Sectional view of the plume region. The ambient, cool air enters from the side at a rate u, while the air inside plume region flows upward at a rate v through each cross-section. (b) Top view of the plume region at a cross-section situated at height z.

where c is the specific heat of air, and is assumed to be a constant here.

Based on the equation of ideal gas and $\frac{d\rho}{\rho}$ is infinitesimal, we have:

$$T(1 - \frac{d\rho}{\rho}) = T + dT. \tag{5}$$

$$\therefore \frac{\Delta T}{\Delta \rho} = \frac{T'}{\rho'} \approx \frac{T'}{\rho}, \tag{6}$$

where T' is ambient temperature.

Since

$$\dot{Q} = \pi r^2 vc\Delta\rho T'. \tag{7}$$

By Newton's second law, take a derivative with respect to z on both side of equation:

$$\frac{dF}{dz} = \frac{d(\dot{m}v)}{dz} = \frac{d}{dz}(\pi r^2 \rho v^2) \tag{8}$$

so

$$\pi r^2 g \Delta \rho = \frac{g\dot{Q}}{vcT'} = \frac{d}{dz}(\pi r^2 \rho v^2)$$
$$\frac{g\dot{Q}}{vcT'} = \frac{d}{dz}(\pi r^2 \rho v^2). \tag{9}$$

In order to solve Equation (9) for the shape and velocity distribution of plume region, a typical solution[1] of $r(z)$ and $v(z)$ is assumed:

$$r = \beta z^m$$
$$v = \lambda z^n. \tag{10}$$

Substitute the above expressions into Equation (9) we have:

$$(2m+n)\beta z^{2m+n-1} = 2kz^{m+n} \tag{11}$$
$$(2m+2n)cT'\pi \beta^2 \rho \lambda^3 z^{2m+2n-1} = g\dot{Q}z^{-n}. \tag{12}$$

So we can get

$$\begin{cases} m = 1 \\ n = -\frac{1}{3} \\ \beta = \frac{6k}{5} \\ \lambda = \left(\frac{25g\dot{Q}}{48k^2 cT'\pi\rho}\right)^{\frac{1}{3}} \end{cases} \tag{13}$$

$$\begin{cases} r = \frac{6}{5}kz \\ v = \left(\frac{25g\dot{Q}}{48k^2 cT'\pi\rho}\right)^{\frac{1}{3}} \cdot z^{-\frac{1}{3}}. \end{cases} \tag{14}$$

From the $v - z$ relation we obtain, we can further get $v - t$ function:

$$v = \left(\frac{100g\dot{Q}}{81k^2 cT'\pi\rho}\right)^{\frac{1}{4}} \cdot t^{\frac{3}{4}}. \tag{15}$$

The constants are taken as

$$g = 9.8m/s^2, k = 0.15, c = 10^3 J/kg \cdot K, T' = 293K, \rho = 1.2kg/m^3.$$

And $\dot{Q}$ is determined by the property of the teabags.

In addition,

$$\Delta T = \frac{\dot{Q}}{\dot{m}c}$$

$$\therefore T = \frac{25}{36\pi\rho k^2}\left(\frac{25gc^2}{48k^2\dot{Q}^2\pi T'\rho}\right)^{-\frac{1}{3}} \cdot z^{-\frac{5}{3}} + T'. \tag{16}$$

The temperature field of candle flame will provide a lift force to the tea bag.

3.2.3. *The Temperature Filed and the Motion*

Then we consider the motion of the tea bag in the temperature field we have calculated. Based on the analysis of forces exerted on the the remanent of tea bag m_1 and the air trapped in the loose and porous structure of the ash m_2 , we have the following equations at thermal equilibrium:

$$\rho'\frac{m_2}{\rho}g - (m_1 + m_2)g = (m_1 + m_2)\frac{d^2h}{dt^2} \tag{17}$$

$$m_2(\frac{\rho'}{\rho} - 1)g = (m_1 + m_2)\frac{d^2h}{dt^2} + m_1g. \tag{18}$$

Introduce $\beta = \frac{1}{V}\frac{dV}{dT}$,

$$V' - V = \beta V'dT. \tag{19}$$

Substitute

$$\frac{V}{V'} = 1 - \beta\Delta T = \frac{\rho'}{\rho}.$$

into Equation (16), where $\Delta T < 0$ for Δz. Then we can see how temperature distribution affects the motion.

$$-m_2g\beta\Delta T = (m_1 + m_2)\ddot{h} + m_1g. \tag{20}$$

From the above equation, if the temperature difference is large (i.e. ΔT is large), the acceleration of tea bag will be large. The above analysis assumes the friction can always hold up the tea bag. If there is no enough static friction, as long as the acceleration of air m_2 is large, high relative velocity between air and tea bag can drive the tea bag up due to viscosity.

From Equation (20) we know clearly that the temperature distribution affects the motion significantly. So the relevant parameters in Equation (16) can influence the motion of the tea bag. Since $\Delta T \propto (\frac{\dot{Q}}{k^2\rho c})^{\frac{2}{3}}$, bigger heat power, smaller coefficient of heat conduction and specific heat capacity of the air are beneficial to the upward motion of the tea bag. Also, the material property of the tea bag is crucial to the motion. The tea bag with the ash has loose and porous structure, so that hot air can be easily trapped.

3.2.4. *The Force Analyze for Different Part of the Motion*

Throughout the whole process of ascending motion, the lift force at the beginning of its rising motion is larger than that during the uniform motion process. The difference between these two stages lies in the time for the flame to change its surrounding temperature field. In our experiments after the ignition of tea bag, there are always several seconds during which the tea bag rests on the ground before its departure. During this time the flame establishes a dramatically-decaying temperature field $T \propto z^{-\frac{5}{3}}$ and the tea bag gains sufficient lift force to produce the acceleration. While the tea bag is rising upward, there is little time for the flame to heat its surrounding air thus the lift force is smaller, only sustaining a uniform motion.

4. Conclusion

The tea bag can take off when the top end of the cylinder is lit in virtue of the hot air around. We theoretically calculate the flow field and the temperature field of a point heat source and obtain $T \propto z^{-\frac{5}{3}}$ along axial direction. This dramatic spatial decay of temperature field creates a distinctive area in which the gas is characterized by a higher temperature and thus lower density than ambience. Such bulk of "hot air" ascends and drags the tea bag upward. Moreover, "hot air" around the flame is replenished through circulation, which sustains a uniform motion until the flame extinguishes. Intrinsic properties of the tea bag, such as its density and shape, as well as the burning heat of candle will affect its motion according to Equation (15). The reason why temperature affects so differently on two forces (the force supports a large acceleration and the force balancing gravity and air resistance to sustain a uniform motion upward), is that establishing the temperature distribution requires a period of time, so that a distinctive "hot air" can be well produced at the beginning of its ascending while there is little time for the flame to create such a temperature difference during its uniform rising motion.

References

1. L. Li. Theoretical analysis and experimental investigation on the vertical jet in horizontal flow. Master's thesis, Tongji University (2004).
2. Z. H. Wen X.D, Liu Z., Reproduction of the temperature field of candle flame using michelson interferometer, *Physics Experiment.* **27**(6), 44–47 (2007).

Chapter 16

2013 Problem 16: Hoops

Lan Chen*, Bo Xiong, Sihui Wang and Wenli Gao

School of Physics, Nanjing University

Jumping of an elastic hoop is a procedure of energy storage, transformation and allocation. The elastic potential energy stored due to deformation is related to the geometric parameters, the Young's modulus Y and the method of pressing. In our work, the efficiency of jumping rather than the absolute value of the height is emphasized.

In considering the energy allocation during a jumping, we define the jumping efficiency η as the ratio between the translational kinetic energy and the total kinetic energy. We proposed the existence of higher vibration modes based on the hoop's equation of motion. The vibration state of a hoop is usually a combination of different modes. The order of vibration modes is found to be a new parameter that determines the ratio of energy allocation, in which $\eta = 2n^2/3n^2 + 1$ rather than a fixed number $\eta = 8/13$ as given by previous studies. The material of the hoop and the corresponding recovery coefficients also affect the maximum jumping height.

1. Introduction

Problem Statement:

An elastic hoop is pressed against a hard surface and then suddenly released. The hoop can jump high in the air. Investigate how the height of the jump depends on the relevant parameters.

The jumping mechanics of an elastic hoop includes the storage, transformation and allocation of energy. In Sec. 2, we will investigate how elastic deformation energy is stored during pressing down, and its dependence on geometrical parameters like width W, thickness d and radius R and materials' parameters like Young's modulus.

---

*E-mail: chenlanphy@gmail.com

Once released, the deformation energy will transform into translational kinetic energy and vibrational energy, potential energy of gravity, and energy loss. E.Yang[1] found the fundamental vibration mode of the hoop and the energy ratio between translational kinetic energy and the total mechanical energy is a constant independent of hoops parameters and initial conditions. In their work, maximizing the height of the jump is reduced to a simple algebraic problem. In Sec. 3, based on Yang's work, we will investigate the problem further and find the existence of higher vibration modes. The vibration state of a hoop is usually a combination of different modes. The order of vibration modes is found to be an new parameter that determine the ratio of energy allocation, which is no longer a constant.

2. Energy Storage

2.1. *Bending Energy of Elastic Hoops*

We regard the hoop's motion as a two-dimensional problem in the vertical plane in which torsion is neglected.[2] Its deformation follows *Hooke's Law* to a good approximation. We consider a piece of hoop under pressure. As shown in Fig. 1, the length of any layer may undergo extension or compression due to deformation. But there exists a *neutral surface*[3] whose length keeps a constant.

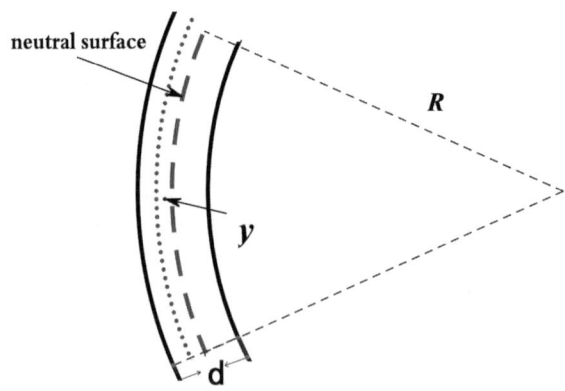

Fig. 1.   Deformation of a layer at a distance y from the neutral surface.

Consider a layer with a thickness dy and a distance y from the neutral surface. Its deformation of in length l_i is:

$$\Delta l_i = (\kappa - \kappa_0) y l_i, \tag{1}$$

where κ is the curvature defined as the reciprocal of curvature radius ($= \frac{1}{R}$) here and κ_0 is the curvature free of loads. Then the bending energy of the layer is:

$$E_{elas} = \sum_i \frac{1}{2} \frac{YW dy}{l_i} \cdot (\Delta l_i)^2, \tag{2}$$

where Y is *Young's modulus*. W is the width of the hoop. Therefore, the total elastic potential energy stored in the hoop is:

$$\begin{aligned} E_{elas} &= \frac{1}{2} YW \cdot \iint_{-\frac{d}{2}}^{\frac{d}{2}} (\kappa - \kappa_0)^2 y^2 \, dy dl \\ &= \frac{1}{24} YW d^3 \int_C (\kappa - \kappa_0)^2 \, dl. \end{aligned} \tag{3}$$

The subscript C means the integral along the whole curve. We can see that E_{elas} is proportional to Y and W. It is evident that a hoop with larger *Young's modulus* Y will store more energy and be able to jump higher. The width W will not influence the final jumping height, because both kinetic energy and potential energy contains the mass m which is proportional to W. The thickness d is much more important, because E_{elas} is proportional to the cubic of d. Having other parameters unchanged, a hoop with large thickness tends to jump higher. The last term in Eq. (3) containing the curvature of the hoop will depend on the deformation of the hoop due to the press and the method of press as described below (see 2.2).

If κ_0 is 0, the total energy[4] is simply proportional to

$$\int_C \kappa^2 \, dl. \tag{4}$$

Another case is a hoop made of a circle of radius R,[5] therefore $\kappa_0 = 1/R$. Generally, the total energy can't be calculated explicitly, unless the shape function of the curve is given.

2.2. *Application to Hoop Under Press*

The deformation δ is defined as the displacement of the top point. We measure the δ of a hoop under certain loads with two pressing methods. One is "flat press", see Fig. 2(a). The other is called "concentrated press", see Fig. 2(b), which is realized by tying a light string to the hoop. The hoop is released by cutting the string suddenly.

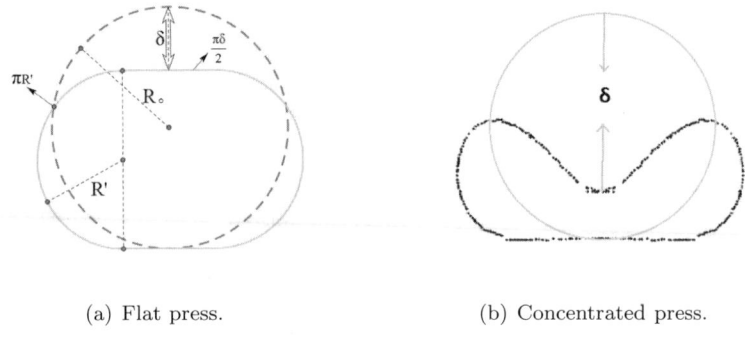

(a) Flat press.   (b) Concentrated press.

Fig. 2.   Different press methods.

2.2.1. *Flat Press*

In the case of flat press, the shape of the hoop can be simplified as two semi-circles with diameter $2R - \delta$, and two straight lines with length $(\frac{\pi\delta}{2})$. So it is convenient to get the integral Eq. (3) as

$$E_{elas} = \frac{1}{24}YWd^3 \cdot \frac{4\pi}{2R - \delta} \qquad (\kappa_0 = 0) \qquad (5)$$

$$or \ \frac{1}{24}YWd^3 \cdot \frac{2\pi\delta}{R(2R - \delta)} \qquad (\kappa_0 = \frac{1}{R}). \qquad (6)$$

Then, the applied load on the hoop can be expressed as a function of deformation δ:

$$F = \frac{\partial E}{\partial \delta} = \frac{1}{24}YWd^3 \cdot \frac{4\pi}{(2R - \delta)^2} \qquad (\kappa_0 = 0 \ or \ \frac{1}{R}). \qquad (7)$$

The expression is the same for $\kappa_0 = 0$ or $1/R$ just like two springs have the same force deformation relation event though their origin positions differ by a initial deviation.

It is more convenient to be expressed as:

$$f = \frac{\pi}{24} \cdot \frac{Yd^3}{R^2} \cdot \frac{1}{(1-\delta')^2}, \tag{8}$$

where $f = F/W$ is the *force intensity*. $\delta' = \delta/2R$ is called *relative deformation* ranging from $0 \sim 1$.

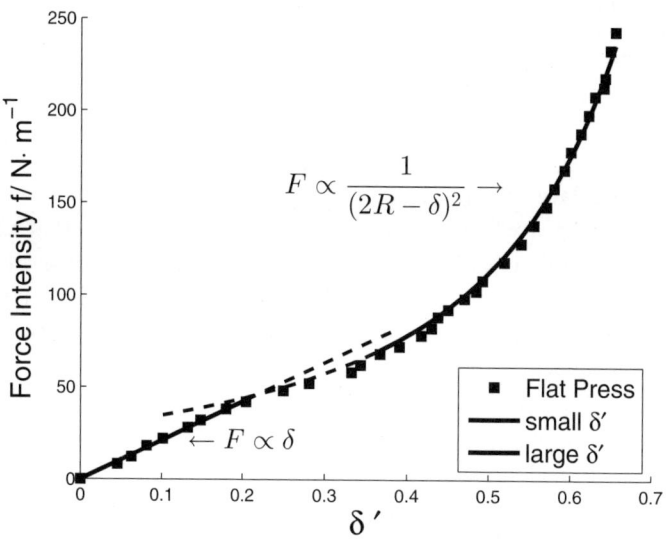

Fig. 3. Force intensity f's dependence upon relative deformation δ' under flat press. The sample is a resin hoop with $R = 6.40cm, d = 0.4mm$. The curve is linear when deformation is small. As the deformation increases, the force on hoop rises rise rapidly as described by Eq. (8).

We measured the applied load F and the deformation δ with a resin hoop with R=6.40cm and d=0.4mm and Y=12.9Gpa.

The data are shown as the squares in Fig. 3. The upper part of the solid curve is the result of Eq. (8). We see that most of the data are close to the theoretical value except those close to origin point when δ' is small. The reason is that in the range of small deformation, the semicircle-and-line simplification is invalid. As a matter of fact, the linear approximation gives a rather satisfactory description in this range, as we can see from Fig. 3.

Based on Eq. (3) and dimensional analysis, we can write the expression for small "δ'" :

$$F = \gamma \frac{YWd^3}{R^3} \delta.$$

i.e.

$$f = 2\gamma \cdot \frac{Yd^3}{R^2} \cdot \delta'. \tag{9}$$

γ is a constant coefficient here.

By linear fitting according to experimental data, coefficient "γ" is determined to be 0.49.

2.2.2. *Concentrated Press*

According to Eq. (8), for flat press, the force and deformation relation is nonlinear in a large range of deformation. More importantly, even the Eq. (8) does not hold when the hoop is pressed too much so that the deformation becomes nonelastic. On the other hand, the concentrated press, though difficult in giving an analytical f δ' relation, has a surprisingly simple experimental result (see Fig. 4). And the hoop can restore its shape after large deformation.

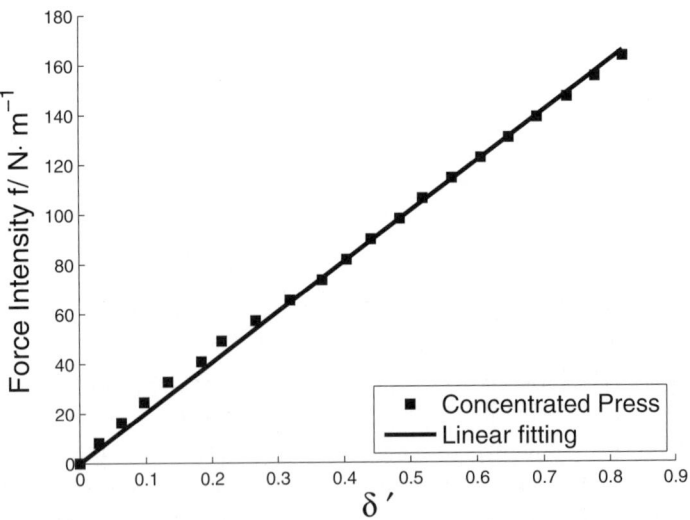

Fig. 4. Force intensity f depending upon relative deformation δ' concentrated press. The sample is a resin hoop with $R = 6.40cm, d = 0.4mm$. The curve is linear for arbitrary deformation.

Based on Eq. (3) and dimensional analysis, we can also suppose that :

$$f = 2\gamma' \cdot \frac{Yd^3}{R^2} \cdot \delta'. \tag{10}$$

Here the fitting result of γ' is 0.47 according to Fig. 4, which is close to γ above. It indicates that flat and concentrated press are similar to each other for small deformation.

Notice that compared to the concentrated press, the hoop is more rigid under flat press, and can store more elastic energy under certain deformation. As a result, the jumping height is also dependent on the method of press. However, considering the linear force relation upon full range of deformation (δ' ranging from 0 to 1), we will do our further research with the method of concentrated press. In this case, we can also deduce that:

$$E_{elas} \propto \delta^2, \tag{11}$$

for $F = \partial E / \partial \delta$.

3. Energy Allocation

3.1. *Vibration Modes*

When a hoop is released, the deformation energy will transform into translational kinetic energy and vibrational energy, potential energy of gravity, and a certain ratio dissipated during the collision with the hard surface. If we focus on the moment when the hoop is just leaving the hard surface, the gravitational energy can be neglected.

The displacements in the axial, circumferential and radial directions are denoted by u, v, w respectively as shown in Fig. 5. Consider the 2-dimensional motion of the hoop, in that $(u = 0)$, we may have the Eqs. (5):

$$\frac{\partial^2 v}{\partial \theta^2} + \frac{\partial w}{\partial \theta} = \rho \frac{(1-\nu^2)R^2}{Y} \frac{\partial^2 v}{\partial t^2} \tag{12}$$

$$\frac{\partial v}{\partial \theta} + \left[1 + k \left(1 + \frac{\partial^2}{\partial \theta^2} \right)^2 \right] w = -\rho \frac{(1-\nu^2)R^2}{Y} \frac{\partial^2 w}{\partial t^2}. \tag{13}$$

The stable solution to Eq. (12), (13) is:

$$v = B \sin n\theta \cos(\omega t + \alpha_0) \tag{14}$$

$$w = C \cos n\theta \cos(\omega t + \alpha_0) \tag{15}$$

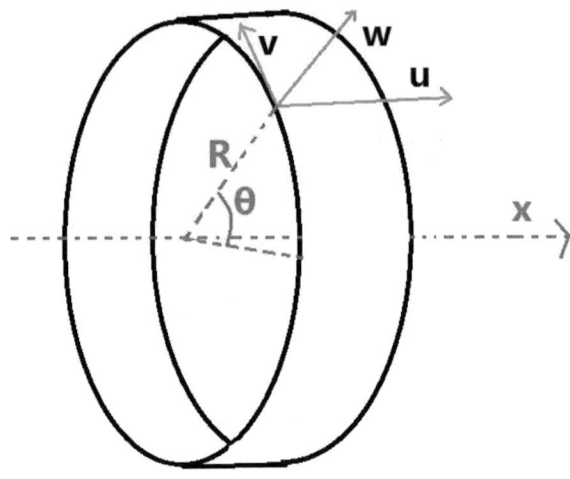

Fig. 5.   Definition of three displacements and u is 0 here.

where the amplitudes B, C satisfy:

$$\begin{pmatrix} n^2 - \Omega^2 & n \\ n & 1 + k(1 - n^2)^2 - \Omega^2 \end{pmatrix} \cdot \begin{bmatrix} B \\ C \end{bmatrix} = \begin{bmatrix} 0 \\ 0 \end{bmatrix}, \tag{16}$$

in which

$$k \equiv \frac{d^2}{12R^2} \approx 0 \quad (\text{as} \quad d << R)$$

$$\Omega^2 = \begin{cases} 0, 1 + k & (n = 0) \\ \frac{1}{2} [1 + n^2 + k(n^2 - 1)^2 \\ \quad \mp \sqrt{[1 + n^2 + k(n^2 - 1)^2]^2 - 4kn^2(n - 1)^2}] & (n \neq 0). \end{cases}$$

In reality $n \geqslant 2$, and $k = 0$, so that the amplitude ratio is:

$$\frac{B}{C} = \frac{n}{\Omega^2 - n^2} = -\frac{1}{n}. \tag{17}$$

See Fig. 6, $2n$ is the standing wave number because $2n$ wave nodes exist for each mode. Only the fundamental mode with $n = 2$ is considered in reference.[1] Generally, multi-mode vibration is inevitable, because the initial condition for a specific mode, say $n = 2$, is strict and hard to control

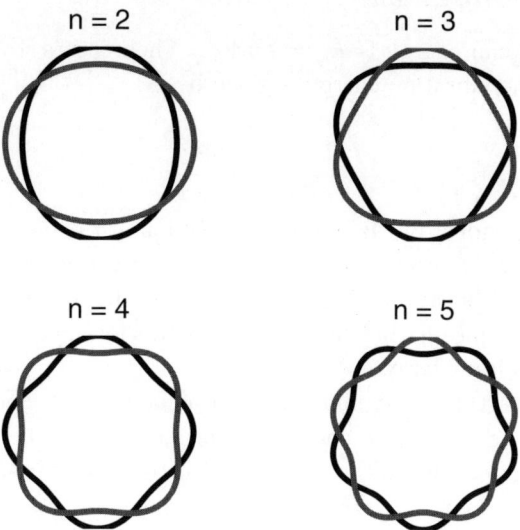

Fig. 6. Vibration modes of different n. There are $2n$ wave nodes for each mode.

in the experiment. For example, an unsymmetrical release would result in odd numbered modes like $n = 3, 5, \dots$.

The initial phase α_0 is taken as $\pi/2$ in our coordinates. So the initial velocities are:

$$\dot{v} = B\omega \sin n\theta \tag{18}$$

$$\dot{w} = -nB\omega \cos n\theta. \tag{19}$$

The lowest point of the hoop must be an antinode of vibration. So in mass center reference frame we get:

$$v\big|_{(\theta = -\frac{\pi}{2})} = V_0.$$

The vibration energy of the hoop with mode number (n) is:

$$\begin{aligned}
E_v &= \frac{1}{2} \int_0^{2\pi} \frac{m}{2\pi} (\dot{v}^2 + \dot{w}^2) \, d\theta \\
&= \frac{n^2 + 1}{4n^2} m V_0^2.
\end{aligned} \tag{20}$$

3.2. *Energy Allocation*

The translational energy is $E_t = \frac{1}{2}mV_0^2$. Then the ratio of translational energy and the vibration energy is given by

$$\frac{E_t}{E_v} = \frac{2n^2}{n^2 + 1} \tag{21}$$

which is independent of the initial deformation of the hoop. It is no longer a constant as $8/5$. Instead, n is an important parameter to determine the kinetic energy allocation. The translational kinetic energy increases for larger n, which means that hoop in high vibration mode can jump higher for it requires less energy.

We consider the energy allocation at the moment the hoop is just taking a leap from surface, so the energy lost in air drag can be neglected. The energy dissipation due to the collision with the hard surface is denoted as E_{coll}. Suppose the recovery coefficient is e, so $E_{coll} = (1 - e^2) \cdot E_{elas}$. From energy conservation we have:

$$E_v + E_t = e^2 \cdot E_{elas}. \tag{22}$$

Combine it with Eq. (21), we can define the efficiency of jumping η as:

$$\eta = \frac{E_t}{e^2 E_{elas}} = \frac{2n^2}{3n^2 + 1}. \tag{23}$$

Fig. 7 shows the efficiency with respect to n. For $n = 2$, the ratio reduces to $\eta = 8/13$, as given by E.Yang.[1] But the efficiency cannot improve much with higher n, because the upper limit is $\eta \to 2/3$, as $n \to +\infty$.

4. Maximum Height

When the hoop reaches the maximum height, the translational energy transforms into the gravitational potential energy (E_h). Neglecting the air friction, we have:

$$E_t = mgH_{max} \tag{24}$$

So the equation of the maximum height with respect to relevant parameters is:

$$H_{max} = \frac{1}{mg} \Big[\frac{2n^2}{3n^2 + 1} \cdot e^2 E_{elas} \Big] \tag{25}$$

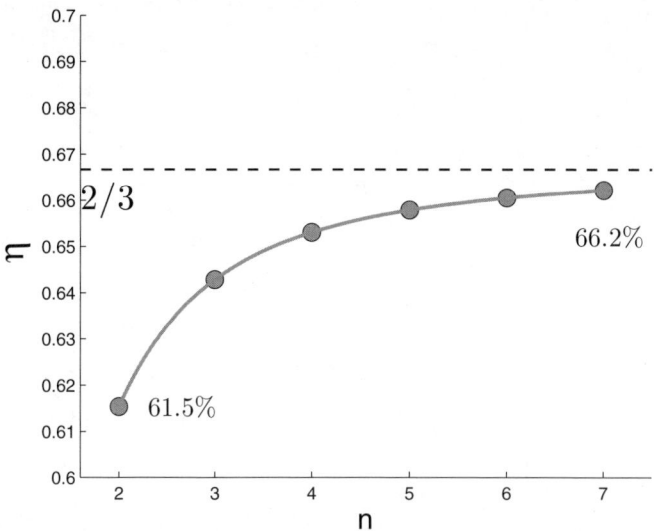

Fig. 7. η changes with n. $\eta \to 2/3$, as $n \to +\infty$. The actual vibration of a hoop is the superposition of different modes.

where:

$$E_{elas} = \begin{cases} \frac{\pi}{6}YWd^3 \cdot \frac{1}{(2R-\delta)} & \text{(Flat \quad press)} \\ \frac{\gamma}{2}\frac{YWd^3}{R}\delta^2 & \text{(Concentrated \quad press)}. \end{cases}$$

e is relevant to the material of hoops, n is an new parameter which is related to the initial condition of the hoop when released.

In our experiment, the relation between maximum height and deformation under concentrated press is shown in Fig. 8. This result means:

$$E_t \propto \delta^2. \tag{26}$$

Let $\eta' = \eta e^2 = \frac{E_t}{E_{elas}}$ which is a constant for a certain hoop. Notice that $E_{eals} \propto \delta^2$ under concentrated press (see Eq. 11), it proves that η' is a constant for a certain hoop. The result for different hoops are listed in Tab. 1. We can see that η' of resin hoops is significantly higher than that of manganese hoops.

Hoops with various parameters are used (see Tab. 1).

Since the initial boundary condition of vibration is unsymmetrical, the vibration of hoops should be generally the superposition of multiple modes,

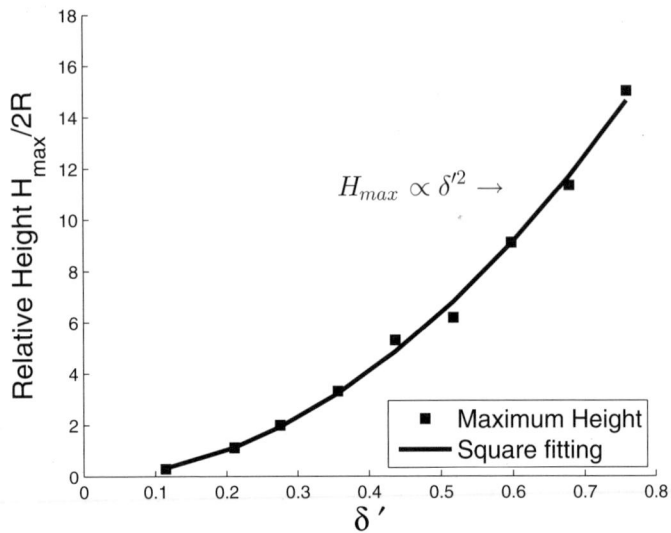

Fig. 8. Maximum height's dependence upon relative deformation δ'.

Table 1. Parameters and coefficients of the hoops.

No.	material	thickness (d/mm)	diameter (2R/cm)	width (W/cm)	mass(m/g)	η' (%)
1	manganese	0.2	8.80	1.46	6.9	47
2	manganese	0.2	10.05	1.58	8.0	41
3	manganese	0.2	10.60	1.50	7.4	45
4	manganese	0.2	11.12	1.51	8.7	41
5	manganese	0.2	11.90	1.30	8.0	49
6	resin	0.4	6.20	3.10	2.7	58
7	resin	0.4	6.45	2.45	2.2	59

rather than only the fundamental mode ($n = 2$) alone. From Eq. (21), η changes between 61.5% and 66.7%, so we can estimate the range of recovery coefficient e as

$$e = \begin{cases} 0.82 \sim 0.85 & \text{(manganese)} \\ 0.93 \sim 0.97 & \text{(resin)}. \end{cases} \qquad (27)$$

5. Conclusion

The problem on maximum height of a jumping hoop includes the storage, transformation and allocation of energy. A general expression shows that elastic potential energy stored under deformation is proportional to its Young's modulus Y and width W, and cubic of its thickness d^3. It also has a complicated dependence on the local curvature due to deformation which is related to the method of pressing. We discussed two pressing methods: the flat press and concentrated press. Experiment shows that the stored elastic energy is simply proportional to δ^2 for concentrated press. Under small deformation, the elastic energy for flat press is almost the same as the result of concentrated press. For flat press under large deformation, the elastic energy can be derived analytically by a simple geometrical model which increases more rapidly with respect to deformation δ. Comparing the two method of pressing, a hoop by flat press has the potential to jump higher owing to more elastic energy under certain deformation. However, considering the linear force relation upon full range of deformation (δ' ranging from 0 to 1), we take concentrated press in the jumping experiment. In this case, $E_{elas} \propto \delta^2$ and $F = \partial E / \partial \delta$. Moreover, the efficiency of jumping rather than the absolute value of the height is more important.

Once released, the elastic energy is transformed into translational kinetic energy and vibrational energy, potential energy of gravity, and energy loss. We proposed the existence of multi-mode vibration, and energy allocation between the translational and vibrational kinetic energy has been reconsidered. In our theory, the ratio between the translational kinetic energy and the total kinetic energy is $\eta = \frac{2n^2}{3n^2+1}$ rather than a fixed number $\eta = 8/13$ as given by the reference.[1] The vibration state of a hoop is usually a combination of different modes. The order of vibration modes is found to be a new parameter that determines the ratio of energy allocation, which is no longer a constant.

The material of the hoop and the corresponding recovery coefficients also affect the maximum jumping height.

References

1. E. Yang and H.-Y. Kim., Jumping hoops, *America Journal of Physics.* **80**(19) (2012).
2. J. G.AMBATI and J.C.K.SHARP, In-plane vibration of annular rings, *Journal of Sound and Vibration.* **47**(3), 415–432 (1976).

3. A. E. H. Love, *A Treatise on the Mathematical Theory of Elasticity*, 4th edn. Dover, New York (1927).
4. F. P.Beer, J. E.Russell Johnston, and J. T.DeWolf, *Mechanics of Materials*, 3th edn. Tsinghua University Press (2003).
5. Wikipedia, Curvature http://en.wikipedia.org/wiki/Curvature.

Chapter 17

2013 Problem 17: Fire Hose

Li Du[*], Sihui Wang and Wenli Gao

School of Physics, Nanjing University

The complicated motion of a liquid-conveying pipe is studied systematically. We derive the linear dynamic equation by introducing the *Intrinsic Coordinate* which simplifies the equation of motion in present literatures. In the experiment part, we measured the frequency of the pipe on a horizontal low friction plane with two parameters controlled: the total length of the pipe and the velocity of the fluid. We also record the shape of the pipe and compare it to the calculated results according to our equation. Satisfying agreement is found between our theory and experiment.

1. Introduction

Problem Statement:

"Consider a hose with a water jet coming from its nozzle. Release the hose and observe the subsequent motion. Determine the parameters that affect this motion."

Interactions between fluid and engineering structures often induces harmful vibrations, some of which can even result in irreversible damage and lead to severe property losses. As a typical case, the famous Tacoma Narrow Bridge collapsed because of aeroelastic flutter in 1940 shortly after opening and aroused wide public concerns. Also, the vibration of fluid-conveying pipes in oil and nuclear industry can significantly influence the reliability of devices. To cope with these engineering problems, a great number of researches have been done in fluid-induced vibrations.

The problem asks us to study the motion of a fire hose, which can be seen as a fluid-conveying pipe with one end clamped and the other end free. We further assume that the hose is placed on a horizontal low friction plane. By introducing the *Intrinsic Coordinate*, the configuration of a pipe

---

[*]E-mail: ldu.nju@hotmail.com

with random shape can be described. Using the kinematic properties under Intrinsic Coordinate, the linear equation of motion can be derived. After studying the solution of the equation, a critical point is found, after which the hose losses stability and begins to vibrate. The vibration frequency and the shape of the pipe is compared between the theory and the experiment.

Nomenclatures

---

e_n	Normal base
e_t	Tangential base
L	Total length of pipe
EI	Flexural rigidity
M	Mass per unit length of fluid
m	Mass per unit length of pipe
U	Flow velocity of fluid
P	Hydraulic pressure
F	Hydrodynamic force
F_i	Inertial force
F_c	Coriolis force
$\mathbf{M}$	Bending momentum
T	Longitudinal tension
Q	Transverse shear force
κ	Curvature
$\beta, u, \Lambda, \Omega$	Dimensionless quantities

---

2. Theory

2.1. *Intrinsic Coordinates and I-C Transform*

Starting from a discrete system, we divide the pipe into N segments. Each segment corresponds to a generalized coordinate (Fig. 1). Thus the pipe is described by $\{\alpha(i,t)\}_{i=1}^{N}$. As $N \to \infty$, the system has infinite degrees of freedom so that the generalized coordinate can be transformed into

$$\lim_{N \to \infty} \{\alpha(i\Delta l, t)\}_{i=1}^{N} = \{\alpha(l,t)\}_{l=1}^{L} \tag{1}$$

where l denotes the length from the origin to a fixed point along the pipe and α denotes the corresponding inclination (Fig. 2).

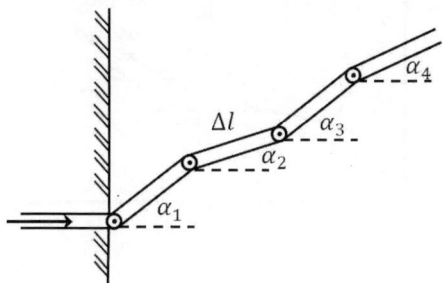

Fig. 1.   The pipe is divided into discrete segments, where $\{\alpha(i\Delta l, t)\}_{i=1}^N$ is the generalized coordinate of the system.

The couple of variables (l, α) are defined as *Intrinsic Coordinates*. It's easy to find the transform between *Intrinsic Coordinate* and Cartesian Coordinate by integrating along the curve starting from the origin. The parametric equation as follow is named to be *I-C transform*.

$$\begin{cases} x = \displaystyle\int_0^l cos\alpha dl \\[2mm] y = \displaystyle\int_0^l sin\alpha dl. \end{cases} \tag{2}$$

It can be noticed here that we only need one variable $\alpha(t)$ to describe the configuration of the pipe. If we can find the motion under *Intrinsic Coordinates* and do the *I-C Transform*, the configuration of the pipe is completely solved.

2.2. *Kinematic Relations*

In this section, some kinematic properties under Intrinsic Coordinates will be studied. The normal and tangential basis (Fig. 2) are defined as follow

$$\begin{cases} e_t^l = cos\alpha \boldsymbol{i} + sin\alpha \boldsymbol{j} \\ e_n^l = -sin\alpha \boldsymbol{i} + cos\alpha \boldsymbol{j}. \end{cases} \tag{3}$$

Under *Intrinsic Coordinates*, the expression of curvature is

$$\kappa = \frac{\partial \alpha}{\partial l}. \tag{4}$$

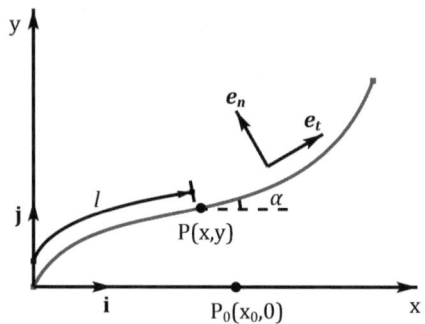

Fig. 2.   The continuous system described under Intrinsic Coordinate which is fixed on the pipe.

The acceleration of a pipe element is studied starting from the discrete system as follow. In the discrete system (Fig. 1), each segment is rotating around its previous node. Set the acceleration at the i'th node to be $\boldsymbol{a}^{(i)}$, then

$$\boldsymbol{a}^{(i+1)} = \boldsymbol{a}^{(i)} + \ddot{\alpha}\Delta l\, \boldsymbol{e}_n^{(i+1)} - \dot{\alpha}^2 \Delta l\, \boldsymbol{e}_t^{(i+1)}. \tag{5}$$

The first term on right-hand-side denotes the acceleration of transportation and the latter two terms denotes the relative acceleration. By summing up Eq. 5, we get

$$\boldsymbol{a}(l) = \lim_{n \to \infty} \sum_{i=1}^{n} (\ddot{\alpha}\, \boldsymbol{e}_n^{(i+1)} - \dot{\alpha}^2\, \boldsymbol{e}_t^{(i+1)})\Delta l. \tag{6}$$

For a continuous system, Eq. 6 can be transformed into integral. By respectively multiplying the tangential and normal basis, we get two components of acceleration as follow

$$\begin{cases} a_n(l_0) = \displaystyle\int_0^{l_0} (\ddot{\alpha}\, \boldsymbol{e}_n^{(l)} \cdot \boldsymbol{e}_n^{(l_0)} - \dot{\alpha}^2\, \boldsymbol{e}_t^{(l)} \cdot \boldsymbol{e}_n^{(l_0)})dl \\[2mm] a_t(l_0) = \displaystyle\int_0^{l_0} (\ddot{\alpha}\, \boldsymbol{e}_n^{(l)} \cdot \boldsymbol{e}_t^{(l_0)} - \dot{\alpha}^2\, \boldsymbol{e}_t^{(l)} \cdot \boldsymbol{e}_t^{(l_0)})dl. \end{cases} \tag{7}$$

Noting the orthogonally of two bases, following differential relations can be found by differentiating both sides by l_0, which is crucial for our study

$$\begin{cases} \dfrac{\partial a_n}{\partial l} = \dfrac{\partial^2 \alpha}{\partial t^2} - a_t \dfrac{\partial \alpha}{\partial l} \\[3mm] \dfrac{\partial a_t}{\partial l} = -(\dfrac{\partial \alpha}{\partial t})^2 + a_n \dfrac{\partial \alpha}{\partial l}. \end{cases} \tag{8}$$

We further assume that deflection angle to be small, i.e. $\alpha \sim O(\varepsilon)$, and neglect higher order terms, thus

$$\begin{cases} \dfrac{\partial a_n}{\partial l} \cong \dfrac{\partial^2 \alpha}{\partial t^2} \\[2mm] \dfrac{\partial a_t}{\partial l} \cong 0. \end{cases} \tag{9}$$

As can be observed, the tangential acceleration is approximated to be zero, which indicates that tangential motion is much smaller than normal motion. So in the following sections, the motion on the tangential direction will be neglected.

2.3. *Equation of Motion*

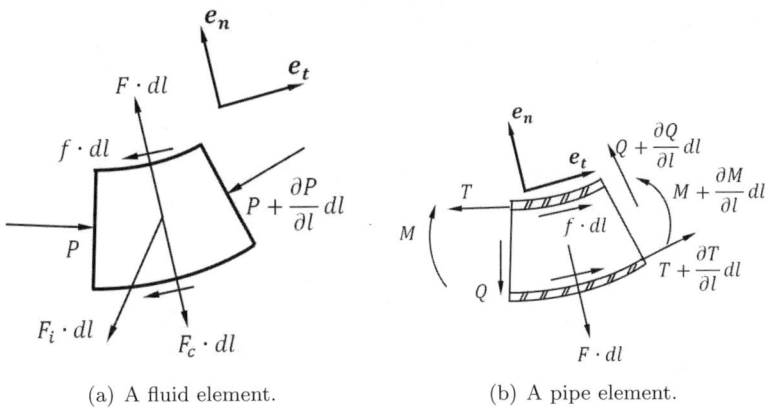

(a) A fluid element.   (b) A pipe element.

Fig. 3.   Free-body diagram.

We consider a uniform and inextensible pipe placed on a horizontal low friction plane. First let's analyze a fluid element inside the pipe. Under Intrinsic Coordinates, the frame is non-inertial with acceleration $a(l)$ and angular velocity $\omega = \dot{\alpha}\mathbf{k}$, thus the Coriolis force

$$\mathbf{F}_c = -2M\dot{\alpha}U\, e_n dl \tag{10}$$

and the inertial force

$$\mathbf{F}_i = -M(a_n e_n + a_t e_t)dl. \tag{11}$$

It can be seen on the free-body diagram that the contributions of hydraulic force, longitudinal tension and transverse shear force on an element are

$$
\begin{cases}
\dfrac{\partial}{\partial l}(-Pe_t^{(l)})dl = -(P\dfrac{\partial \alpha}{\partial l}e_n^{(l)} + \dfrac{\partial P}{\partial l}e_t^{(l)})dl \\
\dfrac{\partial}{\partial l}(Te_t^{(l)})dl = (T\dfrac{\partial \alpha}{\partial l}e_n^{(l)} + \dfrac{\partial P}{\partial l}e_t^{(l)})dl \\
\dfrac{\partial}{\partial l}(Qe_n^{(l)})dl = (\dfrac{\partial Q}{\partial l}e_n^{(l)} - Q\dfrac{\partial \alpha}{\partial l}e_t^{(l)})dl.
\end{cases}
\tag{12}
$$

Applying Newton's second law to the pipe and the fluid on normal direction yields

$$
\begin{cases}
F - 2MU\dfrac{\partial \alpha}{\partial t} - Ma_n - P\dfrac{\partial \alpha}{\partial l} = MU^2\dfrac{\partial \alpha}{\partial l} \\
\dfrac{\partial Q}{\partial l} + T\dfrac{\partial \alpha}{\partial l} - F = ma_n
\end{cases}
\tag{13}
$$

and momentum balance leads to

$$
\frac{\partial \mathbf{M}}{\partial l}dl + \mathbf{e}_t \times \mathbf{Q}dl = 0.
\tag{14}
$$

We assume that our hose satisfies the Euler-Bernoulli beam theory, which means the following relation between the curvature and the momentum holds

$$
\mathbf{M} = EI \cdot \kappa.
\tag{15}
$$

Eliminating F from Eq. 13, we get

$$
EI\frac{\partial^3 \alpha}{\partial l^3} + MU^2\frac{\partial \alpha}{\partial l} + 2MU\frac{\partial \alpha}{\partial t} + (T-P)\frac{\partial \alpha}{\partial l} + (M+m)a_n = 0.
\tag{16}
$$

Since $(T-P) \sim O(\varepsilon)$, the fourth term can be neglected. We differentiate both sides by l and use Eq. 9. The final equation of motion can be derived

$$
EI\frac{\partial^4 \alpha}{\partial l^4} + MU^2\frac{\partial^2 \alpha}{\partial l^2} + 2MU\frac{\partial^2 \alpha}{\partial l \partial t} + (M+m)\frac{\partial^2 \alpha}{\partial t^2} = 0.
\tag{17}
$$

For the clamped and the free ends of the pipe, following boundary conditions are adopted

$$
\alpha\mid_{l=0} = \frac{\partial \alpha}{\partial l}\mid_{l=0} = \frac{\partial^2 \alpha}{\partial l^2}\mid_{l=1} = \frac{\partial^3 \alpha}{\partial l^3}\mid_{l=1} = 0.
\tag{18}
$$

2.4. *Solution*

Let's assume the solution in the form of

$$\alpha = Re[Y(l)e^{i\omega t}] \tag{19}$$

where ω denotes the complex frequency. The imaginary part of ω dominates the amplitude over time, and the real part dominates the vibration frequency. Substitute the expression into Eq. 17 and we get

$$EI\frac{d^4Y}{dl^4} + MU^2\frac{d^2Y}{dl^2} + 2MUi\omega\frac{dY}{dl} - (M+m)\omega^2 Y = 0. \tag{20}$$

By taking the trial solution $Y = Ae^{i\lambda l}$, and defining the following dimensionless quantities for convenience

$$\beta = \frac{M}{M+m}, \quad u = (\frac{M}{EI})^{1/2}LU, \quad \Lambda = \lambda L, \quad \Omega = (\frac{M+m}{M})^{1/2}\frac{L}{U}\omega \tag{21}$$

the corresponding eigen equation is then

$$\frac{\Lambda^4}{u^2} - \Lambda^2 - 2\beta^{1/2}\Omega\Lambda - \Omega = 0. \tag{22}$$

Now each mode can be written as

$$\alpha = Re[\sum_{j=1}^{4} A_j e^{i\lambda_j l}e^{i\omega t}]. \tag{23}$$

The restriction of boundary conditions brings about the following relations

$$\begin{cases} \sum_{j=1}^{4} A_j = 0 \\ \sum_{j=1}^{4} \lambda_j A_j = 0 \\ \sum_{j=1}^{4} \lambda_j^2 e^{i\lambda_j} A_j = 0 \\ \sum_{j=1}^{4} \lambda_j^3 e^{i\lambda_j} A_j = 0. \end{cases} \tag{24}$$

For nontrivial solution, the determinant of coefficient must vanish, thus

$$\begin{vmatrix} 1 & 1 & 1 & 1 \\ \lambda_1 & \lambda_2 & \lambda_3 & \lambda_4 \\ \lambda_1^2 e^{i\lambda_1} & \lambda_2^2 e^{i\lambda_2} & \lambda_3^2 e^{i\lambda_2} & \lambda_4^2 e^{i\lambda_2} \\ \lambda_1^3 e^{i\lambda_1} & \lambda_2^3 e^{i\lambda_2} & \lambda_3^3 e^{i\lambda_2} & \lambda_4^3 e^{i\lambda_2} \end{vmatrix} = 0. \tag{25}$$

To conclude, using Eq. 22 and Eq. 25, we can solve ω and λ. So the solution of the linear equation of motion can be figured out.

We construct an algorithm as follow (Fig. 4). With different flow velocity input, the complex frequencies can be solved and compared to experimental result.

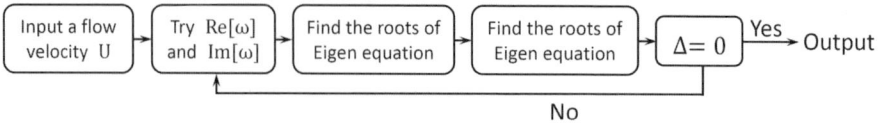

Fig. 4.   Flow chart of the algorithm.

3. Experiment

3.1. *Parameters of the Hose*

We place a rubber hose on a horizontal low friction plane lubricated by liquid soap. The parameters are: inner diameter of the hose $\phi = 5.25mm$, outer diameter $\phi_0 = 7.47mm$, length of the hose $L = 683mm$, total mass of the hose $m_0 = 12.2g$, density of water $\rho = 1000kg/m^3$. The flexural rigidity is $EI \cong 9.10 \times 10^{-5}(N)$.

$$I = \int y^2 ds. \tag{26}$$

3.2. *Experimental Results*

In the experiment, we use a camera to record the motion of the pipe. A set of XOY axis is marked on the plane, and one end of the hose is tightly clamped at the origin. Water flow is pumped into the hose and is controlled by a valve. The flow velocity can be read from a watermeter.

Using Eq. 23 and *I-C Transform*, the configuration of the pipe can be calculated. However, in linear dynamics, the amplitude goes to infinity exponentially which is contrary to the fact. This contradiction is due to the ignorance of nonlinear effect in our theory. As the amplitude becomes considerably large, nonlinear factors will take place and stabilize the amplitude. Here, as a simple model, we just neglect the real part of complex frequency and sketch the shape of the pipe in one half period (Fig. 5), and very similar patterns can be found by theory and experiment.

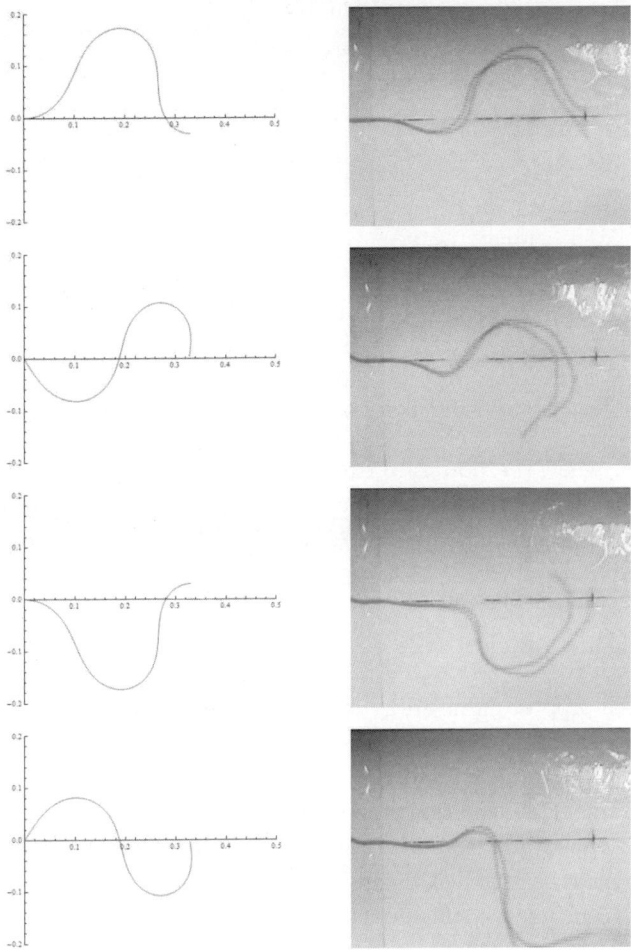

Fig. 5. Comparisons between computational results and real cases of pipe's configurations. The interval between each picture is 1/8 period.

By varying the flow rate of the injecting water, we study the responding frequencies of the hose. In Fig. 6, the squares are the experimental data, the line connecting the dots are the theoretical result. In the figure, the theoretical value of the frequency corresponding to the flow rate below the critical value is taken as zero. In this region, the imaginary part of the complex frequency is positive and the amplitude of vibration is decaying.

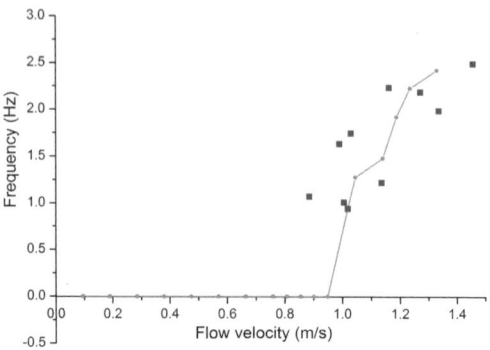

Fig. 6.   Vibration frequency with flow rate. The region before critical velocity is cut off.

Afterwards, when the flow rate exceeds the critical value, the pipe loses stability and begins to vibrate. Satisfying agreement is found in Fig 6, where the theoretical curve fits the data well, and all the data are converged after the critical velocity.

4. Conclusion

Our research proposes a new way for solving the fire hose problem. Unlike previous works, the linear governing equation of motion is derived under the *Intrinsic Coordinates*, and the corresponding *I-C Transform* is found. By solving the eigne equation, the critical flow rate is found at which the pipe loses stability and begins to vibrate. The critical condition and the vibrating frequency is studied theoretically, and compared with experiment. We conduct our experiment on a horizontal low friction board. The flexural rigidity and other parameters are all directly measured and satisfying agreement is found between theory and experiment.

Through our study, we conclude that the most essential property of the fire hose system is determined by the dimensionless mass β – it directly decides the dynamic behaviors of the linear system. But it's hard to tell qualitatively that how β influences the system without a careful computation. Still, we may conclude that – the longer the total length L becomes, the lower the frequency is; the higher the flow velocity U becomes, the higher the frequency is.

Chapter 18

2013 Problem 3: Bouncing Ball

Bo Xiong[*], Li Du, Sihui Wang and Wenli Gao

School of Physics, Nanjing University

In this solution, we study the rebound of a liquid-filled Ping-Pong ball after a free-fall motion. We classify the collision into "rigid-like motion" or "liquid-involved" motion. The most significant parameter is the amount of water. The rebounding height is suppressed most as the amount of water is about half of the total volume, exhibiting a typical "liquid-involved" motion. As the amount of water increases further, the rebounding height gradually recovers, and the ball becomes rigid again. We build a theoretical model to interpret the phenomenon. The model describes the formation of the flow field during the collision stage based on the momentum propagation and flux conservation of the liquid. An effective mass is introduced to describe the confinement effect on water by the sphere. Our model successfully predicts the bouncing height with respect to the amount of water. Releasing height is also an important parameter in determining the "nature of collision". As the releasing height increases, the whole system tends to become more and more "rigid". We classify the nature of collision above a certain releasing height as "rigid-like" regardless of the amount of water inside the ball.

1. Introduction

Problem Statement:

If you hold a Ping-Pong ball above the ground and release it, it bounces. The nature of the collision changes if the ball contains liquid.Investigate how the nature of collision depends on amount of liquid inside the ball and other relevant parameters.

In this solution, we study the rebound of a liquid-filled Ping-Pong ball after a free-fall motion. The process is divided into three stages:

---

[*]E-mail: 1130214891@qq.com

- **Free-fall stage.** The liquid-filled Ping-Pong ball falls down vertically from a certain high. We assume the Ping-Pong ball is released without initial disturbance, thus the surface of liquid maintains horizontal during the free-fall stage.
- **Collision stage.** The Ping-Pong ball collides with the ground. Intense impact occurs between the ground, the Ping-Pong ball and the liquid. At the end of this stage, the surface of liquid is still horizontal, but a flow velocity field is formed due to the impact.
- **Rebounding stage.** The Ping-Pong ball leaves the ground. Interaction between the liquid and the Ping-Pong ball persists because of the flow field that already exists.

According to our analysis, energy loss during the free-fall and rebounding stages can be neglected, especially when the viscosity of the liquid is small. Thus the 'nature of collision' is mainly associated with the dynamic behavior during the collision stage, including the formation of the flow field and the energy dissipation during the collision.

The total energy loss during the whole process can be reflected by the maximum height the Ping-Pong ball is able to reach after rebounding. From the maximum jumping height, we will classify the collision into "rigid-like" motion in which the ball behaves like a rigid body; or "liquid-involved" motion in which energy is dissipated significantly.

The most significant parameter is the amount of water. We will investigate how the bouncing height changes with amount of water inside the ball. We will build a theoretical model to interpret the phenomenon. The model describes the formation of the flow field during the collision stage based on the momentum propagation and flux conservation of the liquid. An effective mass m'_{ball} is introduced to describe the confinement effect on water by the sphere. Our model qualitatively predicts the bouncing height with respect to the amount of water. Satisfying agreement will be found between our theoretical and experimental rebounding height.

Another parameter, the releasing height is also found important in determining the "nature of collision".

2. Experiment

In the experiment, we use a Ping-Pang ball infused with water to study how the amount of liquid affects the nature of collision. Half of the ball is painted black for tracking the motion of the barycenter. A transparent plastic ball is also used to observe the motion of the liquid during and after the collision. A high-speed camera is used to record the motion of the ball and the water inside. The white screen and the light behind it is used to illuminate the ball, as shown in Fig. 1.

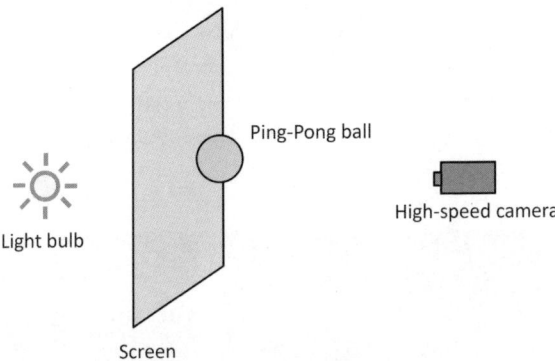

Fig. 1.   Experimental setup to record the motion of liquid inside the Ping-Pong ball.

If you hold the ball and release with your hand, you will encounter a significant uncertainty of rebounding height under the same conditions. An ideal release of the ball needs to avoid initial disturbance as much as possible. We construct a device to release the Ping-Pong ball evenly, see Fig. 2. A chamber (the chamber is made of an empty bottle with lots of holes punched on the side uniformly) is connected to a vacuum cleaner. When we turn on the vacuum, low pressure is produced inside the chamber to fix the Ping-Pong ball. The ball is released as we turn off the vacuum.

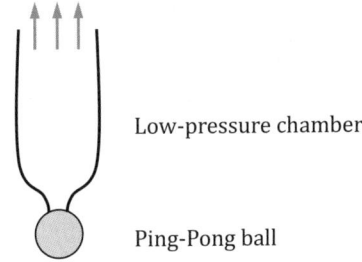

Fig. 2.   The releasing device consists of a low pressure chamber connected to a vacuum cleaner.

Table 1.   Paraments of the ball.

mass (g)	radius of the sphere(cm)	the coefficient of restitution
0.27	2.0	0.86

We take photos with high-speed camera to observe the water level before and after collision, see Fig. 2. The water level is nearly horizontal before collision, as we expected for an "ideal release". Afterwards, the water rises along the edge of the sphere, indicating the formation of a velocity field immediately after collision.

Fig. 3.   The water level before and after collision. The interval between each figure is 0.03*s*.

3. Bouncing Height and Amount of Water

We have measured the parameters of the ball as listed in Table 1.

Now we change the volume of the water and test how the bouncing height changes with the amount of water. The amount of water is measured as the volume percentage of the empty sphere. In Fig. 5, the triangles indicate the experimental result. The releasing height is set as 50.00*cm*.

Fig. 4. The highest point that the Ping-Pong ball can reach with 0%, 50% and 80% of water inside the ball. It can be observed that the rebounding height is greatly suppressed as the water take up roughly half of the volume.

The squares are the theoretical result according to our model than will be given in Section 4.

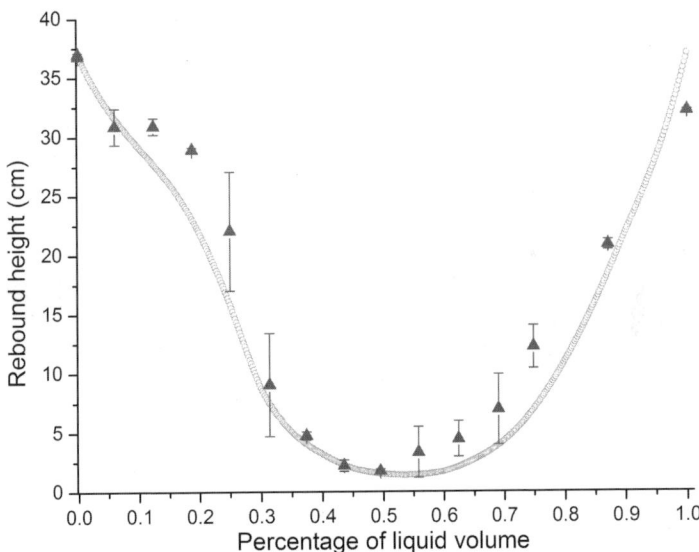

Fig. 5. The bouncing height changes as the amount of water increases. The triangles indicate the experimental result, and the red line is the theoretical result.

It is shown that as the amount of water increases, the bouncing height decrease at first. Then it reaches a minimum at 40 ~ 60%. Afterwards, the height begins to recover. In the end, when the sphere is fully filled, the bouncing height is almost the same as an empty ball. So the explanation to the trend of the bouncing height is substantial in understanding the "nature of the collision".

4. Theoretical Model

As we have pointed out, the motion of the Ping-Pong ball is divided into three stages: the free-fall stage, the collision stage and the rebound stage. We assume an ideal release so that after the free-falling stage, the water level maintains horizontal. The ball and the water inside the ball have the same velocity v_0 downwards before collision happens. We will concentrate on the stage of collision in which intense impact occurs between the ground, the Ping-Pong ball and the liquid.

Collision is usually described by the coefficient of restitution e in which details of interaction are ignored. To understand how water influence the nature of collision, we have to go through more details. We consider the interaction between the ground and the ball, between the ball and the layers of water according to the *time sequence*. We suppose

(1) During the collision, the spherical Ping-Pong ball together with part of water first hits the ground, gains momentum and bounces back, then interacts with the rest of water inside.
(2) Interaction inside the liquid propagates as a longitudinal wave at constant speed (the speed of sound).
(3) Total momentum of the ball and water is conserved during the momentum propagation.
(4) The liquid is incompressible, so that the flux is conserved.

We will introduce an effective mass m'_{ball} to describe the confinement effect on water by the sphere. When the water is below half of the sphere, there is no confinement effect, so that the effective mass equals the mass of the ball itself, $m'_{ball} = m_{ball}$. When the water is above half of the sphere, as shown in Fig. 6(b), m'_{ball} includes the mass of the ball and the that of the water marked in the figure, because the water in this area is confined by the boundary of the ball under gravity.With these hypothesis, we will be able to build the velocity field distribution inside the water after the collision. Hence the total momentum of the system can be found and final height of the ball will be determined.

Assuming the distance between the surface of water and the center of the Ping-Pong ball is d when the amount of water is less than half of the volume, then the mass of the water between two spherical shells with radii r and R (Fig. 6(a)) is defined as

$$m(r) = \frac{1}{3}\pi\rho_0[(R-d)^2 \cdot (2R+d) - (r-d)^2 \cdot (2r+d)] \qquad (1)$$

we have a similar but very long and complex expression for $m(r)$ as the amount of water is more than half of the volume (see Fig. 6(b)), so it is omitted here.

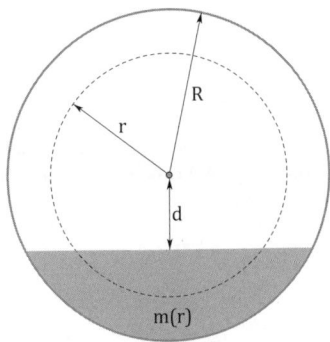

(a) A Ping-Pong ball partially filled with water. The definition of $m(r)$ is marked in the figure when the amount of water is less than half of the total volume.

(b) Definition of the effective mass m'_{ball} and the definition of $m(r)$ when the amount of water is more than half of the total volume.

According to hypothesis 1, the moment the Ping-Pong ball hits the ground and bounces back, the momentum gained by the "ball" before interacting with the water is

$$p_0 = e \cdot m'_{ball} \cdot v_0. \qquad (2)$$

Note that the ball's momentum is upward, whereas water's momentum is still downward (Fig. 6(a)). Then interaction inside the liquid occurs, so as to reverse the momentum of water layer by layer, as shown in Fig. 6(b). According to hypothesis 2, the interaction propagates as a longitudinal wave at the speed of sound. Since the initial wavefront is the sphere of Ping-Pong, we suggest that the wavefront maintains spherical. So we divide the liquid elements according to the *time sequence* into a series of homocentric spheres. The elements in each liquid layer undergo an impact simultaneously. Thus the complicated interactions inside the liquid can be attributed to the momentum spread between liquid layers (Fig. 6).

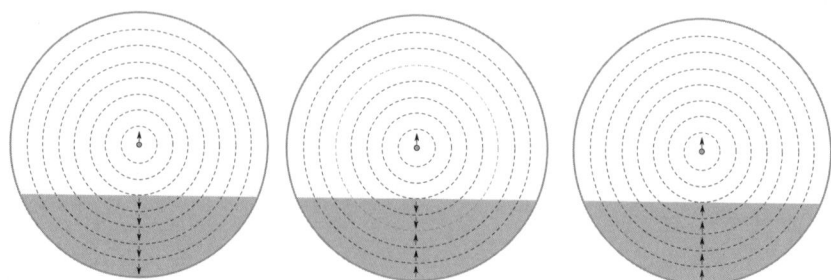

Fig. 6. Process of momentum exchange during the collision stage. The arrows marked on each layer indicate the direction of momentum of the layer. (a) At the beginning of collision stage, Ping-Pong ball has a momentum upward, whereas liquid has a momentum downward. (b) Interactions inside the liquid can be seen as the propagation of longitudinal wave. (c) At the end of the collision, all liquid layers have momentums upward.

The momentum exchange between liquid layers is inelastic, so the velocity of the Ping-Pong when wavefront is on the sphere of radius r is

$$v(r) = \frac{(1+e) \cdot m'_{ball} \cdot v_0}{m'_{ball} + m(r)}. \tag{3}$$

To calculate the velocity distribution, we study a segment in a layer (Fig. 4). In the Ping-Pong ball frame, the flux on the upper interface is

$$\Phi(r) = \rho \iint [v(r) + v_0] \cdot sin\theta \cdot d\sigma. \tag{4}$$

The flux on the lower interface is

$$\Phi(r + dr) = \rho \iint [v(r + dr) + v_0] \cdot sin\theta \cdot d\sigma. \tag{5}$$

According to the continuity equation, we have $d\Phi = v_t \cdot 2\pi r \cos\theta dr$, from which the tangential velocity at the edge of the segment is determined. In addition, after inelastic interaction the normal velocity approaches zero, thus the velocity field inside the liquid is found

$$\begin{cases} v_t(r, \theta) = \dfrac{2\pi e \cdot m'_{ball} \cdot v_0 r^2 (r - d) \cos\theta}{(m'_{ball} + m(r))^2} \\ v_n(r, \theta) = 0. \end{cases} \tag{6}$$

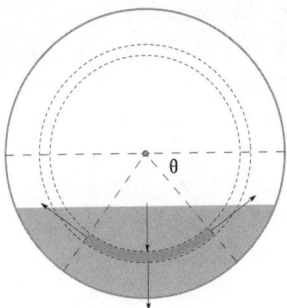

Fig. 7.   The flux on a liquid segment is conserved.

The numerical result for the velocity field distribution after the collision is obtained by Mathematica, see Fig. 8. We can also calculate the z-component of the liquid's total momentum

$$P_z = \rho \iiint v_t \cdot cos\theta \cdot dV. \qquad (7)$$

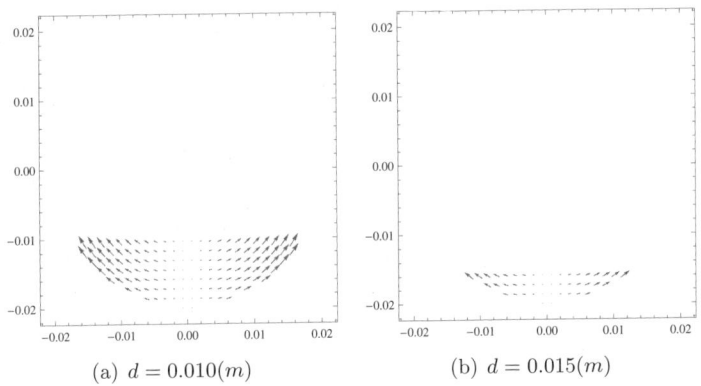

(a) $d = 0.010(m)$ (b) $d = 0.015(m)$

Fig. 8.   Velocity field inside the liquid after the collision stage.

Thus the velocity of the system (center of mass) after the collision stage is

$$v_z = \frac{P_z + m'_{ball} \cdot ev_0}{m'_{ball} + m(d)}. \qquad (8)$$

After the collision, the Ping-Pong ball leaves the ground. In this stage, interaction between the liquid and the Ping-Pong ball persists. But we can ignore the energy dissipation, since the interaction is apparently much weaker than that during the collision stage. The motion of the system (center of mass) is simply determined by the gravity, so that the bouncing height can be finally estimated.

We make a numerical calculation according to our model. The parameters of the Ping-Pong ball are the same as in Table 1, the releasing height $50.00cm$. The theoretical results are indicated by the red line, and the experimental data are indicated by blue triangles in Fig. 5. The agreement between the theory and experimental results supports our model and the assumptions we made.

5. Discussion and Conclusion

The nature of collision is highly related to the amount of water: the rebounding height is suppressed most as the amount of water is about half of the total volume. However, as the amount of water increases further, the rebounding height gradually recovers. We may classify the collision into "rigid-like" motion in which the ball behaves much like a rigid body; or "liquid-involved" motion in which energy is dissipated significantly due to interaction between the sphere and the liquid. The valley area near half-filled ball in Fig. 5 is typical "liquid-involved" motion. The empty or fully-filled ball is typical "rigid-like". Whether it behave like "rigid body" or "liquid" depends on how much the ball-and-water system can move together like a single object. When the water level is below half of the total volume, the water has large degree of freedom to move inside the sphere. So more water will cause more energy dissipation and lower bouncing height. As the water level exceed half of the total volume, part of the water is confined by the spherical boundary, the water begins to lose degree of freedom inside the sphere. In a fully-filled ball sphere, the water lose all its mobility and the ball-and-water recovers to a "rigid body" again! In our model, we introduce an effective mass $m'(ball)$ to describe the confinement effect on water by the sphere.

Other parameters are also important in determine the nature of the collision. As shown in Fig. 5, we compare the bouncing height from different releasing height. The triangles, squares and circles indicate releasing heights of $50cm$, $70cm$ and $90cm$ respectively.

In the figure, we find that when the releasing height is raised above $70cm$, the rebounding height almost remains unchanged with the amount of water. In other words, as the releasing height increases, the whole system tends to become more and more "rigid". This is probably related to the

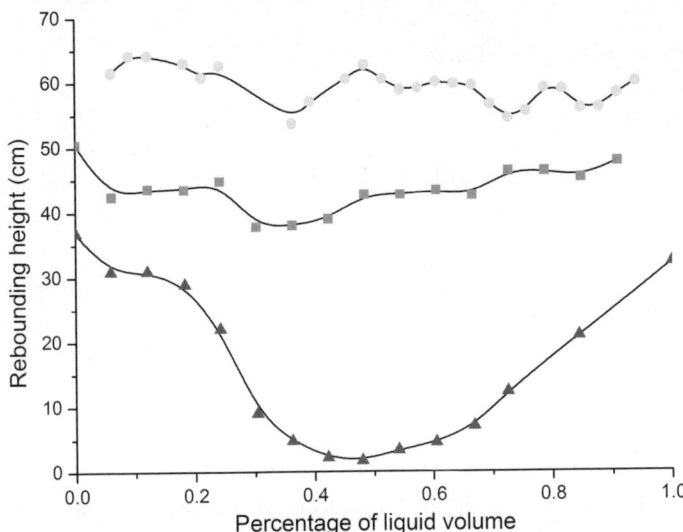

Fig. 9. Comparison of bouncing height for different releasing height. The triangles, squares and circles indicate releasing heights of $50cm$, $70cm$ and $90cm$ respectively.

change in collision time. For sufficiently large releasing height, the collision time τ becomes smaller than the characteristic time required to build a velocity field in the fluid, consequently the macroscopic water flow can not be formed. Instead, it only absorbs energy in the form of vibrational energy, like a rigid object dose. So we may classify the nature of collision above a certain releasing height, say $70cm$ for our experimental condition, as "rigid-like" regardless of the amount of water inside the ball.

We observed that when the ball bounces for a second time, the height becomes much lower. It is related to the separation of the water and sphere. When the ball bounces for second time, the liquid and the ball are detached. This makes the crash between ball and ground and the crash between ball and liquid happen at different time. So that the liquid suppresses the motion of the ball to a great extent.

Appendix A. IYPT Problems of 2012 and 2013

A.1. IYPT 2013 Problems

1. Gaussian Cannon

A sequence of identical steel balls includes a strong magnet and lies in a nonmagnetic channel. Another steel ball is rolled towards them and collides with the end ball. The ball at the opposite end of the sequence is ejected at a surprisingly high velocity. Optimize the magnet's position for the greatest effect.

2. Cutting the air

When a piece of thread (e.g., nylon) is whirled around with a small mass attached to its free end, a distinct noise is emitted. Study the origin of this noise and the relevant parameters.

3. String of beads

A long string of beads is released from a beaker by pulling a sufficiently long part of the chain over the edge of the beaker. Due to gravity the speed of the string increases. At a certain moment the string no longer touches the edge of the beaker (see picture). Investigate and explain the phenomenon.

Newton's beads.
Image courtesy of Hans Jordens (2010)

4. Fluid bridge

If a high voltage is applied to a fluid (e.g. deionized water) in two beakers, which are in contact, a fluid bridge may be formed. Investigate the phenomenon. (High voltages must only be used under appropriate supervision - check local rules.)

5. Bright waves

Illuminate a water tank. When there are waves on the water surface, you can see bright and dark patterns on the bottom of the tank. Study the relation between the waves and the pattern.

6. Woodpecker toy

A woodpecker toy (see picture) exhibits an oscillatory motion. Investigate and explain the motion of the toy.

7. Drawing pins

A drawing pin (thumbtack) floating on the surface of water near another floating object is subject to an attractive force. Investigate and explain the phenomenon. Is it possible to achieve a repulsive force by a similar mechanism?

8. Bubbles

Is it possible to float on water when there are a large number of bubbles present? Study how the buoyancy of an object depends on the presence of bubbles.

9. Magnet and coin

Place a coin vertically on a magnet. Incline the coin relative to the magnet and then release it. The coin may fall down onto the magnet or revert to its vertical position. Study and explain the coin's motion.

10. Rocking bottle

Fill a bottle with some liquid. Lay it down on a horizontal surface and give it a push. The bottle may first move forward and then oscillate before it comes to rest. Investigate the bottle's motion.

11. Flat flow

Fill a thin gap between two large transparent horizontal parallel plates with a liquid and make a little hole in the centre of one of the plates. Investigate the flow in such a cell, if a different liquid is injected through the hole.

12. Lanterns

Paper lanterns float using a candle. Design and make a lantern powered by a single tea-light that takes the shortest time (from lighting the candle) to float up a vertical height of 2.5m. Investigate the influence of the relevant parameters. (Please take care not to create a risk of fire!)

13. Misty glass

Breathe on a cold glass surface so that water vapour condenses on it. Look at a white lamp through the misted glass and you will see coloured rings appear outside a central fuzzy white spot. Explain the phenomenon.

14. Granular splash

If a steel ball is dropped onto a bed of dry sand, a "splash" will be observed that may be followed by the ejection of a vertical column of sand. Reproduce and explain this phenomenon.

15. Frustrating golf ball

It often happens that a golf ball escapes from the hole an instant after it has been putted into it. Explain this phenomenon and investigate the conditions under which it can be observed.

16. Rising bubble

A vertical tube is filled with a viscous fluid. On the bottom of the tube, there is a large air bubble. Study the bubble rising from the bottom to the surface.

17. Ball in foam

A small, light ball is placed inside soap foam. The size of the ball should be comparable to the size of the foam bubbles. Investigate the ball's motion as a function of the relevant parameters.

A.2. IYPT 2013 Problems

1. Invent yourself

It is more difficult to bend a paper sheet, if it is folded accordion style or rolled into a tube. Using a single A4 sheet and a small amount of glue, if required, construct a bridge spanning a gap of 280 mm. Introduce parameters to describe the strength of your bridge, and optimise some or all of them.

2. Elastic space

The dynamics and apparent interactions of massive balls rolling on a stretched horizontal membrane are often used to illustrate gravitation. Investigate the system further. Is it possible to define and measure the apparent gravitational constant in such a world?

3. Bouncing ball

If you hold a Ping-Pong ball above the ground and release it, it bounces. The nature of the collision changes if the ball contains liquid. Investigate how the nature of the collision depends on the amount of liquid inside the ball and other relevant parameters.

4. Soliton

A chain of similar pendula is mounted equidistantly along a horizontal axis, with adjacent pendula being connected with light strings. Each pendulum can rotate about the axis but can not move sideways (see figure). Investigate the propagation of a deflection along such a chain. What is the speed for a solitary wave, when each pendulum undergoes an entire 360 revolution?

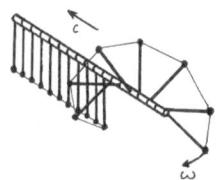

5. Levitation

A light ball (e.g. a Ping-Pong ball) can be supported on an upward airstream. The airstream can be tilted yet still support the ball. Investigate the effect and optimise the system to produce the maximum angle of tilt that results in a stable ball position.

6. Coloured plastic

In bright light, a transparent plastic object (e.g. a blank CD case) can sometimes shine in various colours (see figure). Study and explain the phenomenon. Ascertain if one also sees the colours when various light sources are used.

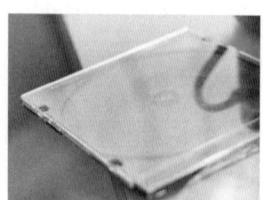

7. Hearing light

Coat one half of the inside of a jar with a layer of soot and drill a hole in its cover (see figure). When light from a light bulb connected to AC hits the jars black wall, a distinct sound can be heard. Explain and investigate the phenomenon.

8. Jet and film

A thin liquid jet impacts on a soap film (see figure). Depending on relevant parameters, the jet can either penetrate through the film or merge with it, producing interesting shapes. Explain and investigate this interaction and the resulting shapes.

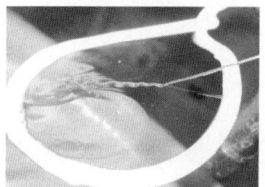

9. Carbon microphone

For many years, a design of microphone has involved the use of carbon granules. Varying pressure on the granules produced by incident sound waves produces an electrical output signal. Investigate the components of such a device and determine its characteristics.

10. Water rise

Fill a saucer up with water and place a candle vertically in the middle of the saucer. The candle is lit and then covered by a transparent beaker. Investigate and explain the further phenomenon.

11. Ball bearing motor

A device called a Ball Bearing Motor uses electrical energy to create rotational motion. On what parameters do the motor efficiency and the velocity of the rotation depend? (Take care when working with high currents!)

12. Helmholtz carousel

Attach Christmas tree balls on a low friction mounting (carousel) such that the hole in each ball points in a tangential direction. If you expose this arrangement to sound of a suitable frequency and intensity, the carousel starts to rotate. Explain this phenomenon and investigate the parameters that result in the maximum rotation speed of the carousel.

13. Honey coils

A thin, downward flow of viscous liquid, such as honey, often turns itself into circular coils. Study and explain this phenomenon.

14. Flying chimney

Make a hollow cylindrical tube from light paper (e.g. from an empty tea bag). When the top end of the cylinder is lit, it takes off. Explain the phenomenon and investigate the parameters that influence the lift-off and dynamics of the cylinder.

15. Meniscus optics

Cut a narrow slit in a thin sheet of opaque material. Immerse the sheet in a liquid such as water. After removing the sheet from the liquid, you will see a liquid film in the slit. Illuminate the slit and study the resulting pattern.

16. Hoops

An elastic hoop is pressed against a hard surface and then suddenly released. The hoop can jump high in the air. Investigate how the height of the jump depends on the relevant parameters.

17. Fire hose

Consider a hose with a water jet coming from its nozzle. Release the hose and observe its subsequent motion. Determine the parameters that affect this motion.

Appendix B. The Regulations of the International Young Physicists' Tournament

B.1. International Young Physicists' Tournament

The International Young Physicists' Tournament (IYPT) is a competition among teams of secondary school students in their ability to solve complicated scientific problems, to present solutions to these problems in a convincing form and to defend them in scientific discussions, called Physics Fights (PF).

B.2. The problems of the IYPT

The 17 problems are formulated by the International Organizing Committee (IOC) and sent to the participating countries not later than in October. These problems may be used in any competition that could lead to selection of a national team. They may be used in International tournaments that involve foreign teams not taking part in IYPT.

B.3. The participants of the IYPT

B.3.1. *The national teams*

Any invited country, as well as the host country, is represented by one team. A country can only take part in the IYPT if it is nominated and accompanied either by the country's IOC representative or by the representative of a candidate IMO.

B.3.2. *The membership of the teams*

A team is composed of five secondary school students. All members of the team must either be citizens of the country they represent, or be enrolled as students in a school of the country they represent. Secondary school graduates can participate in the IYPT in the year of their graduation. The participation of university students is not allowed. The LOC may allow participation of teams of four or three students. The composition of the team cannot be changed during the Tournament. The team is headed by a Captain who is the official representative of the team during the PFs.

B.3.3. *The team is accompanied by one or two team leaders*

B.4. The Jury

The Jury is nominated and organized by the LOC in cooperation with EC. The Jury consists of at least five members, if possible from different countries. Team leaders, at least one from each team, are included in the Jury. The team leaders cannot be members of the Jury in the PF where their teams participate and should not, if possible, grade the same team more than twice.

B.5. The agenda of the IYPT

The IYPT is carried out in a period determined by the LOC (from May to July). All teams participate in five Selective PFs. Selective PFs are carried out according to a fixed schedule as detailed in the attachment to these Regulations. Numbers are ascribed to teams by lot. The best teams participate in the Final PF.

The host country provides a cultural program for the participants.

B.6. The Physics Fight regulations

Three or four teams participate in a PF, depending on the total number of teams. In the course of a PF the members of a team communicate only with each other. Before the beginning of a PF, the Jury and the teams are introduced.

The PF is carried out in three (or four) Stages. In each Stage, a team plays one of the three (four) roles: Reporter, Opponent, Reviewer (Observer). In the subsequent Stages of the PF, the teams change their roles according to the schemes:

Three Team PF				Four Team PF				
Stage	1	2	3	Stage	1	2	3	4
Team 1	Rep	Rev	Opp	Team 1	Rep	Obs	Rev	Opp
Team 2	Opp	Rep	Rev	Team 2	Opp	Rep	Obs	Rev
Team 3	Rev	Opp	Rep	Team 3	Rev	Opp	Rep	Obs
				Team 4	Obs	Rev	Opp s	Rep

B.7. The Stage regulations

The performance order in the Stage of a PF:	Time in minutes
The Opponent challenges the Reporter for the problem	1
The Reporter accepts or rejects the challenge	1
Preparation of the Reporter	5
Presentation of the report	12
Questions of the Opponent to the Reporter and answers of the Reporter	2
Preparation of the Opponen	3
The Opponent takes the floor, maximum 4 min. and discussion between the Reporter and the Opponent	14
The Opponent summarizes the discussion	1
Questions of the Reviewer to the Reporter and the Opponent and answers to the questions	3
Preparation of the Reviewer	2
The Reviewer takes the floor	4
Concluding remarks of the Reporter	2
Questions of the Jury	5

In the Final PF the procedure of challenge is omitted.

The official language of the IYPT is English.

B.8. The team performance in the Stages

The Reporter presents the essence of the solution to the problem, attracting the attention of the audience to the main physical ideas and conclusions.

The Opponent puts questions to the Reporter and criticizes the report, pointing to possible inaccuracy and errors in the understanding of the problem and in the solution. The Opponent analyses the advantages and drawbacks of both the solution and the presentation of the Reporter. The discussion of the Opponent should not become a presentation of his/her own solution. In the discussion, the solution presented by the Reporter is discussed.

The Reviewer presents a short estimation of the presentations of Reporter and Opponent.

The Observer does not participate actively in the PF.

During one PF only one member of a team takes the floor as Reporter, Opponent or Reviewer; other members of the team are allowed to make brief remarks or to help with the presentation technically. No member of a team may take the floor more than twice during one Selective PF or, as Reporter, more than three times in total during all Selective PFs. During the Final PF any team member can take the floor only once.

The LOC must inform about the devices available for presentations not later than two months before the IYPT.

B.9. The rules of problem-challenge and rejection

(1) All problems presented in the same PF must be different.
(2) *Selective PF* The Opponent may challenge the Reporter on any problem with the exception for a problem that:

    (a) was rejected by the Reporter earlier;
    (b) was presented by the Reporter earlier;
    (c) was opposed by the Opponent earlier;
    (d) was presented by the Opponent earlier.

If there are less than five problems left to challenge, the bans d), c), b), a) are successively removed, in that order.

During the Selective PFs the Reporter may reject the challenge of three different problems in total without penalty. For every subsequent rejection the coefficient of the Reporter (see section X) is decreased by 0.2. This reduction continues to apply during the following selective PFs.

The following special rules apply to the last Selective PF:

- The procedure of challenge is omitted. All teams may choose the problem to present. The only exception is that a team may not present a problem, which they presented earlier in the Selective Fights, and all problems presented in one group must be different. In case teams of one group choose the same problem, priority is given to the team with the higher TSP (see section XI).
- Teams must choose their problems for the last Selective Fight as soon as possible after the results of the preceding Selective Fight are official. The choice must be made public immediately.
- The problem which a team presents in this PF may not be presented again in the Final PF by the same team.

(3) *Final PF* Within four hours after the announcement of the results of the Selective PFs the teams participating in the Final choose their problems. In case teams choose the same problem, priority is given

according to the order of presentation in the Final (see section XII). The choice should be made public immediately.

B.10. The grading

After each stage the Jury grades the teams, taking into account all presentations of the members of the team, questions and answers to the questions, and participation in the discussion. Each Jury member shows integer marks from 1 to 10. The mean of the highest and the lowest marks is counted as one mark which is then added to the remaining marks. This sum is used to calculate the mean mark for the team. The mean marks are multiplied by various coefficients: 3.0 or less (see section IX) for the Reporter, 2.0 for the Opponent, 1.0 for the Reviewer and then transformed into points.

B.11. The resulting parameters

(1) *For a team in the PF* The sum of points (SP) is the sum of mean marks, multiplied by the corresponding coefficients and rounded to one decimal.

(2) *For a team in the Tournament* The total sum of points (TSP) equals the sum of SP of the team in all Selective PFs. The number of fights won (FW) is the number of Selective PFs, in which a team received the highest SP from all three or four teams participating in the same PFs.

B.12. The Final

The three teams having the highest TSP in the Selective PFs participate in the Final. In case teams have equal TSP, their participation in the Final is decided by FW. If team(s) winning all their Selective PFs (FW=5) did not reach the Final by TSP, the best of them (determined by TSP) takes part in the final as fourth team.

The order of presentation in the Final is determined by position by entering the final: the higher the position, the lower the number in the scheme of section VI.

B.13. The final team ranking of the IYPT

Students in the top half (rounded up) of participating teams receive medals. The students of the team winning the Final are awarded the winners cup.

If two or three teams have the same SP result in the Final, the winner is nominated according to the highest TSP, in case of equality by FW. All teams participating in the final are awarded 1st place certificates and gold medals. The five best teams not participating in the final are awarded 2nd place certificates and silver medals. 3rd place certificates and bronze medals are awarded to students in all other teams finishing in the top half. All other students receive certificates of participation. Team leaders obtain certificates indicating the ranking of their team.

B.14. The status of the regulations of the IYPT

The regulations are established by the IOC and may be changed only by the IOC.

Author Index

Subject Index

Towards Gender Equality

Students in a city classroom
(John Robinson/South Photographs/
africanpictures.net).

Students mingle outside the classroom
(Guy Stubbs/Independent Contributors/
africanpictures.net).

Towards Gender Equality
South African Schools during the HIV and AIDS Epidemic

Robert Morrell
Debbie Epstein
Elaine Unterhalter
Deevia Bhana
Relebohile Moletsane

UNIVERSITY OF KWAZULU-NATAL PRESS

Published in 2009 by University of KwaZulu-Natal Press
Private Bag X01
Scottsville 3209
South Africa
E-mail: books@ukzn.ac.za
Website: www.ukznpress.co.za

ISBN: 978-1-86914-175-2

Managing editor: Sally Hines
Editor: Alison Lockhart
Typesetter: Patricia Comrie
Indexer: Judith Shier
Cover design: M Design
Cover photographs: John Robinson/South Photographs/africanpictures.net

Printed and bound by Interpak Books, Pietermaritzburg

Contents

Abbreviations

AIDS	acquired immune deficiency syndrome
ANC	African National Congress
ARVs	antiretrovirals
COSAS	Congress of South African Students
DFID	Department for International Development
EFA	Education For All
GETT	Gender Equity Task Team
GEU	Gender Education Unit
GFP	Gender Focal Points
GRB	gender-responsive budgeting
HEARD	Health Economics and HIV/AIDS Research Division
HIV	human immunodeficiency virus
HSRC	Human Sciences Research Council
IFP	Inkatha Freedom Party
LO	Life Orientation
LRC	learner representative council
MDGs	Millennium Development Goals
NAPTOSA	National Professional Teachers' Organisation of South Africa
NGO	non-governmental organisation
NPPHCN	National Progressive Primary Health Care Network
NRF	National Research Foundation
OBE	outcomes-based education
RCL	Representative Council of Learners
SADTU	South African Democratic Teachers' Union

SAP	South African Police
UDF	United Democratic Front
UNAIDS	The Joint United Nations Programme on HIV/AIDS
UNESCO	United Nations Educational, Scientific, and Cultural Organization
UNICEF	United Nations Children's Fund
WID	Women in Development
WHO	World Health Organization

Acknowledgements

The research on which this book is based enjoyed the generous financial support of the British Council, which funded a Higher Education Link from 2000 to 2005. This link was initially between the University of Natal (Durban) and the Institute of Education, University of London, with Goldsmiths College, University of London and then Cardiff University School of Social Sciences joining the link when Debbie Epstein moved from the Institute to first one and then the other institution. During this time, the British Council, through the kind offices of its Durban director, Tony Reilly, also provided additional research funding. Subsequent to Tony Reilly's departure, Barry Masoga continued the support of the British Council, enabling the final phase of research to be undertaken. Robert Morrell's research and writing was also supported with funding from the Ford Foundation and the National Research Foundation (NRF).

This book could not have been written without the generosity and co-operation of the principals, teachers and learners in the schools we worked with. They agreed to be involved, to be interviewed, and to allow us to see their classes in operation. Our presence inevitably impacted on their lives and we wish to record our thanks to all the schools and their staff and students for making us welcome and supporting this research project.

We were ably assisted in the research process by a number of people, some of whom were our students. We would like to extend our appreciation to Nicolette Catelle, Claire Gaillard-Thurston, Vijay Hamlall, Alex Kent, Sakhamuzi Khuzwayo, Nokuthula Masuku, Mxolisi Mchunu and Mark Thorpe.

In the early stages of the project, we received valuable support from Linda Chisholm, Clive Harber and Claudia Mitchell. Lynne Dalrymple of DramAide

kindly made her organisation available as the vehicle to carry a gender-equality intervention into some of the schools.

Robert Morrell would like to record his thanks to the community of gender scholars, in South Africa and beyond, who have contributed to his thinking and work in many ways. Specifically, he would like to thank Cathy Burns, Raewyn Connell, Dori Posel, Jenny Robinson and Imraan Valodia. No research work proceeds in a social or emotional vacuum. We generally need the support of a lot of people to complete a big project, to get through the many challenges that inevitably stand in the way. He would thus also like to thank Penny Morrell, Geoff Schreiner, Keith Breckenridge, Alan Rycroft, Mike Hart, Alan Whiteside, Doug Hindson and Sandra Swart for their unstinting and constant support. He would also like to thank his wife, Monica, his children, Tamarin and Ashleigh, his mother, Bridget, and his siblings, the aforementioned Penny and his twin brother Christopher, for being the family that he needed and depended on. Finally, he would like to pay tribute to his father, Richard Morrell (1930–2007) for teaching him about tenacity and truth.

Debbie Epstein would like to thank Rebecca Boden for her constant support, making cups of tea, cooking dinners for invading co-authors and generally being there, 'reading and commenting on drafts at various stages of the process', and Sally (the dog) for forcing her to take walks, when staying at the computer, while tempting, would have led to a breakdown. She would also like to thank her colleagues, especially those in the Sexualities and Gender Research Group, at Cardiff University School of Social Sciences for intellectual, moral and emotional support over the past five years and, in particular, Barbara Adam, Katy Greenland, Gabrielle Ivinson, Joanna Latimer, Emma Renold and Valerie Walkerdine. She is particularly grateful for the support of Huw Beynon who, as director of the School, facilitated her work and her ability to travel to and from South Africa. During these trips, Costas Criticos provided splendid accommodation in Durban and went far beyond the duties of landlord, sorting out computing, Wi-Fi, and everything else that was needed.

Elaine Unterhalter would like to thank students in the MA course in Education, Gender and International Development at the Institute of Education, University of London, who commented in stimulating and critically engaged discussions on research findings and publications from the project over a number of years.

Additional thanks to those who wrote dissertations and essays on some of the themes explored in the book for the ways they helped to deepen her insight. Colleagues and friends at the Institute of Education provided general encouragement with the project. The special help from Sally Power, Diana Leonard, Debbie Gaitskell, Jenny Parkes and Peter Aggleton is particularly valued. She is also deeply appreciative of the outstanding research assistance of Amy North and Helen Poulsen. She has benefited from conversations over a number of years with colleagues and friends associated with the 'Beyond Access: Gender, Education and Development' project, who helped to set the trends in South Africa in a wider context: Sheila Aikman, Rajee Rajagopalan, Janet Raynor, Elspeth Page, Fiona Leach, Madeleine Arnot, Ingrid Robeyns, David Clarke, Dhianaraj Chetty, Akanksha Marphatia, Nyokabi Kamau and Sujata Khandekar. She is profoundly thankful for the kindness of her family: Richard, Joe, Rosa, Oliver, Sophie, Beryl, Karrie and David who, singly and collectively, provided essential encouragement and good humour throughout this project.

Deevia Bhana would like to thank her late mother and father, Savitha and Parbhoo Bhana.

Relebohile Moletsane would like to thank her late mother and father, Mpono and Monyake Moletsane, for supporting her to be the person that she is.

Introductory Accounts

Preamble

On a sunny morning in May 2000 we made a visit to Lilian Ngoyi Secondary School[1] in Durban. Robert Morrell went to sit among a group of Grade 10 boys. There were five of them, all lively and keen to talk about themselves and the goings-on in the school. He sensed they enjoyed having somebody to talk to, somebody to listen. They were exulting in being allowed to miss class. Around them the business of the school went on, with a few teachers walking purposefully across the uneven grass between the low-lying buildings and occasionally a pupil, neat in a dark green blazer, darting from a class towards the administrative block on an errand.

This was one of the introductory meetings we had planned. Rob was to talk to some male teachers and students to get a sense of gender relations in schools. Our idea was to hear how the lived experience in this township high school meshed with the secondary literature we had been reading. The boys ranged in age from sixteen to nineteen. As they spoke, they warmed to the task. They talked about their work –Tone worked on weekends packing bags in a local supermarket, another rode taxis and collected fares. They described their parents, many of whom were unemployed, and their girlfriends. While there was some bravado, there was also sadness and anxiety. Some were disconsolate about girlfriends who were now going with (dating) someone else, many were anxious about whether they would find another to whom they could, in their words, 'propose love' or ask to go out. There was anger too – about girlfriends who had been unfaithful or the experience of being dumped. The topic of violence in the school was broached. Their stories touched on being attacked at bus stops by gangs, about violent rivalries with other boys for the affections of girls. They said little about school, other than

1

that it was 'okay'. They made light of the danger of violence, although one admitted that the previous week he had run away from knife-wielding men intent on taking his cellphone. Just as the discussion was coming to an end, there was a commotion. The boys walked casually away. Initially Rob did not pay much attention, as break was beginning and children were streaming out of classes. He gathered his notes and chatted with one of the teachers, who pointed to a taut group of figures: plain-clothes police had arrived and arrested a learner for being in possession of an illegal handgun. We had often heard about school violence and read about it in newspapers, but seldom seen it. This was a close reminder of how violence was an all-too-common experience for teachers and pupils in the poorly resourced township schools in which we had begun our work.

On another early visit, Debbie (Epstein) and Lebo (Moletsane) sat in an empty classroom with a group of Grade 10 girls. After the initial introductions, we asked the girls to talk about relationships between boys and girls in the school, their everyday experiences of being girls, their hopes and fears. Between the two researchers there was a frisson of tension about what the girls might or might not say about gender, sexuality, violence and HIV. This was the focus of our research project, but we knew these issues were difficult to broach. On the other hand, we knew that they were likely to be part of the girls' lives and hoped that our initial questions and gentle probes would enable them to open up. Tentatively at first, the girls talked about their need for education and their wishes to stay at school and become 'something' or 'somebody'. Gradually the girls became more confident and started telling us private and sensitive things about themselves. Some of them spoke about sexual violence, saying it was their biggest fear. As the conversation went on, with the young women telling us quite intimate things, one of the quieter girls suddenly burst out with a story that we perceived as quite disconnected in the way she narrated it. She was full of emotion as she told us about her neighbour, an older man. She said he had raped her for several years since she was six years old. Until the day of this conversation, she had never told anyone about the multiple rapes, not even her mother, because she knew that nobody would believe her. There was a stunned silence. Lebo and Debbie did their best to hold the girl safely in the interview space, refraining from asking further questions, but making sure that the girl knew they were not judging her and believed her story. We reminded the others in the group about their promises of confidentiality. At the end of the

discussion, they spoke privately with the girl, asking her what support we could give her and whether she wished us to obtain some counselling for her. Afterwards, we approached a counselling service connected to the university and arranged for her to have some counselling. We come back to this incident in Chapter 4.

These examples give a sense of the drama, the pain and the difficulties of the linked research projects on which this book reports. In our research, we looked at ways of understanding how transformation in relation to gender might take place, particularly in the context of the HIV and AIDS pandemic.[2] We were interested in how different interventions manifested themselves (or failed to do so) in changing gender relationships in schools, how the institutionalisation of change did or did not take place and the kinds of interactions/interactive approaches that might facilitate change. While both the nature and demography of HIV and AIDS and the prevalence and directions of gender-linked violence provide us with important ways of understanding the state of gender (in)equality in wider South African society and, more specifically, in schools, we considered it important also to look at the everyday, mundane practices, beliefs, discourses and institutional arrangements of particular institutions, such as schools, and this provided us with the substance for this book. These are the ideas we have used to help make sense of the images of the boy with the gun and the girl whose memories suddenly overwhelmed her.

Introduction

This book explores the development of gender equality in South African schools in the post-apartheid period during which people worked to put the hopes and aspirations of the liberation struggle into practice. Through detailed work in two secondary and two primary schools in Durban, we are concerned to examine how gender relations and gendered structures of power have shifted and the extent to which gender identities have come to be constructed differently. The period during which the research was conducted was marked by the ravages of the HIV and AIDS epidemic, uneven distribution of income and wealth, and high levels of gender-based violence. Thus the context for our examination of gender equality in education, moving from policy to practice, is marked by many difficulties over and above those entailed in transforming gender inequality in schools.

We seek to develop a nuanced, theoretical understanding of how the goal of gender equality is understood and experienced by school communities in one province in South Africa. We try to indicate where progress has been made, where major challenges remain and why this is the case. The story is not one of unqualified success. It shows that there are opportunities for making important strides towards gender equality in a society in the process of a rapid transition. By the same token, however, there are major obstacles, including poverty, under-resourced schools, relations within families and identities framed by ideas about appropriate, but inegalitarian forms of masculinity and femininity.

We examine these issues using a number of different analytical approaches. First, we seek to explore why there is a divergence of views about the state of gender equality in schools in South Africa, with some people seeing considerable advances, while others are highly critical of a lack of achievement. Second, we look at the HIV and AIDS epidemic and its relationship with the struggle for gender equality in education. Third, we present and analyse data from a six-year study, which allow us to understand how gender equality is embedded and undermined institutionally, and how it is lived by individuals in their personal social relations and their interactions with institutions. Through an analysis of this data, we offer an understanding of teacher and learner responses to interventions that have promoted gender equality, as well as their responses to ongoing patterns of inequality, violence and HIV and AIDS. Our analysis of developments within a few primary and secondary schools allows us to come to some conclusions about the progress made so far in achieving gender equality.

A brief history of the research

In 1999 we were granted funding by the British Council to undertake a study titled (misleadingly in terms of later developments) 'Life and Death in Secondary Schools in KwaZulu-Natal: AIDS, Violence, Poverty and Gender'. Over the next six years, through a Higher Education Links scheme that brought together academics working in South Africa and the United Kingdom, the British Council and a number of other funders provided generous support to the project, which evolved, expanded and led us to make many unexpected connections. By the time the project ended, we had gone in a number of directions. We had worked with and monitored an intervention on gender and HIV introduced by a theatre-in-education non-

governmental organisation (NGO) called DramAide. We had placed postgraduate students to work as ethnographers in two Durban township secondary schools and drawn on the work of a number of postgraduate research projects being conducted in primary schools. We had dovetailed the project with other work on masculinities in schools funded by the Ford Foundation and on the Millennium Development Goals (MDGs) funded by the Department for International Development (DFID) in the United Kingdom. In the process we collected hundreds of hours of taped interviews and piles of survey returns. We had developed good working relationships with the staff in both secondary schools, presented the experiences in the schools to government officials in the provincial education headquarters in Pietermaritzburg and to the national Department of Education in Pretoria. We reflected on the experiences in our teaching in three universities, in papers we presented at seminars and conferences and in our discussions with international organisations. We thus drew on many academic projects in order to help us to understand the mosaic of data.

The shape of the book
The issues of gender inequality, violence, poverty and education that interested us at the outset remained core concerns throughout our research, although we dropped the initial ghoulish reference to 'death' and focused, instead, on issues of coping, care and the relationships that people made. Inevitably, our thoughts matured. Our confidence that we could make a difference, at least at the school level, is less strong as we bring the project to a close than it was at the beginning. Our appreciation of the many challenges that lie ahead has grown. We have struggled to understand their complexity and connections. Gender equality, however one understands it, is elusive and we realise that it is not something that can be achieved in the short term.

Chapter 2 presents a framework for our approach to thinking about gender equality and education and some of the effects of violence and the HIV epidemic on these understandings. This analysis, which tries to distinguish between meanings of equality associated with different approaches to gender, equality and understandings of the HIV epidemic and violence, frames the way in which we have interpreted both the history of shifts in relation to the political economy of gender relations in South Africa (see Chapter 3) and the data from the four schools.

Our discussion of equality, violence and approaches to HIV and AIDS draws on the distinctions formulated by Elaine Unterhalter (2007) between interventions, institutions and interactions.

In South Africa, a variety of agencies (governmental and non-governmental) have been involved in promoting change, often in the form of one or other kind of intervention, whether in relation to HIV, sexual behaviour or the wider issues of gender equality. Such initiatives inevitably enter the discursive fields of schools and interact with them, producing new, organically developed understandings and practices. These are lived out by individuals in social relationships – hence the importance of understanding how gender, sexuality and HIV are played out in schools and elsewhere. If, as many contemporary theorists argue, identities (in this case masculinities and femininities) are as much about doing and becoming as about being, the importance of investigating the reiterated gendered practices of boys and girls, women and men in schools and other contexts cannot be underestimated. Our focus for exploring identities has been to examine the discourses through which these are expressed, that is, the ways in which people speak about identities, social relationships, interventions and institutional changes. We are interested in the ways that ideas are connected together, the kinds of distinctions or dichotomies that are made, the forms of knowledge or practice that are legitimated and those that are demonised. We examine these discursive practices, while taking account of the social structures and interactions that form them. Thus, this is a largely conventional, interpretativist work of analysis. We do not use a thorough-going post-structuralist framework in which the discourse itself constitutes the social relations.

In Chapter 3 we review the broad sweep of the history of the struggle for gender equality in South African education in the context of efforts to change a political economy shaped by race, class and ethnic divisions, as much as by gender inequality. We document shifts in policy and practice regarding gender and education, looking closely at interventions, institutionalisation and interactions. We also look at who participated in campaigns for gender equality in education, what alliances were built and how this affected both the nature of the demands made and the research conducted.

Chapter 4 is devoted to a description of the methods that we used and the schools in which we worked. In collecting and interpreting the data, we tried to

6

get beyond an examination of policy, social structures and surface impressions. Following the ethnographic tradition of obtaining rich data, our analysis has been informed by a deep familiarity with the schools that constituted our research sites. During the six years in which we worked in these schools, we came to know a lot, via observation, interview and interaction, about the schools, the learners and the teachers. As our discussion of our methodology reveals, we have worked to look at the multilayered processes entailed in struggles to put gender equality in education in place.

Chapters 5 and 6 explore school-based interventions, the first via drama-in-education provided by an NGO in the two secondary schools and the second via Life Orientation lessons, as advised by the provincial government in KwaZulu-Natal in the two primary schools. Chapter 7 returns to the secondary schools, to discuss older learners' understandings of gender, sexuality and the HIV and AIDS pandemic, while Chapter 8 explores younger children's gendered identities and ethics of care in relation to HIV and AIDS. Chapters 9 and 10 explore teacher masculinities and femininities respectively, considering how these are constituted and how they impact on what happens in schools. In Chapter 11, our conclusion, we pull together the threads of the book, reflecting on both difficulties and achievements in relation to instituting gender equality in schools and how this has played out in a time of HIV and AIDS. In reading the book, some readers might prefer to skip straight to the empirical chapters where we explore our data, and return to the theoretical framework and historical content at a later date.

People make themselves, but not in conditions of their own choosing. The young people whose lives we analyse in this book are making themselves in conditions of change, danger, stress and possibilities. It is these contradictory threads we explore and we evaluate the ways in which gender relations and identities have changed and how they have been affected by (and in turn affected) both the HIV and AIDS pandemic and national gender equality policy. We identify some of the forces that stand in the way of change and suggest why progress towards eradicating gender inequalities in schooling has been slow. We try to show that gender equality policies have made a difference and that initiatives from central and provincial governments and the work of teachers and school managers are important. Even if a new, brave world in which girls can comfortably achieve their potential and boys are free from the crushing expectations of having to be

strong, virile providers has not come into existence with the end of apartheid, and even if there are many worrying ways in which gender equality is currently ignored, and students and teachers draw too easily on stereotyped views, we believe that knowledge, commitment and many significant actions by individuals and groups are helping to make a different world possible. It will probably not announce itself with posters, training sessions or large rhetoric, but it will be present in small shifts in status, in the forms that arguments take, and in the interactions between students and teachers and their families over many years, some of which we have tried to document here, in order that we can further support their development.

Notes

1. The names of all the schools, teachers and children have been changed to suitable pseudonyms throughout this book.
2. In this book we follow current practice and talk about HIV (human immunodeficiency virus) and AIDS (acquired immune deficiency syndrome) as separate conditions. HIV identifies a virus that steadily reduces the ability of a person's immune system to ward off illness. Antiretroviral drugs (ARVs) can limit some of the most harmful effects of the virus on the immune system. Without the application of ARVs, however, HIV generally leads to AIDS, a condition which leads, often via the medium of diseases such as tuberculosis, to death. For most of the time we worked on this book the practice was to refer to the connection between the two conditions as HIV/AIDS. This was, in part, a political statement and a challenge to views associated with a group of AIDS denialists who refused to acknowledge that the two conditions were linked. Former president, Thabo Mbeki, associated himself with these views and emphasised poverty, rather than unprotected sexual intercourse, as the cause of HIV and AIDS.

Elusive Equality

Theoretical Engagements

Introduction

Gender equality in education is not easy to define, even though the term appears in many national and international policy declarations. The South African Constitution sets equality as a key dimension of the Bill of Rights, noting that 'equality includes the full and equal enjoyment of all rights and freedoms' and committing the state to take legislative and other measures 'to protect or advance persons, or categories of persons, disadvantaged by unfair discrimination'. The Constitution further states: 'The state may not unfairly discriminate directly or indirectly against anyone on one or more grounds, including race, gender, sex, pregnancy, marital status, ethnic or social origin, colour, sexual orientation, age, disability, religion, conscience, belief, culture, language and birth.'

Thus the Constitution commits the state to addressing unfair discrimination in schools and redressing gender inequalities. The South African Schools Act of 1996 aimed to provide equal educational opportunities for all learners, to prevent discrimination and to accommodate diverse needs. One of its major objectives was to provide a free and basic education to everybody up to the age of sixteen. In addition, it banned corporal punishment and prohibited the exclusion of pregnant girls from schools.

These documents provide a legislative vision of gender and race equality in schools. However, realising equality has been an immensely difficult project, particularly in everyday lives marked by very different histories and continuing sharp socio-economic inequalities. The gap between policy formulation and

enactment is a key theme in the sociology of education (see, for example, Ball 1994, 2008; Morley 1999) and is noted in many countries in relation to gender policy and education (Ames 2005; Kenway et al. 1997; Leggett 2005; Vavrus 2003). This book is also concerned with this disjuncture. However, we have attempted to go beyond merely noting the gaps between aspiration and realisation. In this chapter we outline some of the theoretical resources we have drawn on to understand how different interpretations can be made of the same policy because of the elasticity of words. Thus we delineate some of the different meanings of gender equality in education. We suggest connections between these and emphases in approaches to violence, HIV and AIDS in schools. The analysis we wish to explore is that the gap between policy and practice is not only between ideal aspiration and messy reality. If the ideal aspiration for equality is expressed rather loosely in very general terms, it obscures or dissolves sharply contrasting views about what constitutes gender equality in education. The practices that follow might select one particular interpretation of policy because of prevailing socio-economic conditions, but it may also be possible to discern particular alignments of actions, so that particular interpretations of gender equality in education are generally accompanied by particular forms of action, with regard to schooling, violence and work on HIV and AIDS.

In this chapter we identify three frameworks through which gender equality in education has been approached and link them with particular ways of interpreting the connection between policy and practice. Narrow interpretations of equality make it difficult to engage with the wide ambition of how to understand, and hence to intervene in, issues related to violence and HIV and AIDS. However, connecting different aspects of work on gender equality makes the practice of confronting and changing relations marked by violence and discrimination particularly demanding.

Understanding gender equality in education

In *Equality of What?* Amartya Sen (1979; see also 1992), Nobel Prize winner for Economics in 1998, famously posed a question as to how we can measure equality. Should we look at it in terms of distribution, justice or comparisons between individuals? We can extend this question to ask how to define and measure gender

equality in education. Are we to look at the distribution of amounts of schooling between groups of girls and boys, or the comparisons between a particular girl and boy with specific learning needs, or justice in relation to explaining the links between education and the development of a more gender-equitable and fairer society?

Sen highlighted how personal heterogeneity means that individuals differ in how they use resources and the outcomes that follow (Sen 1992: 27–28). He identified four ways in which diversity was significant and should be seen as 'a fundamental aspect of our interest in equality' (xi). Firstly, people differ in their personal characteristics. For example, some are fast learners in particular areas and some are slow. Secondly, people differ in their external circumstances and a society such as South Africa, with its long history of race and class division, makes this graphically clear. Thirdly, people differ in how they convert resources into valued outcomes. Sen's example is that a pregnant or lactating woman needs more food than a woman not in either of these states to remain healthy. Adapting this example to schooling, it is clear that a South African girl, growing up in an impoverished rural family, which has strongly held views that women do not need much schooling, who misses crucial years of learning language or mathematics because of heavy demands to take care of siblings and housework, will need more or different tuition to classmates or girls from other parts of South Africa who have had continuous schooling or those from families that do not have to make such heavy demands for household labour. Sen's fourth form of personal diversity relates to people's different views on what constitutes the good. Thus, for example, people may have very different ideas of what constitutes good sexual relations between teenagers. Our diverse views on the nature of the good in sex, religion and social relations have key consequences for how we think about establishing gender and other forms of equality.

Sen suggests that in thinking about equality and taking diversity seriously, we cannot simply compare outcomes, as we cannot know the reasons and diverse conditions associated with a particular outcome. For example, if two people make the same choice to leave school before completing matriculation, we cannot know, simply by comparing matric statistics (outcomes), what lies behind that choice. Assuming that failure to complete matric is only about conditions in school would

11

yield too narrow an analysis. It would suggest that a boy who leaves school early because he is not doing well, realises he has few job prospects with relatively low qualifications and has strong family expectations to fend for himself after the age of sixteen is the same as a girl who becomes pregnant, experiences scorn for what is seen as her sexual licentiousness and has no one in her family to help with her child because of the long hours her adult relatives work and the low family income. On a very limited and distorted reading of gender equality, there is an equality of outcomes. But this is clearly inadequate because of the very real differences in the circumstances that yielded those outcomes.

Sen suggests that in thinking about equality, we should not make our comparison in the space of amounts of inputs, such as years at school, or in the space of outcomes, such as matric results. We should make our evaluations of equality in relation to the freedoms people have to achieve valuable objectives, that is the capability to function, or the real opportunities people have to achieve valued objectives, that is valued combinations of alternatives. In developing public policy linked to this form of equality, we should seek to build the conditions so that morally irrelevant factors, such as being born a boy or a girl into a poor family, do not constrain capabilities or the range of valued alternatives from which a person can choose. Choosing to forego the last year of schooling to contribute to household income or to care for a small child should not carry with it exclusions from further educational opportunities and severe limitations on future income, health or housing. Gender equality is thus an aspect of building the equality of capabilities.

Many standard sociological accounts look at the connections between the structures of schooling and the outcomes, and work on the difficulty of ensuring equality across class divisions in South Africa (see, for example, Chisholm 2004). This is a particularly fruitful approach that can be extended to look at equality across gender, race and class divisions, in ways that draw on Sen's concern with equality of capabilities. In her earlier work, Elaine Unterhalter (2005) distinguishes between four different frameworks to understand gender equality in education. The table below draws out how each framework uses a different understanding of gender, of equality in education and suggests particular emphases in policy. These policy directions are developed from Unterhalter's later work on gender equality and global social justice (2007).

Table 2.1 Frameworks for understanding gender.

Gender means	Gender equality means	Policy emphasis
Girls or boys	Equal amounts (parity)	*Interventions* to ensure parity
Constructed social relations of power	Transformed structures to redress power inequalities	Building *institutions* to transform power inequalities
Discourses of appropriate or resistant femininities or masculinities	Equality of esteem or recognition for diverse identities	*Interactions*: Encouraging cultures of participation, critique and affirmation of diverse identities
Plural concept, entailing both an intersecting structured positioning and a shifting form of agency and identity	Plural notion of equalities, includes freedoms to achieve valuable objectives and varied combinations of real alternatives	*Empowerment*: *Interventions* + *institutions* + *interactions*

The typology highlights the four ways in which gender has been written about in education. In the first approach, gender is understood as a noun (girls, boys, women, men). It is linked with Women in Development (WID) and policies to get girls into school, and to get women into work or into politics. Thus equality is measured as equal numbers of girls and boys in school, sometimes referred to as gender parity. Equal amounts of expenditure on the education of women and men, or the employment of equal numbers of teachers would be other instances of this form of equality. This approach to equality is often associated with what Unterhalter has termed 'interventions', that is a 'bounded form of action . . . to prevent extreme suffering or want' (2007: 35). Interventions are designed to achieve 'what works' to secure the desired effects of equal numbers. For example, stipend programmes to encourage girls to stay on in school, so that the same number of girls and boys complete secondary school (Raynor and Wesson 2006) is an instance of an intervention. The introduction of an awareness-building campaign on HIV or gender equality for one day in school and its presentation to equal numbers of boys and girls would also be an example of intervention, as would the distribution of free condoms to girls and boys. Often interventions do not look at the wider issues of

13

gendered social division and the ways in which gendered structures of power will form the responses to interventions both in schools and beyond.

The second approach understands gender as socially constructed, that is the socio-economic and political relations between women and men are shaped by social structures, such as laws, norms or legitimated forms of action, regarding levels of pay or assumptions about who takes care of children. In this context, 'gender' can be seen as an adjective – it is about certain actions and behaviours and gender inequality is an attribute of particular structures of power. For example, a gendered curriculum may exclude girls from studying maths and science, presenting these as areas of forbidden knowledge that they are not clever enough to grasp. A gendered education budget might spend large amounts on higher education, where few women study, and virtually nothing on early childhood education and adult education, despite large numbers of women requiring investment at these levels. Gender equality in this approach is understood as the transformation of these unjust structures, termed 'institutionalisation' by Unterhalter. Institutional policy approaches to gender inequality 'address elements of want and inadequate provision' and attempt to remove gender inequalities that detract from the quality of education (Unterhalter 2007: 36). Institutional approaches focus on building routinised processes of institutional reform for gender equality, for example, through making changes in curriculum so that it becomes oriented to gender equality, and these shifts are sustained and supported by attention to teacher education and the introduction of gender equality in employment and expenditure in schools. This requires recognising different forms of contribution, auditing how resources are distributed and why, careful consideration of who participates in decision-making, and whether gender-equitable outcomes result. This may require giving considerable attention to affirmative action and acknowledging the conditions which have produced, for example, the exclusion of women from senior levels of management or the low numbers of girls entered for a particular examination.

In the third approach, the meanings of gender and equality highlight how discourse fixes particular identities and forms of social relationship as appropriate for men and women and 'gender' can be seen as an adverb – it is not just doing, but doing in particular ways. Thus, for example, humility and silence may be considered the correct forms of behaviour for women, while toughness and

aggression might be considered appropriate for men. Emphasis on particular gendered identities generally entails selecting some qualities for praise, stigmatising others, and constructing hierarchies of value. Gender equality here demands the overturning of these hierarchies and boundaries of esteem, opening up a wide range of identities and social relationships as valuable for women and men, subjecting all identities to discussion, critique and change. It also requires challenging and changing the structures that maintain horizontal or group inequalities and subject certain groups to forms of control or cultural contempt, while asserting the value of other groups (Stewart 2008). Unterhalter associates policy that supports this form of equality with what she calls 'interactions'. This policy approach identifies multiple sites for enactment, takes human diversity as a central concern in the implementation of equality and is concerned with supporting cultures of critique, participation and connection across different organisational settings (Unterhalter 2007: 140–53; see also 2006). Thus a government seeking to support gender equality in education using this approach might draw in a very wide range of organisations to comment on and critique policy at national, regional and local levels and would provide conditions and support for questioning all aspects of intersecting raced, classed and gendered identities and the development of more plural notions through the curriculum, for example, of what is good, what is needed by particular gendered groups at particular times, and who is important in taking decisions on school knowledge.

The fourth approach is in some ways an attempt to connect the previous three. Here, gender is understood as a multidimensional concept, entailing both an intersection of structured social divisions of race, class and ethnicity and shifting forms of agency and identity through which each performs in particular ways, and in which active choices, based on particular forms of reasoning, are a key component. In different settings and for differently situated individuals the determinants of structure or the openings of agency might be more or less significant. This multidimensional formation of gender shapes the articulation of capabilities or what each has reason to value. Equality is plural, but entails the establishment of the conditions of justice in which constraints on capabilities can be removed and valued actions can be realised. The policy approach entailed sees gender linked with a strong rights agenda, involving practices that develop strategies, arguments and actions for women and girls claiming rights within

15

education and in all the social and cultural sectors that bear on education. It is also concerned to establish gender equitable conditions that are not formal statements of equal inputs, but go considerably beyond this to develop a wide commitment to equality among women and men. Thus a key concern would be providing and supporting openings for women and men from a range of different backgrounds to participate as equals in the political economy and social and cultural life. This approach draws attention to the intersections of, for example, gender, ethnicity, race, poverty and disability. This kind of transformative approach would require policy to support building and sustaining the interactions between civil society, new social movements – especially women's movements – the state and education and cultural practitioners. It would entail bringing together the women's movement's concern with violence against women, adult education, poverty and sexual and reproductive rights, including strategies on HIV; the Education For All (EFA) movement's interest in access to school, inclusion, quality and gender equality for girls and boys; the broad children's rights movement's concern with health, nutrition and the quality of care and concerns associated with participatory development that draw attention to the need to create space for dialogue, discussion, critique and action of women and men as equals.

Despite some misgivings about nomenclature, we have called this policy approach 'empowerment'[1] and see it as encompassing and connecting in particular ways the three areas Unterhalter identified as interventions, institutions and interactions (2007). Empowerment approaches are the most far-reaching in their attempt to engage gender equity and expand the notion of equality. They involve private organisations and states, obligations to act in cases of inequity and commitment to seeking gender equality for all (not only citizens or fee-paying pupils) in life, not only 'in school'. These policy approaches to gender equality in education would require projects to develop gender equality in multi-sectoral social policy that bears on, for example, health, housing, adult education, employment, and legal changes with regard to rape and sexual violence. These changes would not only be associated with state initiatives, but would also engage communities and individuals, who might be extremely critical of state initiatives, and see shifts in social relations in state and non-state settings. Empowerment approaches would also take seriously the multidimensional identities that people

16

bring into education and shift through educative processes. The implication of empowerment is that people are concerned both with their own well-being and the well-being of others and the social environment (Kabeer 1999; Koggel 2006; Ibrahim and Alkire 2007).

In this book we highlight how official policy on gender has tended to focus on interventions and illustrate how discourses of appropriate identities limit institutional aims to transform unjust structures in schools. The attempt to make connections between gender equality in education policy and other areas of social transformation has been particularly hampered by the impact of the high levels of violence in South Africa (Altbeker 2007; Gibson and Hardon 2005) and the effects of the HIV and AIDS epidemic (Abdool Karim and Abdool Karim 2005; Poku, Whiteside and Sandkjaer 2007; Walker, Reid and Cornell 2004). In interpreting our data, we have had to take account of these social conditions and the challenges they present to our analysis of equality. This has led us to extend the framework, with regard to the different meanings of gender equality, to consider some discussions of gender-based violence in schools, drawing on comparative studies in other parts of Africa (Dunne 2008; Leach and Mitchell 2006; Parkes 2009). We have also considered how analyses of gender, schooling, HIV and AIDS (Aikman, Unterhalter and Boler 2008; Baylies and Bujra 2000; Boler and Archer 2008; Vavrus 2003) require us to think about gender equality and education differently.

Both the nature and demography of HIV and AIDS and the prevalence and directions of gender violence provide us with important ways of understanding the state of gender (in)equality in the wider society and, more specifically in schools. However, we think it is important also to look at everyday practices, beliefs, discourses and institutional arrangements – that is, the gender regimes – of particular institutions. The taxonomy we have developed attempts analytically to sort out approaches to thinking about violence and HIV that focus on particular actions, those that identify social relations and structures as key features of the explanation, and those that identify particular discourses and forms of identification as central to understanding the complexity of the situation. Each way of specifying the problem of violence, HIV and AIDS is also associated with a particular emphasis in policy and advocacy for change and these can be aligned with the different formations of gender equality discussed above.[2]

17

Table 2.2 A taxonomy of ways of understanding gender, violence and HIV.

Gender means	Violence and gender means	Policy approach to violence entails	Vulnerability to HIV means	Policy approach to HIV in schools entails
Girls/boys	An aberration; generally associated with the physical actions of boys	*Interventions* to correct behaviours by improved information, punish perpetrators or protect victims	Essentialist qualities of girls at risk; often implies boys engaged in predatory sexual behaviours	*Interventions* to give out information, comdoms; punish transgressive behaviours or protect girls from risk
Gendered structures of power	An outcome of adverse and unequal gendered social structures	Transform social structures and *institutions*	An outcome of unequal gendered power relations	Transform structures, *institutions* and social relations
Gendered discourses and identities	Enacted within/through everyday relationships as a relational feature of hegemonic and subordinated masculinities	*Interactions*, which give opportunities for dialogue, reflection and transformation	Enacted within/through everyday relationships as a relational feature of hegemonic and subordinated masculinities and particular formations of femininity	*Interactions*, which give opportunities for dialogue, reflection and transformation
A form of social action for empowerment	Multidimensional: The curtailment and denial of freedoms through a combination of actions, structures, identities and discourses	*Empowerment*, which expands freedoms (*interventions* + *institutions* + *interactions*)	Multidimensional: Gender inequalities, poverty, stigma, poor policy, combination actions, structures, identities and discourses	*Empowerment*, which expands freedoms (*interventions* + *institutions* + *interactions*)

We can understand the phenomenon of violence in a number of ways. It can be seen as a series of acts of psychopathology, where individuals may lose control or set out to intend to cause harm (Parkes 2008). It may be seen as the outcome of unequal and unjust structural conditions, between women and men or the poorest

and the richest. In this analysis, it may also be interpreted as one result of poor policy delivery. Thus the argument sometimes runs that boys who have been poorly taught in under-resourced schools, have no job prospects on completing matric, and live in townships, where there are few facilities, turn to crime and violence, partly for income, partly for social networks, and partly for the lack of alternative aspirations and that the lack of adequate policing or a well-functioning criminal justice system exacerbates this (Altbeker 2007: 69–83; Sloth-Nielsen 2007). A third interpretation of gender-based violence is that it is enacted in more everyday encounters and can be understood as a relational feature of hegemonic and subordinated masculinities, enacted in conditions where these are not challenged or redressed. Psychologists have objected to the use of the concept of hegemonic masculinity, suggesting that it misunderstands gendered subjectivity and volition, makes individuals disappear and forces the unified subject onto analyses that rather reflect multiple identities, discontinuities and fluidity (Hollway and Jefferson 2000; Wetherell and Edley 1999). There have been concessions to these critiques, but not to the point of altogether abandoning the concept of hegemonic masculinity and concern with the institutions which produce it (Connell and Messerschmidt 2005). One of the key tensions in utilising the concept of hegemonic masculinity is its applicability to sociological approaches to gender (where the analysis of groups is in focus) and psychological approaches where the focus is on individuals. This third approach to thinking about gender and violence is attentive to the ways in which individuals make sense of their lives and perform their masculinities or femininities and that these are in constant interaction with the gendered realities around them.

These realities bear the imprint of societal inequalities and therefore gender identities and agencies themselves reflect, but not in a simple or linear way, the fault lines of society. Among the most obvious of these fault lines are race and class. As sociological factors, they are both powerfully implicated in the ways that masculine identities are formed. Among young people, working-class experiences and prospects profoundly shape gender identity, even as traditional class systems evolve in the context of globalisation (Gibson and Lindegaard 2007; Morrell 2007b; Ratele et al. 2007; Salo 2003) and ethnic and original factors play out in different forms of masculinity (Mac an Ghaill 1994; Marriott 1996; Westwood, 1990).

19

These three broad approaches to looking at gender and violence throw light on how we can understand the gender dynamics of HIV and AIDS in schools. Thus, working with a meaning of gender linked with descriptive ideas about girls and boys, the epidemic is often portrayed simplistically and in essentialised terms, which stress the vulnerability of women and the viciousness of men (Unterhalter, Aikman and Boler 2008, 17–18). A different form of analysis, which brings out the structural dimensions of the social relationships associated with the epidemic, can be associated with the turn in feminist theorising from thinking about women to thinking about gender as relational, formed in the interactions between women and men (Connell 2002). Power relations are complicated by issues of class, sexuality, race/ethnicity, (dis)ability and nation, to name but a few. This relational and intersectional understanding of gender identified in the second group of ideas is primarily focused on the structures that support and maintain unequal power, particularly as this bears on vulnerability to HIV and AIDS and care for those infected and affected and the ways in which these can or cannot be transformed. A third approach draws out how the progress of the HIV and AIDS epidemic has both formed and been formed by the discourses and forms of identification through which people reflect on their gendered notions of self, social and cultural relationships, and the interconnections between the private realm of the family, or intimate partner relationships and the public realms of school, work, community or religious organisation, and the languages and discursive forms in which these identifications are expressed.

For each different meaning of gender-based violence and the gender dynamics of HIV and AIDS, different policy and practice positions can be discerned. Thus interventions are generally associated with views of violence that focus on the acts of individuals and the essentialised identities that put boys and girls at risk of HIV. Interventions may be seen as a limited form of action to correct behaviours by improved information, the punishment of perpetrators or the protection of victims. Interventions are often associated with a limited notion of equality that emphasises sameness, for example, equal amounts of an intervention (say information leaflets on HIV) for boys and girls. Ideas about gender are thus either descriptive – gender is only about biological sexual differences – or may tend to be highly essentialised, suggesting, for example, that all girls or boys in a certain time or place need a particular form of the intervention. Thus, for instance, all

boys are assumed to be keen on sport and interventions in which star footballers make declarations about safe sex or their concern with HIV are an appropriate means to engage them to change behaviour (Duffet 2006). Essentialist ideas about gender often accompany interventions about violence, suggesting it can only be understood in terms of aberrant behaviour of individuals (themselves often essentialised) and that correction can be brought about through punishment and/or improved flows of information or policy (Parkes 2008).

A number of assumptions underpin much of this writing on HIV-prevention policies and interventions in schools. First, actors in educational settings – learners, teachers, managers – are located outside particular education and social histories. In many of the policy directives, they are addressed as though they have no pasts or, if a past is invoked, it is assumed to be an unproblematic affiliation – for example, the struggle against apartheid in South Africa. Much of the policy writing with regard to schools suggests that children and adults have no social relationships beyond the school and will act mechanically in response to techno-rational instruction (Sears 1992: 8) about safe sex or appropriate school organisation. Thus, for example, Jean Baxen and Anders Breidlid (2004), reviewing a range of education interventions in sub-Saharan Africa, conclude that many interventions ignored the complex social, cultural and discursive fields that shaped their reception and failed to take account of the social relations that shaped teachers' lives. Thus while interventions might be tailored to particular groups, generally they see gender in largely essentialised terms and believe that responses to violence or the threat of HIV can be guided in particular directions by information, changed attitudes and supportive social relations.

In contrast to interventions, institutional work on gender, violence, HIV and AIDS does address unequal gender power relations. Thus, for example, long-term strategies for institutional change will be concerned with the locus of decision-making, the distribution of money, discussions about sex and sexuality and assessments of the consequences of action by teachers and learners (Aikman, Unterhalter and Boler 2008). The development of national and regional education sector plans with a focus on gender and HIV have been an instance of this, although as David Clarke (2008) suggests, much of the institutionalising work has failed to place gender issues centre stage. Attempts to look at horizontal inequalities and

21

their implication in perpetuating violence (Stewart 2008) provide another instance of how an institutional approach seeks to address the social conditions of violence.

However, in much policy writing that focuses on schools as institutions, they are portrayed as if they were not sites of power and struggle, but rather places of rational instruction, engagement and action. The gender dynamics of institutions are overlooked, even if some account is taken of the complex process of making meaning that is entailed in developing understandings about HIV in learners, teachers and teacher educators. Thus, for example, despite being rich in detail on the process of change, neither Jean Stuart's (2006) account of working with visual arts to develop an understanding about the HIV epidemic with teacher educators in South Africa, nor Jackson Amone and Paul Bukuluki's (2004) analysis of how the epidemic impacts on school governance in Uganda, draw out the gender dynamics entailed. By contrast, the very rich, ethnographic accounts evident in studies by Frances Vavrus (2003) in Tanzania and Doris Kakuru (2006) in Uganda show the very complex gendered interrelationships established in schools in the context of rural poverty and the HIV epidemic, but they do not suggest how change might happen.

A third group of writers about change focus on interactions and draw out how gender is fluid and relational. Here, the emphasis with regard to gender equality is on facilitating dialogue between differently articulated masculinities and femininities, and different forms of the relationship with regard to gender, class or race. Work on HIV and AIDS taking this perspective has noted forms of personal transformation (Manchester 2004), shifts in ideas of masculinity (Verma et al. 2003) and transformations of young men in prison for violent crimes (Dobash et al. 2000). This perspective, however, is generally little used in mainstream policy or practice in terms of the HIV and AIDS epidemic or gender-based violence in schools (Epstein 2007; Leach and Mitchell 2006).

Elusive equality
The scars, traumas and corrosive social divisions associated with the apartheid education system have not been easily transformed. Why, despite well-conceptualised policy, wide support among the electorate and sustained economic growth since 1994, have equality and equity been so difficult to bring about? How do we understand the simultaneous and apparently contradictory developments,

which we set out in some detail in the next chapter: high numbers of girls completing school with excellent grades (in many areas, outperforming boys); widening labour market access for women, with unparalleled opportunities available at professional and managerial levels for the best qualified; escalating levels of violence, much of it hinging on questions of gender; and the ravages of HIV and AIDS, where gender identities and power relations in relation to sex and sexuality figure prominently?

Catherine Odora Hoppers (2005) has suggested that the connections may be causal. As girls do better at school and make inroads into higher earning jobs, they elicit a backlash of gender-based violence, with HIV and AIDS amongst the consequences. Others (for example, Gouws 2005; Hames and Koen 2006; Hassim 2006) suggest that *all* the initiatives for gender equality – from changing family law to shifting the balance of power in Parliament and civil society organisations – have had mixed outcomes because of the difficulties of overcoming deeply entrenched gender divisions. This would suggest that education, despite the policy attention to equality, is unlikely to be an exception.

However, to group together a very wide range of gender-equality initiatives in social policy and say all are likely to be difficult to effect seems to give the complexity of change too wide an explanatory remit. It also suggests that the effects of the difficulties of achieving comprehensive gender equality in post-apartheid South Africa have been the same for all women, but this is not the case. Some women have flourished, but many, generally the most disadvantaged, continue to be locked in immobile and impoverished social positions. While the argument about the backlash against women's success may make too tight a connection between women's advancement, sexual harassment and brutal behaviours, it does deal with aspects of the question of violence, which the idea of the complexity of change reduces to no more than one of many factors. A third line of explanation stresses the significance of HIV and AIDS in curtailing equality initiatives in South Africa, but there are two strikingly different perspectives on this. Some have argued globally that the pandemic has 'reinforce[d] socio-culturally constructed inequalities of gender, social status, race and sexuality' (Patton 1994: 2). By contrast, Edward Kirumira (2004) has made the case that the pandemic had created conditions that promote gender equality.

The concurrent emergence of achievement for some women, side-by-side with heightened exposure to violence and HIV, may not be the outcome of a generalised

slippage between gender equality policy and on-the-ground practice. It may equally be due to an emphasis with regard to gender equality that stresses, for example, only outcomes or opportunities, without addressing simultaneously unjust structures and assumptions about violence, gender, HIV and AIDS.

The argument we are making is that using only one approach to gender equality in education, violence or HIV and AIDS is too attenuated. What is required is the intersection of all three approaches that we separated out in the taxonomies above. We do not subscribe to a simplistic idea that gender is always a risk factor for girls. We consider gender largely as a relational concept that binds boys and girls, men and women together in conditions marked by inequalities in power. This approach allows us to avoid essentialist arguments about males and females and allows us to show how gender is constructed and enacted and, in the process, is a locus of struggle, accommodation and change. Whether change is in the direction of expanded meanings of gender equity and equality or not is the focus of this book. We are concerned to look at political economy, institutions and the social interactions in which identities form, reform and incline towards or against a politics of gender equality in education. We show how the multidimensionality of gender, violence and vulnerability to HIV require plural notions of equality and policies and practices that combine interventions, institutional development and inter-actions. Change does not happen suddenly and interventions aimed at bringing about change – whether in relation to HIV, sexual behaviour or wider issues of gender equality – come into the discursive fields of schools and interact with them, producing new understandings and practices lived out by individuals in social relationships.

The theoretical resources we draw on emerge from discussions of equality, diversity, conditions of structured inequality, and the multiplicity of actions people take. We are concerned with what Jane Kenway et al. have called, in the Australian context, 'the mysterious gap between hope and happening' (1997: 1). We have deployed ideas about the elusiveness of equality in relation to data collected over six years in several schools using a wide range of methods. As the chapters that follow show, the practices of teachers, learners, officials and researchers working to try to change gender inequality, understand conditions of violence and support work on HIV prevention, together with developing concern for treatment and care, were not a smooth roll-out of carefully planned strategies, but neither were they a

haphazard collection of disparate events. We see interventions and interactions as being facilitated and constrained by particular forms of institutions. All three forms of policy and practice have had particular histories in South Africa, and to provide this wider context for the detailed analysis of the data collected in schools, the next chapter provides a background analysis of those developments.

Notes

1. We are cognisant of the many critiques, not least within feminist scholarship, that have been made of the term 'empowerment'. As Frank Jacob (1996: 453) puts it: 'The irony of empowerment is that people in a position of authority and power who advocate the empowerment of the individual and community, rarely conceive the empowerment of others in terms of themselves relinquishing their own privileged position of power.' However, the word is used here to indicate progressive and transformational processes. In our usage, we are following the aspirations for empowerment as a concept associated with the work of Naila Kabeer (1999) and Ruth Lister (2004).
2. In developing our taxonomy, we have developed ideas in conversation with Jenny Parkes. There is thus some overlap between the categories we use and those identified in her paper on conceptualising violence (Parkes 2008). We gratefully acknowledge the dialogues that have allowed our work in adjacent areas to develop.

CHAPTER 3

Addressing Gender Equality in South Africa
1994–2008

Introduction

In this chapter we review research on gender and education in South Africa and place particular emphasis on how policy has been interpreted, how it has been implemented and how it has impacted on gender relations. In the post-apartheid period since 1994, gender equality has been one of the goals of education policy. However, in practice, the tendency has been to use quite limited meanings of gender equality, which has stunted the reach of gender transformation.

Gender equality work has found its strongest expression and its greatest challenges in the endemic violence in schools and in the effects on learners, teachers and gender relations of the HIV and AIDS pandemic. Apart from the sheer scale of violence and HIV and AIDS, outlined below, both phenomena are highly gendered and are, in many cases, 'caused' or at least exacerbated by gender inequalities. For this reason alone these issues have attracted gendered attention in the form of laws, policy statements and interventions. To understand violence and HIV and AIDS from a gendered point of view requires a historical perspective, and a brief history of the last three decades of schooling in South Africa is thus provided in the first section of this chapter. In this section we draw attention to how race and gender are mutually implicated and constitutive. No analysis of South Africa can ignore racial inequality. Under apartheid, race was the primary axis of inequality and, to a large extent, it continues to be so. In terms of a crude but nevertheless illuminating race and gender hierarchy, it has been African women and girls who have found themselves at the bottom of this hierarchy. All post-apartheid policies have sought primarily to address racial inequality, while at the same time paying attention to gender imbalances and injustices.

In the second section of this chapter, we examine various aspects of gender inequality, providing a sense of its scale and contours. We also explore different understandings of the concept and show how these generate different views about the extent, depth and seriousness of gender inequality. In the third section we focus specifically on violence and HIV and AIDS. We explore how gender has played out at different levels and in different ways. We examine the quantitative measures, generally national aggregated figures, of gender inequality in the areas of violence and HIV and AIDS. We then focus on how gender is structured and experienced within institutions, how it is experienced by teachers and learners on a day-to-day basis. We also examine the ways in which identities interact with and themselves are implicated in the creation of gender regimes. In the fourth section we analyse the various ways in which gender inequality has been addressed in the schooling system. This section is organised to reflect the different approaches to dealing with gender inequality that were outlined in Chapter 2. We therefore sequentially describe interventions, institutional approaches and interactions, showing the intent and limits of each of these approaches.

When gender is understood as a noun, the focus tends to be on the equalisation of resources. When gender is understood as causing inequality, efforts tend to concentrate on creating institutional mechanisms to monitor and regulate gender disparities. In South Africa this has taken the form of mainstreaming gender work in government departments in an attempt to ensure that it is not marginalised and that gender is not treated as a one-off problem. When gender is understood as being a complex, fluid and dynamic relationship that can produce inequality (and also create equality) the challenge is to focus on the interactions of those who constitute school communities and on the gender regimes of the schools themselves, as it is only in this way that new equitable identities will be created and sustained.

Gender inequality in education under apartheid

Education policy was a central part of apartheid. The experiences and skills that children gained from schooling were determined by 'race' (Kallaway 1984). African children had a curriculum focused on agriculture and they were taught versions of history that stressed their subordinate role. Their teachers were not as well trained as white, Coloured and Indian teachers (Hartshorne 1992). Their schools were poorly resourced – very few had sports fields and many lacked the most basic

facilities, including toilets and electricity, let alone books. Higher education opportunities were severely limited, whereas for white matriculants,[1] university entrance was easily accessible and cheap. As can be seen in Table 3.1, in 1976–77, state spending on pupils reflected gross inequalities.

Table 3.1   Relative financing (in rand) of schools for different racialised groups in 1976–77.

African	Coloured	Indian	White
R48.55 (7%)	R157.59 (24.1%)	R219.96 (33.6%)	R654.00 (100%)

Source: Blignaut 1979: 48

Note: The figure of R654.00 in the last column is given the value of 100% to permit a comparison of the monetary allocations for other race groups (Africans, Coloured and Indians).

In turn, this led to very different ratios of teachers to pupils.

Table 3.2   Teacher–pupil ratios in1978.

African	Coloured	Indian	White
1:48.4	1:28.3	1:26.8	1:20.2

Source: Blignaut 1979: 65

Gender inequalities were also a feature of apartheid education. The number of girls attending and completing secondary school did not equal that of boys until the 1980s (Unterhalter 1991). Women teachers were paid less than male teachers. They were seldom promoted to management positions in school and were not eligible for maternity leave. Unmarried teachers who fell pregnant were forced to resign their posts (Kotecha 1994).

Understanding gender (in)equality in South Africa

Amongst the education systems in Africa, South Africa's is unusual because it has virtual parity of access for boys and girls (Unterhalter 1991). This was an ironic effect of Bantu Education, a system introduced in 1955 to ensure a separate and inferior education for Africans. But it was also a system that gave universal access

29

to state education – for reasons of social control as much as anything else – and this resulted in gender parity in access for children of all colours (Hyslop 1999).

A second significant feature of South Africa's schooling system is the numerical dominance of women in the ranks of teachers. In 2004, 65 per cent of more than 370 000 teachers were women (Moorosi 2006b), a percentage that has altered little since the 1980s (Truscott 1994). Within the teaching profession, however, female teachers are significantly under-represented in senior positions, as shown in Table 3.3.

Table 3.3 Percentages of women teachers at different ranks in 2004.

Rank	Number	Percentage
1 (most junior)	±290 000	70
2 (head of dept)	±50 000	61
3 (deputy-principal)	±27 000	38
4 (principal)	±8 700	26

Source: Moorosi 2006b: 15

The situation in KwaZulu-Natal in the same period broadly reflected national trends, although efforts to increase the number of women in management positions have had an impact. In 2004, 69 per cent of teachers were women, while 41 per cent of principal positions and 41 per cent of deputy-principal positions were held by women (Moorosi 2006b: 17).

A surprising feature of the system is the academic performance of boys. Looking elsewhere in the world, but particularly in Africa, one would expect to find boys doing much better than girls. But this is not exactly the case. In the 2002 matriculation exams, Helen Perry and Brahm Fleisch found that 'young women are performing substantially better than their male counterparts' (2006: 108). The number of girls who take matric exceeds the number of boys and although in 2002 fractionally more boys obtained matric endorsements (permitting entrance into higher education institutions) the proportion of girls obtaining merits and distinctions far exceeds the proportion of boys (61 per cent of matriculants with distinction in 2002 were girls) (Perry and Fleisch 2006: 114). This is a trend that has continued to 2007. Even in the areas of maths and science, globally the preserve of boys, girls are catching up, although they haven't caught up altogether (Kahn 2006).

It should be added, however, that in 2002 girls attending the worst-resourced schools (often in townships and in the countryside) fared worst, worse than their male colleagues.

At the other end of the educational scale, there has been the reversal of a continent-wide gender pattern in which girls have tended to leave school in large numbers and substantially before boys. The number of boys now dropping out exceeds that of girls, as can be seen from Figure 3.1.

Figure 3.1   Number of male and female learners by grade, 2000.

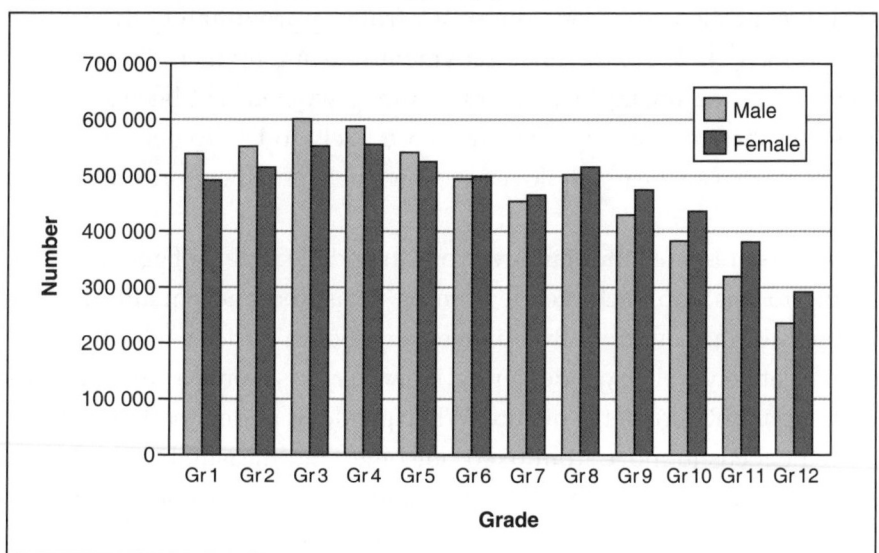

Source: Department of Education Annual School Survey, 2000, in Perry and Fleisch 2006

These improvements in girls' educational achievements are mirrored in the workforce. From 1995 to 2003, the number of women in employment rose from 3 785 000 to 5 194 000 and their share of the workforce rose from 39.3 per cent to 44.7 per cent (Casale and Posel 2005: 22). At the top end of the market, the number of African women in professional positions (legislators, senior officials and managers) grew from 69 000 in 1995 to 130 000 in 2003 (an increase of about 90 per cent). In the same period, the figures for white women were 77 000 to 175 000 (or an almost 130 per cent increase). The move into better-paid positions has meant that more women are richer: whereas in 1995 close to 20 per cent of those in the

31

top-earnings decile was women, by 2003, this had risen to 33 per cent (24). Yet things are not all rosy. More than a million women are still employed as domestic workers and they earn only a quarter of what women earn on average. And income is still unevenly spread between men and women. In 2003 men earned 64 per cent of the total income pie, although this was down from the 70 per cent that they earned in 1995 (26). Unemployment rates among women have risen sharply (and more sharply than for men). The female unemployment rate (the percentage of all women who were seeking paid work) rose from 38 per cent in 1995 to 49 per cent in 2003. Almost five million women were unemployed. By comparison, male unemployment also rose considerably – from approximately 1.7 million to 3.7 million men: 'In 2003, the unemployment rate among men (36 percent) was still lower than the unemployment rate among women had been in 1995. The unemployed therefore are significantly more likely to be women than men – in 2003, 58 percent of all the unemployed were women (and therefore only 42 percent were men)' (22).

The trends presented thus far flow from an understanding of gender as a noun. According to this approach, gender can be measured and equality gauged by technologies of (ac)counting. By this measure, South Africa's education system performs relatively well in gender terms. However, this is only a part, and possibly the least important part, of the picture. When gender is understood as an adjective, it allows the focus to move from fixed outcomes and measurables to the way in which gender is arranged. The concept, 'gender regime', developed by Raewyn Connell in 1987 described the arrangement of gender relations within organisations and drew attention to gender hierarchies, relationships of power and inequality. Using this concept, authors have often viewed schools as 'gender factories' and masculinising institutions (see Heward 1988; Mac an Ghaill 1994). That is, their curricula, teaching and disciplining styles contribute to the ways boys and girls understand and produce their gender identities. The constructions of masculinity in the schools that service South Africa's poor (largely African) population are strongly marked by poverty and harsh learning conditions. In some schools there is an almost Darwinian acceptance that the strong will tyrannise the weak. Struggles over food, territory and women mark many daily experiences, though peace-loving, often church-going, boys offer different ways of relating to their peers (Bhana 2005). Boys without the opportunity to affirm their masculinity by performance in sport

or attaining high marks often seek hyper-heterosexual avenues to express themselves. This may lead to early sexual debut, sexual violence and increased HIV risk for the boys themselves and for girls (Jewkes et al. 2007; Leach 2002; Niehaus 2000; Sathiparsad 2005; Vetten and Bhana 2001; Walsh and Mitchell 2006; Wood and Jewkes 2001).

Particularly during their teenage years, boys tend to seek peer approval for their actions. At its most benign, this fuels a nationwide obsession with sport, particularly soccer for African boys and rugby for white boys although integrated schooling has begun to shift sporting preferences (Alegi 2004; Grundlingh, Odendaal and Spies 1995). But sport can also generate or be the location for the development of homophobia and hierarchical distinctions based on strong bodies and race (Bhana 2008) and within schools there are racialised antagonistic groupings that can generate gangs and racialised violence (Hamlall 2003).

While discourses of gender equality and human rights have created a language for new forms of harmonious gender relatedness (Morrell 2007a), it remains the case that relations between boys and girls tend to be sexualised, with friendships being all too rare (Reid and Walker 2005). In coeducational contexts this results in high levels of sexual harassment (Human Rights Watch 2001; Subedar 2003) with consequences for the school experiences of girls. It remains the case, particularly in rural settings, that boys often believe that girls are inferior, that boys have a right to discipline them and a right to have their proposals of love accepted (Sathiparsad 2005; Zakwe 2005; see also Chapters 6 and 8).

The experience of school for many girls, particularly those in township and under-resourced schools, is often traumatic. Subjected to sexual harassment, corporal punishment and unsympathetic teachers, girls often leave school with low levels of self-esteem and with little ability to challenge the social forces that lock them into domestic subordination and limit their ability to break into the public world of paid work and public recognition (Unterhalter 1999).

While homophobia remains widespread in schools, constitutional support for gay rights and an active and public gay rights movement (Gevisser and Cameron 1994) mean that more space is being opened up for gay issues to be discussed, if not exactly leading to tolerance of gays in schools. Nazir Carrim (2009) documents the ways in which young gay men have sometimes been able to come out, but also shows how fragile some of the affirmations of tolerance and human rights by schools have been.

33

Violence and HIV

If we follow the same approach to the examination of violence and HIV as applied to gender above, i.e. looking at numerical indicators, we can rapidly come to the conclusion that South Africa is one of the most violent societies in the world and also one of the most grievously afflicted by the AIDS pandemic.

South Africa has experienced a great deal of violence, particularly in the political arena after the 1976 Soweto student uprisings. Political violence was a common occurrence in township schools involving 'the comrades' (mostly young black men) in violent confrontations with apartheid forces or, in KwaZulu-Natal, with Zulu nationalist forces, organised under the banner of Inkatha. But anti-apartheid opposition was only one form of violence. Schools were historically violent places where corporal punishment was frequently used and sexual harassment and gender-based violence were common (Marks 2001; Morrell 1998). Post-apartheid, as Anthony Altbeker (2007) puts it in the title of his book, South Africa is still 'a country at war with itself'. In a population of 47 million people, 50 people are murdered every day (which translates to between 18 000 and 20 000 murders annually); 500 000 are victims of assault and attempted murder; 200 000 are robbed and 55 000 raped (Altbeker 2007: 37–38).

Although there are no databases for violence in schools, we know that they are not exempt from the violence that characterises other public and private spheres. While all schools suffer some form of violence (Human Rights Watch 2001), the intensity and frequency of violence are concentrated in schools that service the poorest communities. The violence takes many forms, including corporal punishment, sexual violence (by male teachers and learners against female students) and gang-related violence (Eliasov and Frank 2000; Niehaus 2000). Human Rights Watch described violence in schools as 'widespread', indicating that no group was free of it. This report indicated that 'for many South African girls, violence and abuse are an inevitable part of the school environment' (2001: 5).

South Africa has one of the highest rape rates in the world. Between April 2005 and March 2006, 54 926 rapes and attempted rapes of women were reported to the police (SAP 2008). This is equivalent to 117.1 per 100 000 population, a rate that is nearly four times higher than the United States (US Department of Justice 2006). In an earlier study, Rachel Jewkes et al. found that 1.6 per cent of a sample of more

than 11 000 women had been raped before the age of fifteen and that the largest group (33 per cent) of perpetrators were teachers (Jewkes et al. 2002). In an Eastern Cape study conducted with 1 370 young men aged 15–26, 20 per cent said that they had been involved in rape or sexual assault (Jewkes et al. 2006). The mean age of the first sexual assault or rape was seventeen, an age at which most South African boys are still at school.

In South Africa, as elsewhere, males are more likely than females to be victims of violence other than rape (Kimmel 2001). In 2003 roughly seven times as many South African men as women died as a result of homicide. Estimates based on the National Injury Mortality Surveillance System (which drew on an area containing approximately 40 per cent of the country's population) suggest that in 2003, 22 000 men and 3 600 women were murdered (Peacock, Khumalo and McNab 2006).

Experience of violence is also age-related. Several studies have shown that young people are significantly more likely to be participants in and/or victims of violence than adults. Lezanne Leoschut and Patrick Burton, for example, showed that one in ten young people they surveyed reported feeling fearful at school (2006: xiii). Almost one in six feared travelling to school and one in five had been threatened or hurt by someone at school (xiv). More than half of those who had been threatened at school indicated that they had had these experiences more than once (xv). Corporal punishment was one of the forms of violence frequently reported (by more than 50 per cent of respondents), while boys were more likely to be victims of violence than girls (67). By race, Coloureds were most likely to have experienced assault: 17.8 per cent compared to 17.0 per cent for Africans, 11.8 per cent for Indians and 9.9 per cent for whites (2006). Ninnette Eliasov and Cheryl Frank (2000) came to similar conclusions in relation to both primary and secondary schools and suggested that crime and violence were more severe in poor, working-class schools.

South Africa has the highest number of people living with HIV and AIDS in the world (UNAIDS 2007: 16). The disease is a leading cause of death in sub-Saharan Africa. While globally infection rates are roughly the same among men and women, in sub-Saharan Africa, females make up 61 per cent of those infected (8). Prevalence rates in South Africa were recorded as 18.8 per cent in 2006, down from 20.9% in 2004 (11). This should be compared with Senegal (with the lowest recorded sub-Saharan rate) of 0.9 per cent, and with Swaziland (33.4 per cent) and Botswana

(24.1 per cent) with the highest levels. The Joint United Nations Programme on HIV/AIDS (UNAIDS) believes that 'HIV infection levels might be levelling off' in South Africa, as antenatal figures dropped from 30 per cent in 2005 to 29 per cent in 2006. But in KwaZulu-Natal these levels are at a staggering 39 per cent (16).

These figures confirm earlier statistics generated for KwaZulu-Natal, the South African province with the highest rates of infection. In 2000 it was estimated that among 15–19-year-olds, 15.64 per cent of African girls were likely to be HIV-positive, compared to 2.58 per cent of African boys (Morrell et al. 2001: 51). In a more recent study, from March to August 2003, a national survey of HIV prevalence among 11 904 15–24-year-olds found that young women were significantly more likely to be infected with HIV than young men (15.5 per cent compared to 4.8 per cent) (Pettifor et al. 2005). The racialised nature of the pandemic can be seen in equivalent figures for white (1.25 per cent and 0.26 per cent) and Indian (1.29 per cent and 0.26 per cent) females and males respectively.

High levels of infection are found among teachers. Overall, 12.7 per cent of all educators are HIV positive (HSRC 2005b). HIV prevalence is highest in the 25–34 age group (21.4 per cent), followed by the 35–44 age group (12.8 per cent). Educators who were 55 and older had the lowest HIV prevalence (3.1 per cent). There are major racial differences in HIV prevalence: Africans had a prevalence of 16.3 per cent, compared to whites, Coloureds and Indians, whose infection rates are less than 1 per cent (HSRC 2005a, 2005b). The proportion of attrition in national teacher numbers due to HIV and AIDS (that is, the proportion of teachers dying of AIDS or giving up teaching because of ill health) has risen from 7 per cent in 1997–98 to 17.7 per cent in 2003–04. In this period, 14 192 educators have died of AIDS. The heaviest toll is on African female educators in the 20–49 age range, with women in KwaZulu-Natal being the worst affected (HEARD 2005: 12–15).

The exact reasons for the greater vulnerability of young (African) women remain unclear, although it is now generally accepted that biology plays a role. In acts of sexual intercourse, more sexual fluid is transmitted from the male to the female than the other way around and there is more surface area (which in turn is more susceptible to being breached) in females' reproductive organs that substantially increases the risk of infection. Such risks are heightened in young girls because the reproductive organs are not mature and are therefore unable to resist infection successfully and are more liable to physical damage, which heightens the risk of HIV infection.

Early sexual debut increases the risk of HIV infection. In South Africa, the average age of first sexual intercourse falls between fourteen and seventeen, although more than half of young men and women are sexually active by the age of sixteen (Harrison 2005: 270). Although there is little difference in the age of sexual debut between boys and girls, girls are rendered more vulnerable to infection because of the physical reasons discussed above. Their risk, however, is also increased by inequalities that impact on the way in which sexual intimacy is negotiated. Women often have less power than men in situations of sexual intimacy and this unequal situation is exacerbated in cases where young girls are sexually intimate with older men. Examples of this inequality would be the inability of women to negotiate condom use[2] and the tendency of men to resort to violence. Where violence is used, the risk of HIV infection is increased (Dunkle et al. 2004) because forceful, unlubricated and unprotected sex is likely to result in physical injury, which provides a portal through which the infection passes.

There are obvious dangers in girls being sexually active at a young age, not least of which is becoming pregnant. In 2003, 9.4 per cent of girls between the ages of 15 and 19 had already become mothers. Figures are not accurate and fluctuate over the years depending on what measures and definitions are used, but every year since 1991 there has been a high rate of teenage motherhood with a strongly racialised distribution. In 2003, the rate for young African women was 10.2 per cent; for Coloureds, 6.4 per cent; for Indians, 2.2 per cent and for whites 0.0 per cent (Health Systems Trust 2007). It appears that many young women regard childbearing as an affirmation of their adult femininity (Preston-Whyte 1993; see also Bullen, Kenway and Hey 2000).

When we examine gender regimes, we move beyond the quantitative indicators to examine the content and nature of gender relations within institutions. We now understand gender as an adjective, as something not so much to be acted on, but as an active relationship, a constitutive part of relationships and social structures. In this sense, gender is dynamic and alive and may even be a cause of inequality. If we examine school violence from this perspective, we see how institutions are implicated in the performance of violence, while also having the capacity to limit or prevent it. This is nowhere more evident than in the continuing use of corporal punishment in schools. Prior to 1996 corporal punishment was inflicted frequently and for many different reasons, including sexual impropriety. For example, when

a boy was discovered to have placed a love letter in a girl's school bag he was forced to lie down on a table in the principal's office and given more than 50 strokes (Niehaus 2000: 393). The impact of corporal punishment on the daily lives of learners is captured by Deevia Bhana, who witnessed punishment being administered in a Durban primary school in 2001:

> Mrs. H. walks in the classroom. Quiet. She picks on two children. They have not done their work. Mrs. H. shouts. She gets the stick from the table. It's a branch from a tree. She hits them on their back and legs. Stick breaks. Teacher gets another stick. This time it is not a branch. They call it a pipe. Mrs. H. continues where she had stopped. They are crying, sobbing quietly. (Bhana 2002: 116)

The gender regimes of some schools legitimate and perpetuate forms of violence, even though schools are generally conceived of as vehicles for emancipatory and democratic values (Harber 2002).

Schools have become a major focus of HIV-prevention work in line with the understanding that what happens in schools is critical to the ability to reduce HIV infections. Schools are understood as institutions that have a major role to play in promoting safe sex and gender equality, and often the campaigns seek to yoke these two goals together. Institutional responses may focus on the sexual vulnerability of young women (Wood, Maforah and Jewkes 1998) and seek to create conditions in which protective measures are put in place. These may include mechanisms to facilitate the reporting of sexual harassment and lessons devoted to teaching about sexual risk. Institutional approaches, however, have limitations and these are examined in some detail in Chapters 5 and 6.

Violence and talk of violence is often constitutive of gender identities. Violence can also be seen as a way of constituting gendered discourses of, for example, tough or loyal masculinity, which are taken up in the formation of identities (Parkes 2007). In a research project in secondary schools in Pietermaritzburg, David Blackbeard and Graham Lindegger (2007) found that adolescent boys battled to meet the normative demands of hegemonic masculinity and adopted various strategies to cope with these demands. Boys dealt with demands for toughness, risk-taking, and emphasised heterosexuality in a number of ways, although they were not always able to develop counter-hegemonic narratives that would produce stable and viable masculine subjectivities. There are basically three reasons for

this: the power of teachers, the power of peer groups and the influence of the school's gender regime, which did not support alternative forms of action (Attwell 2002).

Interactional approaches to gender equality need to impact on peer groups, which generally establish gender norms for boys and girls, and to hold a school population to gender-equitable norms. This is often an exceedingly difficult goal to achieve (Luyt and Foster 2001; Ratele 2001). Many schools have gender regimes that promote what Jane Kenway and Lindsay Fitzclarence (1997) have called 'poisonous pedagogies'. These make it difficult for boys and girls to develop subject positions that are at odds with hegemonic standards. In some schools, however, counter-hegemonic discourses are emerging. In the formerly white, well-resourced schools, for example, there is evidence that 'new man' versions of masculinity exist (Attwell 2002; Morrell 2007b; Richter 2007; Richter and Morrell 2006). These include a willingness to care for siblings and parents and the desire to seek stable working lives in domestically equitable circumstances. In the more violent and resource-poor schools that serve black learners, counter-hegemonic discourses also exist, but they are more marginalised and less likely to challenge hegemonic school identities (Bhana 2005).

Psychological studies in school and youth settings show that certain groups are marginalised. Misogynistic and homophobic discourses sideline girls and gay boys (Wells and Polders 2006). Yet these studies also show that gender identities are changing, especially in the face of the AIDS pandemic (Walker 2005). Although poverty remains a very strong force that shapes identities, there are interventions and contexts that provide opportunities for children to make choices that are healthier for themselves and for society (Ramphele 2002).

Gender identities are shaped by the AIDS pandemic which, as it matures, is changing the dynamics of intimacy. Ethnographic work in Mandeni shows how masculine ideals that validated many girlfriends (and which therefore endangered the sexual health of males and their sexual partners) have begun to shift as the disease has begun to take its toll. The positive connotations of *isoka* (that is, a man having many partners) are being challenged and a question mark is now being placed against the equation that being a real man requires having many girlfriends (Hunter 2004). The pandemic is also encouraging more communication between young people, including in the negotiation of intimacy (O'Sullivan et al. 2006).

New discourses of sexuality are empowering girls to express their agency, which in contexts of poverty they often do in instrumental ways. They may, for example, barter their sexuality for favours of one kind or another, often with older men with money who are called 'Sugar Daddies' (Hunter 2002). Mark Hunter sees this process as one in which girls powerfully assert and deploy their sexuality, rather than a further expression of inequality. We would argue that both dynamics may be simultaneously in play: that is, the ways in which girls assert and deploy their sexuality, however powerfully, are constrained by the gender inequalities within which they are able to act. However, it is important to acknowledge that sexuality is a key aspect of girls' identities, even in the early years. Deevia Bhana et al. (2006) and Relebohile Moletsane (2007) have shown how the play of primary school girls attending school in an impoverished area reflects their knowledge of sex, their initiation into a heterosexual world of desire and the way in which the dangers of this world are normalised. Older girls are often very conscious of their sexuality and though some may still regard some form of surrender to men as the most appropriate model for adult behaviour, some flaunt their sexuality, challenging school authority and returning the sexualised gaze of boys (Gaillard-Thurston 2008).

Gender transformation: Work and outcomes
Interventions
There have been two distinct types of gender transformation interventions (understood as initiatives that mobilise a parity view of gender equality). The most powerful of these have been the state's legislative programme, its resource allocation policy and the attendant establishment of gender machinery (a bureaucracy designed to work for and monitor progress towards gender equality). The second type of intervention undertaken both by the state and by non-governmental organisations (NGOs), often with project-specific funding, operated at curriculum and school level to spread the message of gender equality and, in the context of the worsening AIDS pandemic, safe sex. Both types of interventions approach gender as a noun, which is to say, a subject upon which actions can be performed.

In line with the Constitution's Bill of Rights (which prohibits discrimination on grounds of gender, sexual orientation, race, religion and so on) a number of laws have been passed prohibiting discriminatory gender practices. These include

practices that discriminate against female teachers (unequal pay and employment conditions) and practices such as corporal punishment, which infringe on the dignity of learners. Apart from creating an equitable legislative framework, enabling legislation such as the Employment Equity Act (1998) created mechanisms by which gender inequities (in the workforce) could be addressed. An effect of this has been to privilege applicants on grounds of race and gender in competition for employment as a teacher or promotion within the education hierarchy. Further legislation guarantees women the right to at least four consecutive months on maternity leave from work or study, although there is no minimum payment for such leave.

Parallel to the development of the new legislative framework, there has been a concerted effort, at the level of resource allocation, to tackle racial and, to a lesser extent, gender inequalities. The goal was to equalise resources across schools, districts and provinces and thus redress the chronic imbalances created along race lines under apartheid. The programme was vigorous and, on its own terms, successful. By the turn of the twenty-first century, obvious financial inequalities had largely been removed (Gustafsson and Patel 2006; Van den Berg 2006). State funding was consciously redirected to black and particularly rural schools. Indeed, Servaas van den Berg argues that 'school education experienced the largest pro-poor shift in the post-transition period' (2006: 50). However, even this mechanical understanding of how to create equality has been tripped up by the system's complexities. Shifting money to poor black schools was not accompanied by a move of the best-qualified and most experienced teachers to these schools. The best-trained teachers continue to be located in the better-resourced, more stable, suburban schools. Nevertheless, there has been a significant redistribution of state expenditure on teachers in public schools, such that the number of teachers in black schools increased from 24 to 31 per 1 000 pupils in the period 1991–2001, while it declined from 59 to 31 in white schools (60). Suburban schools have been able largely to offset this decline in state expenditure by using fees from parents to employ an additional 12 teachers per 1 000 pupils, which is one reason, though not the only one, why suburban schools continue to outperform black and particularly rural schools (Reddy 2006).

At the national level, budgeting may have been equalised, but often this process was not gender sensitive. Shortly after the achievement of democratic government,

two major problems were identified in a critique of the budgeting process. First, departmental policies did not respond strongly to women's interests, but even where policies were appropriate, 'the budget does not guide resource allocation accordingly (De Bruyn and Seidman-Makgetla 1997: 60). The Women's Budget Initiative lobbied for the rights of women and children to be prioritised in government spending (Budlender 2000; Zulu 1998) and analysed the impact of government expenditure and revenue on women and girls, as compared to men and boys. It was an attempt, also, to strengthen the state's capacity to use gender analysis in planning, evaluating and monitoring the budget and increase women's participation in the budgetary process. Although research demonstrated the impact of the Women's Budget Initiative on women's economic fortunes (Valodia 1998) no research has been done on the effect of gender-responsive budgeting (GRB) on schooling. While the impact of GRB on gender inequalities has been limited, partly because of the failure to link rights-based and budget work (Budlender 2005: 4), GRB can claim credit for the establishment of the Child Support Grant (for poor young mothers), in the roll-out of antiretrovirals (ARVs) (to prevent mother-to-child transmission of HIV) and in increasing grants to micro-economic enterprises (a market niche dominated by women). The GRB approach has ensured that in some provinces gender is considered when budget allocations are made (Budlender 2005: 11). Gender-based violence received some budget share as a result of GRB lobbying.

Africans affected by poverty get a poorer education, which is compounded by socio-economic conditions. Very large numbers of families have only one wage-earner working in the lowest paid sectors. Millions of children live in extended families that include a number of unemployed or under-employed adults, as well as dependents from family members who are either ill or deceased, often as a result of AIDS (Anderson, Case and Lam 2001). In recognition of the need for extended state support, a means-tested cash grant to guardians of children up to the age of twelve (Child Support Grant) was introduced from 1998. In 2003 it was awarded to 1.9 million caregivers (Lund 2006) and by April 2007 was reaching 8 million children (Lund 2008: 77). The grant had some dramatic effects, particularly in rural areas (Shisana et al. 2005) but is not in itself enough to offset the difficulties of low and uncertain income and health, which have made educational attainment for the poorest children so elusive.

Once a legislative framework had been created and monies equitably distributed, gender equality was to be monitored and sustained by the gender machinery. This was established in terms of the report of the Gender Equity Task Team (GETT), which was appointed by Minister Sibusiso Bhengu in 1996 and which submitted its report the following year (Wolpe, Quinlan and Martinez 1997). The report covered every aspect of education from early childhood development to adult education, but naturally placed its major emphasis on schooling. It made an astonishing 400 recommendations that, if implemented, would effectively have produced a gender revolution in the education system.

GETT identified gender violence as the major obstacle to achieving gender equality in education. The report noted that violence took many forms, some of which originated in the turbulent political struggles that had ensued after the 1976 Soweto student uprising. It expressed grave concern about the ways in which girls, particularly African girls in impoverished contexts, were affected by violence. Among its recommendations, therefore, were steps to create safe school environments, to stop sexual harassment and in general to protect girls.

Its chief recommendation was the establishment of a Gender Education Unit (GEU), which would 'ensure that gender equity is a systematic consideration' (Wolpe, Quinlan and Martinez 1997: 237). The GEU would have to challenge inequalities at a technical level (that is to say, where they were enshrined in various practices and policies) as well as at an ideological level, where normative assumptions would have to be raised and debated. GETT set out a detailed list of functions for the GEU. It was to involve itself in all policy formulation. It would have to monitor departmental activities from a gender perspective and arrange training programmes for personnel, especially senior management. It could also commission research on specific priorities to ensure that the activities of the department were not gender blind.

The GEU was established in 1999. It was supposed to be the motor of gender equality. It was expected to institutionalise gender equality by ensuring that laws were heeded and new policies were passed. One of its tasks was to co-ordinate the work of Gender Focal Points (GFPs). These offices exist at provincial, district and, in some cases, school level. Officers are charged with cascading national gender-equality goals down to lower levels. While little research has been done on the GFPs, the general impression is that they have not succeeded in implementing

gender-equality policy. Another primary function was to monitor and support the gender aspects of employment equality (which means, in part, to assist the process that places women in promoted posts in order to rectify the skewed profile of management posts). The GEU has produced training materials and hosted workshops, which have largely focused on improving girls' education and on ending sexual violence (Wolpe 2005).

The philosophy that has been used by government with regard to institutionalising gender equality has been 'mainstreaming', 'the shorthand term in use for approximately 10 years by activists working inside institutions to orient the values, policies, organizational processes and forms of evaluation so that these take due account of gender equality' (Unterhalter 2007: 130). This approach demands that people be informed about gender issues and that they incorporate this information into their lives and work (Kleintjes et al. 2005: 2). Mainstreaming is effected by infusing the curriculum with gender sensitivity, making each teacher and school manager gender sensitive and by metaphorically (and often literally) having 'gender' as a standing item on the agenda. Mainstreaming assumes that the goal of gender equality can only be approached by shaping the institutions that have the power to effect gender transformation. According to Catherine Odora Hoppers, the GEU's conception of mainstreaming was that it

> consists of an ideological and an institutional component, both reinforcing each other. The ideological component refers to the key *theories*, *paradigms* and *assumptions* about development or about society at any given time. The institutional component refers to the *organizations* and *people* who make key decisions. Because it is the ideas and practices of the mainstream that determine who gets what, and provides a rationale for the allocation of opportunities in society, it is very important that both the ideological and the institutional dimensions of the mainstream are clearly identified in order that integration leads to the desired goals. (Odora Hoppers 2005: 67; italics in original)

It is doubtful that these dimensions have either been clearly identified or enacted.

Mainstreaming as a strategy has limitations. It is inattentive to issues of difference and dissent. Departments that are responsible for gender mainstreaming are often marginalised and the strategy frequently fails to alter existing power

relations since it is often understood in quite technical terms, adrift from the broader feminist goals that underpin gender equality (Meintjes 2004). '[F]ormal institutional commitment to "do something" about gender mainstreaming does not always translate into practice that legitimates the significance of gender equality' (Unterhalter 2007: 132).

When GETT issued its report, the full impact of AIDS on education and gender equality was not yet appreciated, although in 1996 HIV prevalence among antenatal attendees was already 14.2 per cent (Gouws and Abdool Karim 2005: 56). The GETT report says virtually nothing about AIDS, although it is very concerned about school violence and teenage pregnancy (both of which are closely tied to HIV risk). The term 'HIV/AIDS' appears in the list of acronyms (Wolpe, Quinlan and Martinez 1997: 17) but there is, astonishingly, no mention of the disease in the executive summary or in the many recommendations.

Within a matter of a few years, the full extent of the epidemic came to public attention. By 1998, HIV prevalence amongst antenatal attendees had risen to 22.8 per cent (Gouws and Abdool Karim 2005: 56). The impact of AIDS on education was devastating in terms of related deaths, morbidity (teachers becoming ill), the growing number of orphans and the escalating family demands made on children of school-going age for care (HEARD 2005). In this context, we now turn to interventions undertaken by NGOs, which were, by their nature, limited in terms of goals and funding and which utilised minimalist definitions of gender equality. These interventions either focused on (teacher and learner) training or on the curriculum.

The Department of Education was slow to respond, but in 1997 it utilised funding from the European Union to implement a teacher-training scheme for HIV prevention (James 2002). Peer education was central to this – teachers would teach one another and learners would do likewise. Peer education has been used in out-of-school youth settings, as well as in schools, but generally without the rates of success claimed for it (Campbell 2003).

A more systematic approach to the epidemic has been via the Life Skills curriculum, introduced as part of the national overhaul of the curriculum called Curriculum 2005. Life Skills offers a slot in school timetables that can be used to discuss issues of sexuality, gender, disease and risk-taking. Life Skills was compulsory from 2000, but the subject only became examinable at matric level in

2008. In theory, it should be the ideal vehicle for teaching and promoting lessons in gender equality. To date, however, achievements have been limited and, depressingly, they have been least successful in schools that serve learners who are most at risk (Moroney 2002). In understaffed township schools the subject is frequently not taught or is taught by untrained and unenthusiastic teachers who lack the necessary teaching materials (see, however, Chapter 10). The situation is even worse in rural schools. And when Life Skills is actually taught, a lack of resources and training mean that a rote, transmission approach is used: for example, a Grade 9 learner interviewed in 2001 described the teaching methods: 'the [Life Skills] programme at the moment is focused on telling us the information but not showing us' (Raniga 2006: 183). This is not, however, inevitable. In Chapter 6, for example, we focus on sex-education lessons, taught as part of Life Skills in two primary schools, which both illustrate and contradict this view.

At the institutional level in schools, gender equality has been pursued in various ways. Women have been promoted into senior positions. Curriculum initiatives, such as the teaching of Life Skills, have ensured that schools address gender bias and teach about gender. National policy also obliges schools to adopt a formal code of conduct, which binds learners to uphold certain values and gives schools power to act against transgressors.

Alongside the formal work within schools have been a host of complementary interventions, often funded and implemented by NGOs, which have addressed various aspects of school life with a view to achieving or contributing to gender equality. We shall discuss one such intervention (DramAide) in Chapter 5. Many of the NGO interventions have sought to bring about changes in sexual attitudes and behaviours, particularly among young people, as part of HIV-prevention strategies. Probably the best known and funded of these is *loveLife*, which runs peer education programmes, has large billboards and uses youth-oriented marketing techniques. It was started in 1999 with the aim of 'positively influenc(ing) adolescent lifestyle' (Parker 2003: 1). Its ambitious goal was to reduce 'the incidence of HIV among 15–20 year-olds by 50% over the next three to five years' (Parker 2003: 1). In a critical review, Warren Parker argues that *loveLife* operates in a competitive way to discredit other interventions and approaches and uses scare tactics (exaggerating the extent of the epidemic and its dangers) to bully youth into accepting its messages (2003). On the other hand, Nancy Lesko (2007) finds

reason for optimism in her research with twenty African *loveLife* peer educators in Durban. She notes that their involvement with *loveLife* has given them the opportunity to deal with a variety of life conflicts centred on sexuality and sexual health. She argues that *loveLife* allows young people to engage constructively with issues of their own agency and dependence.

Interventions based on peer educators in schools have been advocated for at least ten years, although their capacity to engage with the complexity of young people's relationships with HIV has been questioned (see, for example, Campbell and MacPhail 2002). Much writing on school-based interventions tends to be either marked by aspiration with regard to the potential that schools offer for providing education about the pandemic or to report on particular local initiatives (Kelly 2002; Rugalema and Khanye 2004). These generally look at who carried out the initiative and how young people responded, but are often bounded by the frame of the intervention itself, rather than posing wider questions concerning gender or violence. Often policy guidance suggests what should be done, rather than reflecting on how and why particular aims or practices are desirable and with little concern with some of the effects. For example, for southern and eastern Africa, the policy guidance is contained in directives to school governing bodies on how to develop an HIV plan for their school, in guidance to teachers on how to work with stigma and counselling (Department of Education 2003) and to education departments on how to plan for high levels of teacher or pupil mortality (Boler 2003).

Some interventions are particularly concerned with gender, for example, the work of Stepping Stones (Jewkes et al. 2007), which requires participants to look critically at gender and their own relationships. But on the whole, there has been little long-term follow-up research on intervention and the extent of attention to the gender dynamics of their work is uneven. Mark Thorpe reported on work with teacher trainers using drama to convey messages about gender and HIV in Mozambique (Thorpe 2005), but many studies merely outline the nature of the work done in schools or in adult education programmes and some of the responses by teachers or learners, without considering issues of gender or debates regarding equality.

An interventions approach to equality has thus, in formal terms at least, corrected some of the worst bias in the system by focusing on the funding of state resources to the poorest and by creating a legal framework supported by gender

machinery to create the conditions for a gender-equitable system. This work has been complemented by NGOs and various state agencies, which have attempted to change conditions in school via curriculum interventions and training. Yet these wide-ranging initiatives have clearly not been able to effect enduring change in schools, nor have they clearly impacted on the life experiences of the majority of the population. Nor is it clear that these initiatives have actually passed into the daily workings of school, in administrative or pedagogical terms. In other words, while resource parity has been achieved, it has not effectively been institutionalised nor has it effected the holistic interactions that are a true measure of transformation.

Institutional change
There is some debate as to whether formal schooling does or does not protect against HIV and AIDS. In articles published worldwide reporting on data collected before 1995, more education was related to higher HIV vulnerability. However, after 1995 more education was either not related to HIV vulnerability or indicated lower levels of vulnerability (Hargreaves and Boler 2006: 25). In studies that looked at the impact of education on a cohort over time, as the epidemic evolved, the link between higher levels of education and higher levels of vulnerability weakened. This, James Hargreaves and Tania Boler conclude, was the result either of the risks of infection declining for the more educated or increasing for the least educated (2006: 26). Five out of seven studies that looked at the age of sexual debut and schooling showed that young women attending school were less likely to have had sex than those not attending school, although they concede that girls who were more sexually active may drop out of school as they become pregnant and that longitudinal studies were needed to see whether school impacts positively on age of sexual debut (Hargreaves and Boler 2006: 27). A number of qualitative and quantitative studies in South African schools suggest that girls do leave school when they become pregnant even though the South African Schools Act (1996) made the expulsion of pregnant schoolgirls unlawful. For example, using data collected in KwaZulu-Natal in 2001, Monica Grant and Kelly Hallman found that 76 per cent of young women (aged 14–24) who were pregnant at the time of dropping out of school, gave this as their reason for doing so, although other reasons, such as financial constraints and childcare responsibilities were also salient (2006: 9, 22). One factor that may prompt girls to leave school after giving birth is the absence of any male support (Loila-Nuahn 2004). On the other hand, large

numbers of girls return to school after giving birth for the first time and appear to delay any further childbearing until their twenties (Kaufman, De Wet and Stadler 2001: 149–50). In 1993, approximately 34 per cent of all girls younger than 24 who had had a child as teenagers were currently attending secondary school (155). This probably has to do with the realisation that improved education is one of the few means of upward social mobility, but it also testifies to ongoing familial support often provided by grandmothers.

Hargreaves and Boler's (2006) conclusion from their systematic review of literature is that the more schooling girls had, the more likely they were to use condoms, suggesting higher levels of schooling are linked with the ability to negotiate safer sex. But Sisana Majeke (2008) suggests that schooling and gender interventions may create heightened awareness of gender rights and thus fuel the frequency of sexual intercourse. The rigour of the systematic review is important for helping to explore the question of the levels of education that might provide protection against the epidemic and for understanding which aspects of HIV vulnerability education might be able to protect against. But the review does not examine the way in which different kinds of school experiences might shape responses to the epidemic, with regard to delaying sexual debut, negotiating safe sex, avoiding coerced sex or helping friends who face these fears.

Ensuring that learners, boys and girls, attend school (at least until compulsory schooling ends at age sixteen) is an obvious goal, but figures that indicate school attendance tell us nothing about what happens in schools. Some indication of the variation of school conditions is provided in a study of some Eastern Cape secondary schools. Anthony Lemon suggests that the very different matric pass rates (from 92.73 per cent in the most affluent school to 0 per cent in the poorest rural school) are the result of a lack of professionalism and commitment among teachers in the rural and township schools manifested in lateness and absenteeism and caused by, among other things, 'poor resources including lack of textbooks, insufficient information and training on the delivery of the new syllabus, and sometimes poor leadership' (2004: 289). Attempts to transform this through programmes, such as *Imbewu*, for example, initiated in 500 schools in the Eastern Cape to rebuild community support for schools and develop teacher capacity, started to have an impact, but sustaining momentum was a considerable challenge (Coombe and Godden 1996; Adonis 2008).

Interactional transformation within schools

Schools are key sites in the creation of gendered (and sexualised) identities (Connell 1987; Epstein and Johnson 1998; Mac an Ghaill 1994; Thorne 1993). Yet while schools have a role to play in making gender, it is the people in schools who most obviously have agency and shape the institutions that establish the conditions in which they act. Examining interactions involves acknowledging difference and complexity among groups. Research on schoolboys now routinely makes distinctions between different groupings of boys and the forms of masculinity that are associated with each group (Connell 1989; Mac an Ghaill 1994; Martino 1999). These analyses generally attempt to locate various configurations of masculinity in relation to gender power and, in turn, as relative to other configurations of masculinity. It can be seen that interactions, critique and social networks are important in the formation of masculinity and similar points can be made with regard to femininity. But the ways in which these are linked with structures and institutions is a matter of considerable discussion.

When the focus of gender equality falls on interactions, it reveals the complex connections between the various elements of the school community, the gender regimes of schools and processes of meaning making. Jean Stuart's (2006) account of working with visual arts to develop an understanding about the AIDS epidemic with teacher educators in South Africa shows how changes in gender identity among learners, teachers and teacher educators are contingent and unpredictable. By contrast, the very rich, ethnographic accounts evident in studies by Frances Vavrus (2003) in Tanzania and Doris Kakuru (2006) in Uganda show the very complex gendered interrelationships established in schools in the context of rural poverty and the HIV epidemic. On the whole, however, the literature on schools and HIV tends to focus on interventions and institutions and does not work with nuances, the complexity of relations and ideas, and the gender dynamics entailed in the formation of identities, institutions and practices.

Evidence of emerging equality

It is easier to see gender change from an interventionist perspective than from any other. This is because research can and does measure various types of outcomes. It is less easy, but still possible, to subject institutional change to some kind of qualitative assessment, for example, to examine how curriculum changes have

been implemented and how training programmes have unfolded. An example of identifiable change comes in the staffing profiles of schools that now reflect the influence of the National Education Policy Act. In six districts of KwaZulu-Natal (see Table 3.4.) the movement of women into senior management positions is evident and marks a significant shift from the 1994 position when, nationally, 58 per cent of principals, 69 per cent of deputies and 50 per cent of heads of department were male (Wolpe, Quinlan and Martinez 1997: 198).

Table 3.4 Levels of seniority by gender in schools in six KwaZulu-Natal districts.

Rank	Male (%)	Female (%)
Level 1	26	74
Head of department	38	62
Deputy-principal	54	46
Principal	60	40

Source: Nair 2003: Appendix 2(A)

A ripple effect of the legislation and its implementation is that women teachers now more readily seek promotion (Van Deventer and Van der Westhuizen 2000) even though they still encounter opposition when they assume management positions (Maharaj 2003; Moorosi 2006a; 2006b).

It is less easy to make sense of the complex and tenuous changes that are to be found at the interactional level. Studies that focus on interactional change show it to be non-linear and contested, such that gender changes that occur or seem imminent, might actually generate responses that undermine or oppose progress towards equality.

In this chapter we have shown that gender inequalities were marked in the past and still exist in the present. The shape of gender inequality was and still remains affected by race and class, where African girls in rural areas remain in the most precarious position. Since 1994 there have been major legal and policy interventions designed to improve the lot of, particularly female, learners and to put in place frameworks and processes that will steadily reduce gender inequalities and improve the experience of learners. While these measures have met with some

51

success, especially from approaches that treat gender as a noun (interventions) and as an adjective (institutions), there are obvious shortcomings, not least in the continuing very high levels of violence in schools. What remains unclear is the extent to which gender-equality efforts have impacted on schools and on learners themselves and how these efforts have been hindered or assisted by conditions created by the AIDS pandemic. For our purposes the search for equality must take account of the relationships between learners, the way in which identities are constructed and meanings made. The complex and contradictory interactions between learners, teachers and managers are where, ultimately, an assessment of gender transformation must focus. It is here where some sense of new gender/ed capability, in Amartya Sen's sense (1999), can be obtained.

Pursuing gender equality is not without its contradictions and the rest of this book will show how gender identities and relations have changed, how they have remained the same and offer some explanation for the incomplete nature of gender transformation.

Notes

1. Matriculation (or matric) is the public school-leaving examination in South Africa. Whether matriculants can go to university depends on how well they do in these examinations.
2. As the Women Risk and AIDS team noted, this is also true in the United Kingdom. See, for example, Holland et al. (1990; 1998).

CHAPTER 4

Setting the Scene
Research Approaches and Sites

Introduction

The research for this book was carried out by a team of people who came together from a number of different backgrounds, life histories and epistemological standpoints, all of which impacted on our research. In this chapter we are concerned with the details of our actual research study, the schools we worked in and the varying qualitative methods we employed. The study involved divergent partnerships and the contributions of different people at contrasting moments. In other words, our qualitative research built on multiple perspectives, sustained and changed by a team of researchers with different racialised and gendered histories and connections to South Africa and KwaZulu-Natal, as well as the contributions made by research assistants and Master's students who worked for relatively short periods, though often very intensively, on the project.

We begin with detailed descriptions of the six schools in our study, contextualising them within the educational system and infrastructure of South Africa (see also Chapter 2). We go on to discuss the research methods we used, illustrating how our multi-methodological approach enabled us to broaden and deepen our understandings of gender and sexuality in differing schools in the context of HIV and AIDS. We show how a range of methods was used to uncover and explore the gender regimes, the different understandings of learners and teachers and their racialised and gendered backgrounds. In doing so, we argue for the importance of detailed qualitative and ethnographic-style research carried out longitudinally and for the use of mixed methods of data collection and analysis.

The schools sites

As we showed in Chapter 2, South African schools continue to be marked by the history of apartheid and separated by race. At one extreme were schools for white children, elaborately equipped and well funded. At the other extreme, schools for black children, whether in the urban townships or the rural areas, were grossly underfunded, employed teachers who were underqualified and had very poor resources, buildings and other facilities. In between were the schools for the other racial groups identified in apartheid law – Indians and Coloureds. As apartheid came to an end, schools started to desegregate, in some cases illegally, and the white National Party government introduced legislation about the conditions under which white schools could admit African, Indian and Coloured learners. This system, known as the Clase system (after the minister of education who was responsible for it) allowed white schools to ballot their (white) parents to decide whether or not to desegregate and to what extent. The different levels of desegregation were classified as models A, B and C, with A being the most restrictive in their admission policies. In 1992 Mr Clase announced that in future all formerly white schools would become 'Model C' schools. These were to be state-aided schools run by a management committee and principal and funded partly by the state and partly by the parents.[1]

As shown in Chapter 3, post-apartheid South Africa has aspirations to provide a fully funded state system of free education for all children and young people, but money is always very tight and some schools are more able to supplement their state-provided income than others. School rolls also continue to be strongly racialised. While there are no all-white schools left in South Africa, a large number of schools in the rural areas, townships and informal settlements continue to be attended only by black learners. In the next section, we describe the schools in which our work took place, including a discussion of how they were racialised and classed. Most of our work took place in two township secondary schools and two primary schools, with additional data collection in three further schools that were a former white (Model C) school, as well as an Indian school and a Coloured school.

The project started in Lilian Ngoyi and Dingiswayo High (that is, secondary) Schools, both located in large townships in the greater Durban area in KwaZulu-Natal. Lilian Ngoyi is in KwaMashu, to the north of the city centre and Dingiswayo

is in Umlazi, to the south. These two schools were chosen for the project because they are functional and were accessible. Some township schools scarcely operate because of factors such as crime and poor management. The schools selected, however, are regarded as among the better schools in the townships. The chief director of education for KwaZulu-Natal at the time and the respective principals consented to the proposed research and the subsequent intervention based on drama. Teachers within the schools expressed willingness to be involved in the project.

KwaMashu is the older of the two townships, established between 1958 and 1965 as a result of the forced relocation of people from the more central area of Cato Manor. Umlazi was constructed a few years later (in the late 1960s). Initially there was little to distinguish one from the other. Both attracted residents looking for job opportunities in the city, both were overcrowded – household sizes in both were seven to eight people per tiny two-bedroom abode (May 1986: 119; Moller et al. 1978: 6), both initially poorly resourced. From the start, both sprawling townships had high levels of crime, which remain a feature to this day (Altbeker 2007; Moller et al. 1978: 19; Ndabandaba 1987: 83). And yet both were also, relative to squatter camps (now generally known as 'informal settlements'), places where people were better off, better resourced and more frequently employed (Freund 1996: 131; May 1986: 31).

Since the schools were built around the same time, it is not surprising that their architectural design is similar.[2] Both consist of a series of long, low, concrete buildings around a courtyard, which serves as a suitable place for school assemblies. Both are surrounded by fencing and access is through a gate, which is protected by a guard. At Lilian Ngoyi, access to the school and its management block from the car park is gained through a covered veranda running the length of the building on the right as one enters. At Dingiswayo, access to the head's office is through a front entrance in the middle of the block facing out to the car park. This means that at Dingiswayo it is not possible to enter the school without going past the administrative and principal's offices, while at Lilian Ngoyi one could go into the courtyard and strike out across it towards the classrooms or beyond. This seemed to symbolise the approaches of the two school principals. While both were very concerned about what was going on in their schools, the principal of Dingiswayo seemed more controlling and, perhaps, more in control. Neither of the schools

had good facilities and, at the time of the research, the technical block at Lilian Ngoyi was unusable in the rainy season due to flooding (though drainage for the school has now been renewed). Classrooms were basic and crowded. Each of the schools showed evidence, in different ways, of the care taken by teachers. For example, at Lilian Ngoyi there was a small vegetable patch and teachers provided food for children at mealtimes, while Dingiswayo had an excellent choir, run by one of the teachers in her free time, which performed not only in assemblies but also, on occasion, externally.[3]

The differences between the two townships lie in their histories. KwaMashu's initial population, while diverse, had a historically uniting experience of communal living in Cato Manor; Umlazi's residents did not. This meant that although both communities were disjointed and lacked cohesion, KwaMashu had a stronger civic, urban tradition, whereas Umlazi was more available to traditional mobilisation, the power of the Inkatha Freedom Party (IFP) and the allure of Zulu nationalism. While these differences were arrested to some extent by the 1977 incorporation of KwaMashu into the KwaZulu homeland, subsequent political developments bear the mark of community involvement and nationalist politics quite different from the ethnic nationalism of the IFP, which is much more obviously at play in Umlazi.

The respective histories of school politics in the two townships are illuminating. KwaMashu students responded much more energetically than did their Umlazi colleagues to local dissatisfaction and national campaigns from 1980 onwards. This generated violent intergenerational conflicts, but these were resolved more in favour of the youth than the elders, whereas in Umlazi, after 1985 particularly, the IFP was ascendant and the democratic traditions of the United Democratic Front (UDF) were muted (Bonnin et al. 1996: 174; Booth 1987; Kentridge 1990: 220). At school level this meant that COSAS (Congress of South African Students), an affiliate of the African National Congress (ANC)-aligned UDF, was able to propagate its policies vigorously in KwaMashu. The prefect system and corporal punishment were singled out for particular attention. Many schools introduced student representative councils. Student militancy and radicalism peaked with demands for 'pass one, pass all' in the late 1980s and some schools, caught up in the politics of ungovernability, effectively ceased to function. By contrast, in Umlazi, schools remained by and large 'orderly'.

The schools in our study reflect this recent township political history and we chose their pseudonyms to reflect the politics of the townships in which they are situated. Lilian Ngoyi High School, in KwaMashu, is named for this woman leader of the ANC, while Dingiswayo High School, in Umlazi, is named for the Zulu king. Lilian Ngoyi was unable to collect school fees until 1994 because of a refusal to pay by students and parents. Subsequently, levels of non-payment have remained high. Absenteeism at the time of the first tranche of fieldwork was sometimes close to 20 per cent. Corporal punishment has been resisted by learners and led to a teacher being chased from school. By contrast, Dingiswayo does not have a history of ungovernability. In the mid-1980s the school principal resisted COSAS's efforts to organise the school and brought order by using corporal punishment and a zero-tolerance approach towards lateness, drug usage and gangster entry into the school grounds. In this school, absenteeism, calculated by a teacher on a day early in September 2001 was at 5.3 per cent. Corporal punishment continues to be used.

The schools were founded in 1978 (Dingiswayo) and 1982 (Lilian Ngoyi) by the KwaZulu Department of Education and Culture. They remain poorly resourced compared to middle-class, former white, suburban schools, though by national standards they would probably be considered to have reasonably good facilities.[4] Both are headed by African, isiZulu-speaking male teachers who have been in office for more than a decade. They are relatively orderly, with school bells signalling the beginning and end of lessons and with learners mostly in class during stipulated teaching periods, although at Lilian Ngoyi, there seems to have been much more latitude extended to (or taken by) learners.

There were also significant differences between the schools. Matriculation results in the two schools fluctuated widely from year to year (see Table 4.1), with those at Dingiswayo apparently declining and those at Lilian Ngoyi improving in the three years from 2004 to 2006 (inclusive). However, it is important not to read too much into these figures. Numbers in each year are too small to provide statistical reliability, for a start, and it is not known what other variables were at play in each case. As far as we were concerned, we found that learners at Lilian Ngoyi seemed markedly more comfortable in English when speaking to the researchers than those at Dingiswayo. This might be accounted for by the proximity of Lilian Ngoyi to English-speaking areas of greater Durban that were primarily inhabited by Indian and Coloured people.

Table 4.1 Numbers and percentages of learners entered for and achieving pass marks in
matriculation examinations in 2004–06 in each school.

a) **Dingiswayo High School**

Year	Passed			Failed			Totals by gender		Totals
	Girls	Boys	Subtotal	Girls	Boys	Subtotal	Girls	Boys	
2004	73 (99%)	59 (97%)	132 (97%)	1 (1%)	2 (3%)	3 (3%)	74 (55%)	61 (45%)	135
2005	92 (76%)	87 (76%)	179 (76%)	29 (24%)	27 (24%)	56 (24%)	121 (51%)	114 (49%)	235
2006	63 (75%)	32 (73%)	95 (74%)	21 (25%)	12 (27%)	33 (26%)	84 (66%)	44 (34%)	128

b) **Lilian Ngoyi High School**

Year	Passed			Failed			Totals by gender		Totals
	Girls	Boys	Subtotal	Girls	Boys	Subtotal	Girls	Boys	
2004	70 (51%)	49 (56%)	119 (53%)	66 (49%)	39 (44%)	105 (47%)	136 (61%)	88 (39%)	224
2005	44 (94%)	54 (51%)	98 (64%)	3 (6%)	7 (6%)	10 (6%)	47 (44%)	61 (56%)	108
2006	120 (73%)	73 (76%)	193 (74%)	45 (27%)	23 (24%)	68 (26%)	165 (63%)	96 (37%)	261

Dingiswayo seemed more orderly in the way in which learners dressed and
behaved. There was an attitude of 'respect' towards the principal that was
associated with substantial social distance between learners and teachers. In Lilian
Ngoyi, in contrast, the environment seemed 'freer'. Teachers were not particularly
authoritarian and hence were able to talk and joke with learners. On the other
hand, learners seemed more inclined to 'do their own thing', including being out
of class during lessons or leaving the school in an unauthorised manner.

In 2003, the team invited Deevia Bhana to join the project, in order to extend
our work to primary schools, where her research was based. This enabled us to
include two primary schools, Bullwood and KwaDabeka, both situated to the north
of Durban. The schools are about twelve kilometres apart, but they are completely
different in terms of their racialised and class context, as are their physical
environments.

KwaDabeka is situated in and serves an informal settlement. In this respect, it serves a population poorer than either of the secondary schools. Homes are generally insubstantial structures, often in the form of tin shacks; poverty is endemic; family structures are weak, with many single-parent (usually female) headed families; and there are a significant number of families in which children head the households. The Shisana et al. study (2005) found that 3 per cent of South African children between the ages of twelve and eighteen headed their households, with responsibilities for physical and emotional care and financial support of their families including both younger children and older, ill parents or grandparents. KwaDabeka includes a large number of such child-headed households. At the same time, the settlement includes networks of closely knit, albeit often migrant, communities. The school itself has remained fairly stable in difficult circumstances, providing much-needed services, such as feeding schemes, for its pupils. Class sizes are large, with at least forty children in each class and attendance by children is good – perhaps because of the provision of food. The children's English is often not very good. Their first language is usually isiZulu.

The school opened in 1988 and was renovated in 1997. The renovations were part of a massive government injection into improving apartheid-constructed black schools. The renovated school provides a structural anomaly to the geography of the area. Perched on the edge of a hilltop, its face-brick structure is different from the hundreds of informal dwellings that spread down the hillsides. The perimeter of the school is surrounded by high, barbed-wire fences and a gate, which is protected by a school guard. However, this does not prevent the school from being a frequent target for theft, vandalism and even violent attacks. The school consists of single-storey, long brick buildings around a courtyard, which, as at Lilian Ngoyi and Dingiswayo, serves as an assembly area. The courtyard is also the central area where children congregate to play at breaks and eat their snacks. At the rear end of the buildings, a small patch of land is used by senior boys of the school to play soccer.

Inside the classrooms children sit in rows at old desks. Missing in the classrooms are the bold colours and general paraphernalia associated with primary classrooms in wealthier communities. For example, the children write and scribble with old worn-out pencils, many share crayons and there are no reading and carpeted areas. Teachers in this situation tend to be very authoritarian and corporal punishment

is widely used, even though it has been officially abolished (see Bhana 2002; Deacon, Morrell and Prinsloo 1999). A chorus of sounds, particularly the words, 'yes, tisha [teacher]', follows any visitor to the junior primary (infants in the United Kingdom) areas of the school.

Bullwood, in stark contrast, is located in a well-established, affluent, formerly white-only suburb with many palatial, well-secured and high-walled homes. The school is extremely well resourced and reflective of the context, with much of its money being raised by the parents – in other words, it is typical of formerly white (Model C) schools. The school is well maintained, situated in beautiful grounds, which have areas of lush shrubbery, lawn, playing fields and a swimming pool. The buildings of the school are reflective of the surroundings. A high fence surrounds the perimeter and an electric gate with an intercom system provides access to all visitors to the school. The principal's office is accessible via a reception area and resembles a home environment with a lounge and other furnishings typically found in a middle-class home. Schools in this style could be found in any middle- or high-income suburb of any city in the United Kingdom – although it is probably better resourced than most such schools. Class sizes are approximately twenty-five pupils. The school remains predominantly white and wealthy and most of the pupils have English as their mother tongue.

Classrooms are furnished with a variety of materials, equipment and the paraphernalia associated with young children in well-resourced schools in South Africa and elsewhere. Desks are arranged in groups of about five or six and while the teacher does have power as an adult (Thorne 1993), there tends to be more negotiation between children and teachers than at KwaDabeka. The sounds of young children talking, whispering or squealing and their constant movement are a daily feature of the classroom context. Before the actual breaktime, the children are allowed to sit outside in groups in the vicinity of their classroom and it is here that they eat their lunches that can sometimes consist of sandwiches, juice, snacks, muffins, nuts and raisins, while they laugh and chat.

Additional data from teachers were gathered from three further secondary schools. Oak High is situated in a leafy suburb close to the centre of Durban. It consists of a large imposing building with a spacious entrance hall, offices and classrooms. It is a single-sex, Model C, formerly white, but now multiracial suburban school and is very well resourced in terms of its physical environment,

including large playing fields, equipment, books and other learning resources and staff. It has a racially mixed learner population of more than one thousand boys. Most are from secure middle-class environments, although there are also boarders (mostly middle-class white boys from rural areas) and some working-class African boys from the townships. Reddy School is a former Indian coeducational school, which has over the last fifteen years become a major provider of schooling to African learners from local townships. It has 1 200 learners, most of whom are now African. However, an important factor in how the school approaches the pastoral needs of learners is the fact that most of the teachers are Indian men and women. The school is quite well resourced and managed and exudes efficiency. Gladstone is a former Coloured school, located in a working-class area of Durban. Most of the learners are Coloured, but there is an increasing intake of African students. Like Reddy, it is adequately, but not lavishly, resourced.

Methods of data collection

A key feature of our study is that we used a variety of methods of data collection and interim analysis over a long period of time. Thus we have the advantage both of longitudinal information and of being able to use a range of lenses provided by the different methods we adopted. In this section, we describe briefly the methods of data collection and analysis that we used, situating them within the different approaches to gender equality in education discussed in Chapter 3.

Our research began in 2000 with a combination of quantitative and qualitative methods designed to provide us with starting points for understanding gender regimes and relationships and the making of meaning about gender, sexuality and HIV and AIDS in Lilian Ngoyi and Dingiswayo. Initially, single-sex focus groups with learners in Grades 10 and 11 were carried out by Relebohile Moletsane, Elaine Unterhalter and Debbie Epstein, with groups of girls at Lilian Ngoyi and with both boys' and girls' groups at Dingiswayo. Robert Morrell interviewed groups of boys at Lilian Ngoyi. We found that both boys and girls at Lilian Ngoyi were more forthcoming and appeared to be more open than those at Dingiswayo. Although Relebohile Moletsane understands and speaks isiZulu, it is not her first language (which is SeSotho). Elaine Unterhalter understands a little isiZulu and Debbie Epstein does not speak isiZulu at all. Consequently, the greater ease that the Lilian Ngoyi learners had in English had a significant impact on what we could learn of

their strategies for talking and thinking about the issues we were interested in. The emotive and highly personal conversations held at Lilian Ngoyi had a major impact on our thinking and approaches as the project continued.

A second early approach to data collection was the administration of a survey of learners' attitudes to and knowledge about sex, HIV and AIDS (see Morrell 2007c). The number of surveys completed was fewer than we had hoped for, which makes generalisation – the usual aim of quantitative data – more difficult. More important from our perspective were the limitations of this method of research in gaining in-depth information and understanding. Consequently, although we certainly believe that the survey was helpful and worth doing, we were clear from the start that we wanted to understand the nuances and complexities that could not become visible through a survey. As Wendy Hollway and Tony Jefferson (2000: 2) point out, quantitative research is able to answer the 'what?' and 'how many?' questions, but not those that begin with 'why?' or 'how?'

The third method of data collection used in the first phase of the research was an intervention focusing not only on HIV, but also on gender equality by a theatre-in-education group, DramAide. Consequently in 2000–01, Mark Thorpe, then a Master's student with Elaine Unterhalter, acted as an ethnographic research assistant, spending time at Lilian Ngoyi and Dingiswayo in order to observe, evaluate and talk to pupils and teachers about the intervention and its impact (Thorpe 2001b). The DramAide intervention, 'Mobilising young men to care: Addressing gender issues in relation to HIV/AIDS', used participatory and drama methodologies in teaching and training of life skills and AIDS-related themes, to learners and a group of teachers in each school (Thorpe 2001b: 2). Among the issues the intervention aimed to address, were 'gender issues, care and support, grieving and the social impact of the HIV/AIDS pandemic' (DramAide 2000: 2). The intervention's primary aim was to encourage 'young men to become involved in health promotion care thus demonstrating personal responsibility for their own behaviour' (9). It also aimed to train a core group of learners in play-making and acting skills, as well as to mobilise both learners and educators to become activists for gender responsibility. This would be achieved by encouraging the participants (girls and boys) to explore and understand the implications of living in a patriarchal society (3).

For the learners, the intervention took the form of fifteen workshops spread over a month. Thirty 'willing' learners in each school were selected to participate

in the workshops conducted by DramAide facilitators and (by default) Mark Thorpe. On the whole, the DramAide intervention in Dingiswayo involved learners who were younger (in Grade 8) than those in Lilian Ngoyi (Grades 9, 10, and 11) and this may well have limited their ability to pursue personal projects of gender equality. Teacher workshops with willing teachers at each site were also conducted on the same topics, using similar drama methodologies.

Such processes of data collection and intervention lay much more within the realm of what we have termed, in the previous chapter, 'interventions' than in either 'institutions' or 'interactions'. Our questions and the DramAide workshops worked more at the level of the individual learners, despite the fact that in our next tranche of data collection – follow-up interviews with teachers and focus groups with learners – we did ask, and were told, about some minor levels of institutional change (for example, at Lilian Ngoyi, responsibility for handing out free condoms was shifted from the principal to the young women who ran the 'gender desk').

Some months after the DramAide intervention Elaine Unterhalter, Debbie Epstein and Relebohile Moletsane conducted interviews with teachers and learners at both schools. If the pre- and post-DramAide data collection can be seen, together with the intervention, as forming the first phase of the project, the second phase can be seen as a process of making connections beyond the research team. In this phase several events took place, none of which directly involved data collection, but all of which had an impact on how we were able to make sense of our data.

First, one teacher at each of our two secondary schools spent a term at the Institute of Education, University of London, to receive research training in the hope that they would be able to assist us in data collection on their return. Both benefited enormously from the experience. However, on their return to their schools, pressures of their regular work combined with a lack of research experience meant that they were able to contribute less to the development of our project than we had hoped. Nevertheless, both provided us with descriptions of their schools, commenting particularly on gender.

Second, we began to work with practitioners and policy-makers in KwaZulu-Natal, providing a combination of mentoring for them and gaining insight into their ways of working for us. Both Robert Morrell and Relebohile Moletsane had meetings with members of the provincial AIDS desk (which involved collaboration

between the Departments of Education and Health), addressed gatherings of senior teachers and gave feedback on policy implementation and the production of AIDS workshop materials. In addition, they met with KwaZulu-Natal members of the Commission on Gender Equality. These meetings and engagements helped to give a sense of the different levels of engagement in the education sector, but did not develop into any long-lasting, mutually beneficial relationships, because we, as academic researchers, were overcommitted and because officials were working with what we termed in Chapter 3 an interventionist or institutional model and we were more interested in trying to work at school level with an interactionist approach.

The third phase of the project picked up on the ethnographic insights gained by Mark Thorpe in his work with DramAide at Lilian Ngoyi. Alex Kent, also a Master's student with Elaine Unterhalter, spent a term carrying out an ethnographic study of Lilian Ngoyi (see Kent 2002; 2004) which contributed significantly to our work. Later, she returned to KwaZulu-Natal and spent time at Dingiswayo. What her work enabled us to do was to interrogate the data we had collected through interviews and focus groups. While these, and the evaluation made by Mark Thorpe, seemed to indicate significant levels of change at Lilian Ngoyi based on the DramAide intervention, Alex Kent's detailed account of the gender regimes of the schools showed us how little had changed in practice. Her account of sexual advances made by a male teacher, the way the principal positioned her as a 'girl', the beauty pageant held by the school and other such incidents recorded in deep ethnographic detail in her research journal, dissertation (2002) and article (2004), demonstrated both the importance of in-depth qualitative work and of longitudinal approaches. We draw heavily on this work in Chapter 6, particularly. Had we left the field after the first phase of our work, we would have come away with a very different impression of what and how transformation could take place.

Similarly, the fact that Deevia Bhana joined our team in 2003, with her experience of ethnography in primary schools, gave us the opportunity for in-depth study at Bullwood and KwaDabeka, both of which she already knew well. In these schools we did not initiate our own interventions, but were interested in what was going on in them and initiatives around gender, sexuality and/or HIV undertaken at the behest of the Department of Education in Life Skills lessons. The very different situations of the two schools led to somewhat different methods

of ethnographic detail. Whilst Deevia Bhana had some knowledge of isiZulu, it was not sufficient to work closely with the children whose knowledge of English varied, so at KwaDabeka a translator was necessary leading to somewhat closed questions. At Bullwood, however, the children were able to freely express their delight about being part of the research, greeting Deevia Bhana enthusiastically and constantly wanting to be heard and seen and to be acknowledged in their own right. She also observed the actual teaching of HIV and AIDS lessons in the two schools.

In April 2002 we organised a colloquium at the University of Natal, entitled 'Instituting gender equality in schools: Working in an HIV/AIDS environment'. The meeting brought together academics, policy-makers, non-governmental organisation (NGO) workers and teachers and focused on the theme of 'gender, sexuality, HIV and AIDS in sub-Saharan Africa'. The event provided an invaluable pause for thought, analysis, theorising and planning in the light of the contributions. Edward Kirumira, a keynote speaker from Makerere University, Uganda, helped to concentrate our thoughts by raising a number of important issues around questions of connections between HIV and AIDS and gender relations in sub-Saharan Africa. In approaching this book, we have been concerned to ask why it is so very difficult to create transformation of gender regimes in schools.

In the fourth phase of data collection, we were particularly concerned to understand the discursive strategies available to secondary school learners for understanding issues of gender and sexuality and, in particular, their hopes and fears in this regard. We asked Grade 10 and 11 students to write essays, for which there would be a prize, in response to the following prompt: 'Write a letter to "Sis Dolly"[5] about a problem in your life. It can be your own problem, or something about other people, but it must be something real in your life. Then write a reply giving Dolly's advice.'

Letters were received from 59 learners in Grades 10 and 11 (the higher grades in the school) written in English and isiZulu, although most were written in English. Ten letters were from Lilian Ngoyi and forty-nine from Dingiswayo. Only ten were written by boys, all in their very late teens. The age of the authors was fifteen to twenty, with the majority being fifteen, sixteen or seventeen.

Problem-page letters, such as those to 'Sis Dolly' in *Drum*, demand a particular genre of writing. Within this genre, there is an identifiable problem, normally of a

personal nature, which is explained and the answer given nearly always appears to offer a solution. The genre generally suggests a dialogue, often between the worried questioner and the problem-solving answer giver. Problems are often, but not exclusively, related to questions of relationships, romance, or their failures, and sexuality, often in its least desired or desirable forms. This genre might not be appropriate to elicit the wider concerns of youth regarding social relationships. But other genres for eliciting children's views – for example the freeform essay, the survey, or the structured discussion in school – used in research projects in other parts of the country all yield material with many similar elements. Children tend to represent ideas in the form of problems with relationships or questions of identity located in family, neighbourhood and national contexts (Barbarin and Richter 2001; Jones 1993; Parkes 2002). The data from this phase forms the substance of Chapter 8.

Having focused most of our attention on learners, in the fifth and sixth phases of data collection, we turned our gaze largely on teachers (discussed in Chapters 9 and 10). The fifth phase consisted of the collection of life histories of male teachers (and learners) by Robert Morrell at Oak High, Dingiswayo and Gladstone High School during 2004–05.[6] This is the basis of Chapter 9 of this book, where we focus on teacher masculinities in the context of their particular schools and the HIV and AIDS epidemic. The most obvious danger of using the life-history method is the temptation to treat the story told as being a simple and unvarnished reflection of the 'truth' of someone's life. In such an approach, little or no note is taken of the role of the interviewer in the creation of meaning and the eliciting of particular stories in the course of the interview. The problem with the alternative approach to life history as narrative and discourse is the tendency to reduce it to text, without reference to the materiality of people's lives and experiences. We note the importance of recognising the conditionality of interviews and how their relationship to reality needs always to be understood as contingent on many factors (the self, the interviewer, the interviewee-interviewer relationship, the various agendas floating in this space and so on). Our approach builds on these insights and treats a personal life as: 'a path through a field of practices which are following a range of collective logics, and are responding to a range of structural conditions which routinely intersect and often contradict each other' (Connell 1987: 222).

Our selection of interviews reflects a mix of race and class factors. We also compare teachers whose pedagogic attitudes can broadly be described as

'progressive' (child-centred, respectful, tolerant) with those whose styles reflect more authoritarian approaches.

The sixth and final phase of data collection took place in 2005. As noted above, we retained a focus on teachers, in particular those engaged in Life Orientation (LO) education, who have responsibility for sexuality and HIV and AIDS education. Part of this data was collected through focus group discussions with LO teachers at Lilian Ngoyi, Dingiswayo and Reddy schools and part of it through Deevia Bhana's ethnographic work in Bullwood and KwaDabeka. At this point we were concerned to uncover the caring work of teachers, on the one hand – discussed in more detail in Chapter 10 – and the ways they approached teaching about HIV and AIDS – discussed in more detail in Chapter 5. As well as talking with LO teachers and observing lessons, we also interviewed a group of enthusiastic peer educators drawn from the learners at Lilian Ngoyi, discussed in Chapter 8.

Dilemmas in the research

Any long and complicated research project, such as this one, will reveal a number of dilemmas during its course. It is in the nature of qualitative research, at least, that researchers must make pragmatic, on-the-spot decisions in the light of the events taking place. As Martyn Hammersley points out, researchers must make practical decisions

> both in planning their work and over its course. It has often been recognised in discussions of ethics that good practical judgements will not usually amount to a straightforward application of rules. One reason for this is that . . . multiple ethical considerations may be involved that pull in opposite directions . . . Equally important is that ethics cannot be separated from other considerations in the practice of research. (2006: 5)

Here we highlight two such moments in our work among the many that we met on a daily basis. The first such dilemma for us was the question of our relationship as relatively affluent, middle-class researchers – and, in the case of Elaine Unterhalter and Debbie Epstein, situated in the wealthy United Kingdom – working in a country with extremes of wealth and poverty. While none of the schools fell into the group of the very poorest in the country, three of them (Lilian Ngoyi,

Dingiswayo and KwaDabeka) served very poor communities and, as described above, included learners with very, very few resources. There are two issues here. First, what difference might it have made that we are all middle class and three of the central team as well as the two ethnographic researchers in the secondary schools were white? To what extent might this have impacted on what both learners and teachers revealed to us, and the levels of deference or resentment they might have felt? We have no evidence that this did make a difference, but it is nevertheless something we need to bear in mind in making our analysis. Second was the question of language. We need to be aware that for much of the time we were speaking English with people for whom it was a second or third language. Issues of translation must be borne in mind here. Third, what should our response have been to the poverty of the schools where we were working? In the case of Lilian Ngoyi and Dingiswayo (but not in the case of KwaDabeka due to differences in our funding arrangements), we were able to make small but, to the schools, significant payments in return for access. While consent to our research was given before the offer of payment was made, there could, nevertheless, have been a constraining influence as a result of such payments, although again we have no evidence that this is the case.

The schools needed and were grateful for the money, but this should not be seen as purely instrumental either from the schools' points of view or from that of the researchers. For the schools, the relationships developed and strengthened with the University of KwaZulu-Natal were and are a resource that goes beyond money. For the researchers, the connections we felt with the schools and our responses to the poverty we saw there have stayed with us and, indeed, some of us have been able to make further personal donations beyond the end of the project. Obviously, this does not solve, in any way, the problems of poverty in South Africa, but it demonstrates the affective nature of research in these areas of poverty, hardship and stress. We were glad to be able to help these schools financially and admired the strenuous efforts of their teachers to care for the learners, while not always agreeing with the direction of their interventions, as discussed in Chapter 10.

We began this book with the second moment we wish to highlight – an account of our first focus group discussion with young women at Lilian Ngoyi. While we knew, in principle, that there was a high rate of sexual violence in the country

from which schools cannot escape, we were, nevertheless, taken aback and shocked by the revelations of sexual abuse and rape. In particular, we became emotionally involved and concerned for a young woman who disclosed to us, for the first time ever, that she had been raped as a young child. We were faced with two issues. First, how should and could we support the young woman? Second, what were we to do with our own emotional reactions? In dealing with the first issue, we were faced with two problems. How could we offer her professional counselling support when neither the project nor the young woman had the financial resources to pay for such help? Our planning of the first phase of the project had not contemplated the need to provide counselling. Nevertheless, as noted in Chapter 1, we were able to call on the resources of the University of KwaZulu-Natal to offer some counselling support to this young woman and, when we visited the school a year later, she rushed up to us to tell us how it had changed her outlook and, consequently, her life since she was now doing much better scholastically and hoping to go to university. This anecdote demonstrates the ease with which even experienced researchers such as ourselves, with significant levels of knowledge about the area we were researching, can fail to anticipate such issues arising in such a stark way (though we would not do so again). Neither were we prepared for the emotional impact of these revelations on ourselves as interviewers. Again, we feel we have learnt a lesson here about the importance of building in the possibility of researchers in such sensitive areas needing a space to 'unload', somewhat similar to that provided for therapists in their supervision. While this is not entirely unproblematic (see Corden et al. 2005), it would have eased the situation for us.

Conclusion

This chapter has situated our research in the South African schools where we worked, described our methods of working and highlighted some dilemmas we faced, while in the next chapter, we turn to the data collected in order to consider the gender regimes of the two primary schools, as reflected in sex education lessons at each of them. Looking back, we realise that although we always had interactionist intentions, our time and material resources were insufficient to fully realise these goals. Furthermore, the frame in which both policy-makers and teachers were working was, generally, institutional or interventionist – and this is reflected in

the rest of the book. In thinking about future work on gender, sexuality, HIV and AIDS, we would want to develop further work that is more located within the space of what we have called interactions. As we argued in Chapter 3, we need an approach to gender equality that draws on all three emphases – interventions, institutional change and interactions. This is no less the case in research than in policy.

Notes
1. For further details see Chisholm, Vally and Motala (1998). See also various articles in Kallaway (2002).
2. A diagram of Lilian Ngoyi can be found in Kent (2004).
3. For example, at the memorial for Professor Ronald Louw, held at the University of KwaZulu-Natal in July 2005.
4. Yet in Dingiswayo School the numbers of learners per class varied from 83 (Grade 9) to 47 (Grade 12) when we visited in 2002. This should be understood in terms of a notional national standard of 35 learners per teacher.
5. 'Sis Dolly' is the agony aunt for the popular magazine, *Drum*, established specifically for African readers in 1951 and still aimed at a black urban readership. A briefing to learners was made with members of the research team present by one teacher at each school who facilitated the competition. It was stressed that the letters could be fictional, but must deal with problems that learners felt to be real, either problems about their personal lives or their social circumstances.
6. The data is drawn from a project funded by the Ford Foundation (Grant No. 1035–0493). The financial support is gratefully acknowledged.

The Challenge of Instituting Gender Equality in Secondary Schools

Introduction

This is the first of two chapters detailing contrasting interventions in the schools, this chapter dealing with the secondary schools and the next with primary schools. The HIV and AIDS and gender equality intervention by DramAide in the two secondary schools in the study exposed the role of school cultures in welcoming, sustaining or undermining initiatives concerning gender equality and HIV and AIDS education. The schools can be seen as sites of cultural, political and social struggle where the meanings of gender equality and responses to HIV were in flux for both teachers and learners. In this context we consider some of the reasons for the approaches we observed to equality and violence. In our assessment we draw on three rounds of data collection, each of which seems to reveal a different part of the picture: the first round in 2000 during and immediately after the implementation of the DramAide intervention (Thorpe 2001b); the second in 2003 in an ethnographic study in the two schools (Kent 2002; 2004) and the third in a follow-up study we carried out in 2004–05 in the two schools.

As noted in Chapter 4, the schools were located in townships with very different histories and exemplified some markedly different features related to these histories. At Lilian Ngoyi, learners and teachers seemed able to mobilise high levels of social capital. They were connected with a number of local community-based organisations outside the school that had been built up over the decades of the anti-apartheid struggle. In addition there was an active branch of the South African Democratic Teachers' Union (SADTU), and a well-supported learner representative council (LRC) at the school. The language of these organisations was strongly

inflected with ideas associated with mass democratic mobilisation in the area, with a stress on democracy, liberation and equality (Kallaway 2002; Pandor 2005). Thus teachers and learners at Lilian Ngoyi tended to utilise an interactionist approach to equality. Major concerns were with transforming the unjust structures of apartheid and mobilising community action. In contrast, at Dingiswayo, the focus was more frequently on efficient school management, ensuring that children were on time and that teachers were in the classroom. Equality here was much more likely to be seen as ensuring the same amounts of time on tasks for each class and equal provision of equipment for girls and boys.

In the section that follows, we address the question: What do these different approaches to equality more generally tell us about the gender regimes in the two schools and the challenge of instituting gender equality? We begin with an exploration of the responses to the DramAide intervention at, first, Lilian Ngoyi and then Dingiswayo, noting the different levels of enthusiasm and facility that greeted the intervention. In the second half of the chapter, we consider rather different findings from later stages of the research. Our argument here is that there is a fragility about the sustainability of gender equality interventions, which needs to be recognised and addressed in the context of education work on HIV and AIDS.

Responses at Lilian Ngoyi

In the first stage of the research, we found Lilian Ngoyi to have high levels of social capital in terms of learners and teachers in a context of commitment to social transformation that were translated into highly articulate views with regard to institutionalising gender equality (see, for example, Moletsane et al. 2002). Thus in response to the DramAide intervention at Lilian Ngoyi, the girls' higher levels of confidence were noted at all stages of the research process, during and after the intervention. Mark Thorpe, in his evaluation of the DramAide intervention at Lilian Ngoyi notes: 'An atmosphere of openness developed and at times students were sharing things, often passionately, about their views, and even their own experiences' (2001b: 9). The willingness of female learners to express their opinions and challenge the views of boys was also noted: 'I was struck by the lack of the "victim" image from some of the girls, such as one saying she went for "status" in a relationship, which was followed by a "whooping" like on an *Oprah Winfrey* show' (2001a: 18).

Teachers at Lilian Ngoyi also spoke about the girls' confidence being enhanced by the intervention and about how the school was providing new openings for girls because of this. For example, after the intervention, a group of girls who had participated volunteered to address a school assembly on similar issues. In addition a girls' football club was started.

In this context, we identified some profound changes in gender discourses among staff at Lilian Ngoyi, of the kind demonstrated in the interview extract with Sicelo, a male teacher:

> Before we closed [for the June holidays] there was one learner who could not write my English paper. We got a report that she was raped on her way to school. As a result she could not come. What we did, we spoke to the principal to kind of refer these learners to the police so that a case can be opened. The nurses and whatever, the proper channels, were followed. So I'm saying as an institution we are responding to that corner whenever there is a need for us to do so . . . I think the effect with the elder, let me call them the elders, is that they are seeing things that they have buried, particularly in this institution. There is always an open talk that 'I've buried my friend last week' or 'my relative that other week'. So that has had a lot of impact on their minds such that when you talk HIV and AIDS all of us are willing to assist no matter whether it's male or female.

In this extract, Sicelo did not see the girl's problem with learning as hers alone and, it seemed, the school reacted as an institution in ways that demonstrated that this was a more general reaction. This was evidenced by the way in which the teacher had responded to her report that she was raped: by working with other teachers ('*we* spoke to the principal'). The rape was not seen as the girl's fault. Rather, the school referred the case to police and health officials and took pride in having used 'proper channels'. The data seemed very clear and seemed to indicate that social capital was being built within and beyond the school: 'as *an institution, we* are responding'. Similarly, Sicelo described the high numbers of AIDS deaths in terms of the losses of friends and relatives (rather than in distancing terms of denial), drawing out the social relationships lost. At the end of the transcript extract, he talked of how 'all of us are willing to assist no matter whether it is male or female'. The implication, here, is that gender equality is to be seen as treating men

and women the same, rather than in terms that carried something of Amartya Sen's notion (discussed in Chapter 3) of the need to address issues of equality through an understanding of need and capability. This more radical view seemed not to have reached the school, despite the prevalent rhetoric of struggle and social transformation.

In a focus-group interview, the Lilian Ngoyi teachers reflected on how issues of gender equality were affecting them in their personal lives, as well as new processes they were putting in place in the community. Thuli commented: 'There is also an impact [of DramAide] because we as teachers when we discuss things people who used to brag about having so many girlfriends, nobody now wants to sleep around. Everybody just sticks to one partner. They've just come to value their relationships now and the feelings of their partners because now they are afraid of these things, of AIDS.'

Sicelo agreed, adding: 'And also dealing with cultural stereotypes. For example, in our culture a man could have more than one girlfriends or women. It has been an acceptable norm, but this workshop was able to deal with that at the level of educators as well as the learners.'

Encouraged by her male colleague's support, Thuli went further: 'Even the stereotypes in terms of using condoms because there were people who were saying, "I will never use that. I won't use the plastic" and so forth. All those things now have changed. And the fact that there are people that they know of who have died of this disease now, it's no longer a myth. It's a reality.'

Thus it seemed that the teachers were making connections between their personal behaviour through discussions in the staffroom and considering shifting views with regard to cultural stereotypes. We considered this a nascent interactionist critique where more critical views about equality might be able to develop.

Nonetheless, even at this stage, in linking these understandings to teaching about HIV and AIDS after the intervention, the teachers did not seem to be clearly addressing gender equality. For example, talking about her guidance lessons, Thuli commented:

I talked to the Grade 10s . . . about AIDS in guidance. I asked them, 'If you found out you are positive what would you do?' One boy said, 'No, I'd go on with my

partner because I know that she is positive as well.' 'Will you use a condom?' He said, 'No, I won't . . . because I know that I have AIDS so what's the use?' But other boys, you know, said, 'No, it's wrong if you don't use a condom even if you know that you are positive. You have to use it to prevent re-infection.' So I saw that [the intervention] had an impact on both boys and girls.

While this response suggested that Thuli had a relatively sophisticated understanding of HIV and AIDS and that she and her learners were concerned with self-preservation, she made no comment and did not draw out links between the importance of gender equality and the use of condoms, although this had been a major theme in the DramAide work. By not questioning the learners' stance on this, she ignored and indeed might have reinforced the lack of respect for dignity and equality in relationships. Her understanding of gender equality seems to be concerned primarily with similar or even-handed treatment of boys and girls.

Nonetheless, the Lilian Ngoyi teachers appeared to be more confident in making connections between gender issues at school and in the community than the Dingiswayo teachers. The teachers at Lilian Ngoyi were particularly proud of changed attitudes to the distribution of condoms:

The effect has been quite great because thereafter [the DramAide intervention] kids have been coming to us demanding condoms . . . which is a positive development. Before the schools closed for June recess, we had had to decide as the staff where to put these condoms so that we place them in toilets or in offices. The strategic points where it will be easier for them to access them because we had identified one weakness, because they are kept by the deputy-principal, that it might happen that there are those learners who may be shy to come forward and request them. (Sicelo)

We saw this as an indication that teachers were picking up the initiative from the DramAide intervention and working to change some social relations in the school, with potentially important consequences for gender equality and the prevention of HIV infections.

Learner responses at Lilian Ngoyi showed similar movements toward interactions and empowerment with regard to gender equality. In a follow-up

group discussion girls were unreservedly positive about the intervention, speaking about the way they had found it personally empowering. Thembi, for example, said: 'I've learned that if . . . you've been abused, like raped . . . break the silence, you have to talk about it. Stand up and talk about and tell everybody that this and this and this and that people have abused me.'

Nandi went on to tell us of successfully helping a friend who was being abused by her stepfather to put him in jail: '[My best friend's stepfather] was abusing her. Like, her mother passed away and she was living with him . . . He was abusing her in different ways, sexual abuse and all that . . . I told her, "No don't stay, break the silence. Just go stand up and tell the police what he is doing to you." So he's locked up.' She attributed her ability to help her friend to the intervention. When asked, 'Do you think that's because of the workshop?', she enthused: 'Yes, yes, yes, 'cause Mark (the participant observer) said if you don't want to do something, don't do it. It's okay for you to say no.'

Similarly, Bongi reported that she was able to stand up to an authoritarian (and abusive) stage manager in a drama club she belonged to:

We used to have arguments with my stage manager, Skiri . . . He used to say to me I'm so young to be in drama. And I said, 'You know what, maybe I'm too much matured than you.' . . . 'Skiri what I don't like is when you talk to us and talk like you talking to the younger kids or younger babies or just harsh to us like that.' He [used to beat] us when he [taught] drama and I told him, confront[ed] him and I said, 'You know Skiri, this is not good . . . We are not [children] anymore. We are matured. We are in high school . . . Listen here Skiri, I'm a girl. I've got my own rule. I've got my own thing on tight.' So this is girls' power.

We were also impressed by the fact that some students were confident enough to criticise the DramAide intervention. For example, one of the girls disapproved of the way the disease was portrayed in the plays that the learners produced and presented to the school. Her character, who was HIV-positive, died. To her, the play suggested that being HIV-positive means instant death, a view she disputed. Identifying with and claiming her feminine agency, she declared: 'I [wanted my character] to continue to live with that disease.' According to her, this would teach people to live with the disease in ways that enhance their own lives and protect

others. This is a powerful statement as KwaMashu, the township where Lilian Ngoyi is located, is the area where in 1999 Gugu Dlamini was stoned to death for declaring that she was HIV-positive. For us, what was evident in all these responses was the girls' developing sense of agency, their ability to challenge discriminatory and abusive practices and to select the appropriate institutional intervention, be it through the police or a family member.

However, the extent to which our initial assumptions that gender equality could be relatively easily instituted by a particular intervention, such as DramAide, were shown to be highly questionable as further data were collected, as we discuss later in this chapter. At Dingiswayo a less evidently transformative language concerning gender equality and the prevention of HIV infections was noted.

Responses at Dingiswayo

As indicated at the beginning of this chapter, our initial analysis of the responses from Dingiswayo suggested that the focus there was frequently on efficient school management and that this often translated into ensuring the same amounts of time on tasks for each class and an assumption that equal provision of equipment for girls and boys indicated gender equality. There were few challenges to existing gendered assumptions. For example, Mark Thorpe found that in the workshops at Dingiswayo, in contrast to Lilian Ngoyi: 'The atmosphere was far more subdued and "respectful" in the old-fashioned educative terms. People didn't speak out of turn, they were slower in responding.' (Thorpe 2001b: 39) Many of the girls, he noted, never answered a question in a group context. Even the most outspoken girl at the school did not challenge the boys in their views. He commented: 'Though it may sound sour to say so, most of her actions in front of the class seemed entirely to impress the boys.' (12)

Dingiswayo teachers' understandings and responses on the nature of HIV and AIDS, like those of their counterparts at Lilian Ngoyi, were based on a sound knowledge of the disease and the need for preventative work. However, unless specifically probed, teacher responses were generally devoid of any gender equality language and acknowledgement of the role played by gender inequality in the pandemic. Thus, for example, the following teachers' comments on DramAide's work at Dingiswayo focused strongly on HIV and AIDS, rather than gender equality issues:

It went very well because it was dealing with the issues which the students are familiar with, AIDS, abuse and things like that. After the training the students became enabled to talk about it to their peers. (Norma)

So it was sort of a reinforcement. It made us to be aware of the fact that we are like an integral part . . . I think the teachers have more time with the kids. I think they are the ones who are supposed to teach the kids on how to, like to prevent this killer disease. (Sandile)

The teachers saw little need to enhance girls' confidence, feeling that the intervention had a more positive impact on girls than on boys, in part because the girls were more mature. As some of them explained:

Thandi: It's always the case, men are backward. They take time, you know . . . I mean girls always take the initiative. Even in church you find many [more] women than men.

Vuyo: Yes, it's a fact.

Unfortunately, these taken-for-granted 'facts' meant that there was a level of acceptance of gender inequality and Norma noted that some staff members took gender inequalities as commonplace: 'Yes, [the DramAide intervention] did deal with sensitive issues and that even here at school there are some teachers who think that women are inferior.' Based on this, she did not think the DramAide intervention would have any lasting impact on her male colleagues' attitudes towards women: 'Men will always think that they are superior. Maybe it will change after some time because these things are new.'

In contrast to the Lilian Ngoyi approach to learning from the intervention that drew on a repertoire of actions linked to talking about social transformation, the Dingiswayo teachers spoke about some of the effects of the intervention in terms of the failures of the intended 'cascading effect', the implication being that any institutionalisation would have to take place through top-down action and courses for teachers. According to Thandi: 'Ja, it made a difference, but unfortunately [only] to those who were exposed to the DramAide . . . because there were no discussions after that. There were no internal workshops after that.' For Thandi, then,

institutionalising messages concerned with gender equality would require training, discussion and deepening the insights from the intervention. In retrospect, her assessment pointed to the difficulties of institutionalising gender equality that the rhetoric at Lilian Ngoyi masked.

In contrast to the Lilian Ngoyi teachers apparently putting the new learning to work as appropriate situations came up, for example, in relation to the girl raped on her way to school, the Dingiswayo teachers waited for management directives (or DramAide and the research team) for further workshops. We concluded that the fatalistic acceptance of gender inequalities by male and female teachers at the school echoed the more conservative cultural ethos of the locale of the school and the teachers' approach to their work, which they saw as professional and technical, rather than as personal and political. Guidance as a subject had been dropped from the curriculum at the school and the teachers all lived some distance from the area in which the school was located. We concluded that in this school, after the DramAide intervention, teachers were making plans for action focusing only on HIV and AIDS and excluding gender equality. This seemed steeped in a moralistic stance, in which 'doing the right thing' was the main lesson being drummed into the learners:

> We had our Miss Dingiswayo [pageant], we were fundraising ... So, we called upon one of the stars from *Yizo Yizo* [a television drama that worked with a number of HIV and AIDS awareness messages] just to tell them and mould them towards the right direction. Telling them that as they [the actors] are playing *Yizo Yizo*, it does not mean that it's not real life. What they are showing is that there are people who are doing such things [and] in the end they end up in jail. (Sandile)

Linking *Yizo Yizo* messages to practices such as a beauty pageant to 'mould' students in 'the right direction' was in direct conflict with the main gender equality messages of the DramAide intervention. However, we acknowledged that involving children through fundraising, a beauty pageant and connection with television stars could be seen as highly motivating, but that this orientation indicated how difficult it was to institutionalise gender equality or build some of the interactionist dynamic. One teacher argued that the management style of the principal had contributed to a stifling of efforts to develop new institutional structures building on some of the

ideas suggested during the intervention. Interviewed a year after the DramAide workshops, she explained:

Nomusa: But it's inactive because he wouldn't allow a meeting where the teacher is not there. But his children would want to have things they discuss in our absence, so he doesn't allow that space because he thinks that is an arena to breed disruption in the school. So he doesn't give them that . . .

Debbie: So it's still quite controlled and . . .

Nomusa: Very controlled. Very controlled. He wants to have structures in place in case somebody comes in to question him, but as to whether they are operational that remains to be seen. They are not in operation.

Debbie: Ja.

Nomusa: But he wants people to say, yes we've got an LRC, but it doesn't operate the way an LRC is supposed to. They are supposed to sit in the governing body council of the school. Not one sits in the governing council of the school, because he does not allow that to happen. I know from experience because I was once chosen to represent teachers in the governing council, and not one meeting would he allow the learners to come and represent the interest of the learners.

Debbie: So . . .

Nomusa: He looks very democratic, but in operation he still feels very threatened by the involvement of learners, so he just bosses them around: 'You are going to do this; you are going to do that. I'm not going to allow this in my school.' That is . . . he calls it . . . 'You are paving the way for destruction to take place', then he will give you examples of schools where learners are allowed to voice their opinions. That those schools are a disaster.

It can be seen that institutionalising gender equality in conditions where the principal was reluctant to give students space to discuss even a limited number of concerns was a difficult proposition.

Learners' responses to the intervention at Dingiswayo confirmed the difficulties of 'pushing against the system'. The girls' responses, in particular, tended to suggest a wish to preserve a 'good girl image', and to mirror their teachers' concern with teaching them morally acceptable behaviour, as Ntombi's comment suggests:' I

liked a lot that some of us were showing other children how to behave because I like people who [respect] each other . . . what I hated about [Thabi's character] was . . . because she wasn't respecting her parents . . . she even became pregnant because of not listening to her parents.'

These girls were more inclined to seek change within existing gender parameters. Rather than challenge the existing gender (and other) regimes in the school, family and community, they felt more confined by their ages (and status as the youngest class in the school) and by cultural norms and taboos against speaking to elders about sex and this was not disrupted by DramAide. Philile commented: 'In the drama I was a teacher, teaching children to respect the teachers at school and [Nonto's character] did not want to respect her teachers.' The Dingiswayo girls generally endorsed abstention from sex and, it seemed, found it difficult to imagine challenging boys in a relational context. As Thobile said: 'Some of the kids have stop[ped] having boyfriends. They just started waiting, [have started] a new life.'

Only one girl reported sharing her experiences with her parents in a positive and affirming way. Her peers were, apparently, unable to do so. This seemed to be because of the status of the participants as the youngest class in the school (Grade 8), as well as due to the imperative to follow cultural taboos against discussing sex with one's elders. 'I didn't [tell others] because they say, "You are a young child, what do you know about HIV?" Because we know about HIV. Don't tell us about sex because you are too young.' (Anna)

From the first round of data from Dingiswayo, we concluded that the approach employed by the school in working with the ideas put forward through the intervention did not develop a sense of personal empowerment among the participants (learners in particular). Instead, the girls seemed to be obeying the rules laid down by their elders in their families and communities, as well as by their teachers for the effective running of the school.

The boys, on the other hand, had similar responses at both schools. For example, the Lilian Ngoyi boys' talk about improved relationships with girlfriends was echoed at Dingiswayo: 'I'm [now] speaking to my girlfriend. Last time I was not using a condom [when having] sex with her, but now I know that it is not good not to use a condom. Now I use a condom if I do it.' (Thulasizwe) Two of the other boys added:

[There] is change about myself and with my girlfriend. You have to use condoms. You have to talk about sex before we do anything about it. We have to talk about it and we have to give each of us an opinion how to use a condom when I like to have a sex with her if she don't like I mustn't force her to do sex. We have to talk to each other and agree [with] one another about using a condom, having sex or not to have sex. (Thulani)

Ja. [My relationship with my girlfriend] improved a lot because I wasn't respecting [her]. If she told me she wanna go to town if I don't agree she [couldn't] go . . . now I know [we] have to respect each other and give her a chance to . . . live her life together (Bongani).

In general, however, the boys at Dingiswayo spoke less about changes in their personal relationships and, like their female counterparts, spoke more about how they had learned to obey rules. For example, responding to a question about the fatal refusal of some learners to heed HIV warnings, two of the Dingiswayo boys had this to say:

It means that if you behave well at school and respect your parents, everything just . . . (Siyabonga)

If you have good friends who are well behaved, you should have good friends who will contribute to your education. (Khanya)

Yet among the boys at the school, there were also heartening signs of reflection and change as the comments below illustrate:

I am confident now and I can decide if something is wrong or right. I am able to make decisions that will make other people happy as well. (Nhlanhla)

Boys like grabbing girls but they don't do that anymore. We treat each other as friends. (Khanya)

Initially, we concluded that Lilian Ngoyi was more able to deploy and utilise the language of gender equality and was more likely to implement change, while

Dingiswayo seemed to appropriate the gender equality aspects of the intervention in more limited ways. However, we acknowledged that other aspects of the Dingiswayo school environment (such as the disciplined and well-organised environment) might provide a stronger base for building on the achievements of the DramAide intervention. Learners at Dingiswayo attended school more regularly than at Lilian Ngoyi and teachers were often better organised. Thus, it seemed that Dingiswayo potentially provided a more stable base for teaching and curriculum development that would increase the chances of success for a Life Skills curriculum within which messages concerning gender equality and HIV and AIDS would fall, after the DramAide intervention concluded. In other words, the stability of the school may have provided some capacity to transform gender relations, should the principal and senior teachers within it become committed to such a process. In the event, the school's management was not interested in transformation of gender relations and the learners were unable – and maybe even unwilling – to engineer such a process from below.

The challenge of institutionalising gender equality in schools
As the course of the fieldwork unfolded, we came to the view that our initial reading of the situation was rather optimistic for Lilian Ngoyi. While a *language* of gender equality was developed within the school, it did not seem to have the necessary stability to change the school culture and enable new forms of practice to help sustain the messages espoused by the DramAide intervention. Equally, the hope that there was potential for management to lay the foundations for gender equality actions at Dingiswayo was not realistic, given the lack of additional interventions in this area and the management's lack of interest. The follow-up ethnographic study (Kent 2002; 2004) and our own fieldwork during 2004–05 indicated that we had not paid sufficient attention to the pointers regarding how difficult this was in both schools for slightly different reasons.

Beyond the rhetoric: Examining gender equality practice in the schools
In contrast to our optimistic initial assessment of the impact of the DramAide intervention in Lilian Ngoyi particularly, our follow-up data collection and analyses uncovered a much bleaker picture. The gender rhetoric that followed immediately after the intervention seemed not to translate into school practice. Alex Kent's

conclusion from her ethnographic study of Lilian Ngoyi was that the HIV and AIDS and gender equality discourses that DramAide and other interventions had developed among teachers and learners in the school had failed to develop gender equality in the school and instead had 'encouraged a form of "literacy" and talk about sex, [HIV and AIDS] and gender . . . [which had] served to conceal compulsory heterosexual and gender discriminatory practices. Under the "disguise" of gender and [HIV and AIDS] literacy, hegemonic masculinities and gender inequalities persist, making schools "dangerous" places for addressing [HIV and AIDS] and gender equity.' (Kent 2002: 3) She interviewed a male teacher, Musa, about what had happened since the DramAide interventions:

> Alex: So what has changed since HIV?
>
> Musa: Not much has changed, because so many students have fell pregnant this year, the majority of them have said, they are still victims of pregnancy and some of them are still victims of [HIV and AIDS].
>
> Alex: Since the project, is anything different?
>
> Musa: Yeah, there's much talk about sex, that's one thing. (Kent 2002: 57)

Here Musa saw the intervention as the problem, rather than the persistence of unequal gender relations. Kent concluded that the intervention had done little to shift the hegemonic masculinities and subordinated femininities in the school, noting that sexual harassment, corporal punishment, male management and the exclusion of pregnant girls from school functioned to reinforce both compulsory heterosexuality and the unequal status of males and females. The regular use of corporal punishment in the school was justified as a mechanism for effective control of large classes and of children's behaviour. It was also justified in terms of cultural traditions. Sipho, a male teacher, explained all three of these justifications:

> As an African I know what it takes, I know how to discipline and when to discipline because some of them are so rude and they are, they bully each other, and if you are not careful as an educator you might get seriously injured by the outsiders, if you are too strict if you give them severe punishment you put yourself in danger, I do not want to hide that. So we have to be extra careful not to forget to stamp your authority as an educator. You must demand respect from them, you know. Show them that you are a parent. (Kent 2002: 51)

Alongside the widespread, normalised use of corporal punishment (despite its illegality), sexual harassment was a pervasive presence. Most often, it was girls who were the targets of such behaviour, but it also encompassed women teachers. Indeed, sexual harassment was seen as both laughable and an acceptable expression of manhood, as a male student member of the LRC declared: 'It's ok to touch a woman like that, or touch her breasts. It's like a style to represent that you've grown, that you see her . . . it's a way of touching a lady.' (Kent 2002: 47)

A second illustrative example of the ways in which the school had abandoned the rhetoric around gender equality relates to Kent's observation that during her fieldwork at Lilian Ngoyi, females (teachers and students) tended to be restricted to what she referred to as 'passive' and 'feminine' roles and this was enforced through policing mechanisms that ensured that girls and women in the school would not have the same privileges of space and privacy as boys and male teachers. The strategies she identifies for maintaining this involved an emphasis on a 'proper' female appearance, including excluding girls who were pregnant from school, in spite of the fact that this practice had been outlawed by the South African Schools Act No. 84 of 1997. In this way, the sexuality of girls was excluded from the school arena (see and compare with O'Flynn and Epstein 2005) Kent quotes an interview with the principal, who refers to an announcement he had made in the morning assembly:

Mr Hlophe: So I was informing them that it was wrong for them to remain here at school while pregnant because should they, eh, during the labour period we don't have the expertise to deal with the problem. There are no ambulances here . . .

Alex: And when is this from, as soon as she becomes pregnant, or in the last month?

Mr Hlophe: No, as soon as it becomes a problem or prominent, when it becomes conspicuous. Because we are having a problem. It is not even good for the image of the school and a very, very bad example to our young ones, Grade 8, they are very young, young, young, young girls there. (Kent 2002: 48)

The general feeling among the teachers was that once a girl becomes pregnant, her right of access to education could no longer be a priority. Rather, the school's

reputation, which was placed at risk by the girl's bad/unbecoming behaviour (becoming pregnant), was more important. One of them went so far as to describe pregnancy as a sickness, which required handling by healthcare professionals: 'A pregnant girl is sick, a sick person, who should be attended by those people who are professional in handling such cases. Sometimes they are dizzy, sometimes they vomit.' (Kent 2002: 48)

Another teacher declared that a pregnant schoolgirl's future is doomed, a position challenged by some of the girls in the study. For example:

> Bianca: If I got pregnant I would keep it and I would come to school. But I will have to, just because . . . (inaudible) I will have to deal with it. Even now we have a girl in Standard 10 [matric] hey, she's pregnant, yah!
>
> Yoliswya: And she still comes to school every day, she is improving every day, with her lessons. (Kent 2002: 48)

The treatment of pregnant girls was one of the most prominent instances of gender bias. Despite the strong statements made about removing girls from the school, no questions were raised about who might have fathered the babies or their responsibilities in this regard. The entire blame was placed on the girls.

In mapping the school space and the ways in which it was used by boys and girls and male and female teachers, Kent concluded that girls, and to a lesser extent, women teachers, were confined to a domestic/inside space, while boys and men teachers were allowed to move freely, to occupy the bulk of the space in the school and to take risks. Men were involved in implementing violent school discipline, such as corporal punishment, dominating management in the case of male teachers, breaking school rules (smoking behind the school buildings) or having multiple girlfriends in the case of boys and male teachers. After the DramAide intervention the promised distribution of condoms had been forgotten:

> A year later, when I carried out my research, the same boxes of 2 000 condoms were, dusted over, sitting in the corner of the staffroom. In a 'safe' place and not readily accessible, although the odd student may have asked for a condom, in a school of over 1 300 learners, one would expect more than ten to be used up in a year. (Kent 2002: 57)

Instead, the male teachers often made jokes about 'flesh-on-flesh' (sex without condoms) and the elected gender equity officer was often openly and grossly sexist. Furthermore, the gender equality discourse was often ridiculed through sexist jokes in the staffroom. Citing the work of Kehily (2001), Kent shows how humour, including sexual jokes, was used as a strategy for the subordination and under-mining of women. For example, a joke was printed out and passed around the staffroom:

> A woman comes home late from a gender meeting. The husband was waiting eagerly. When she arrived she said: *'Baba Kuthiwe Kube yi 50/50 nani: nipheke, niwashe, nigeze nabantwana.'* (Now we are 50/50 you must do the cooking, washing and look after the baby)

> The man replied: *'Manje amasende baninikile yini?'* (Did they also give you testicles so we can be equal?) (Kent 2004)

We had thought that the guidance teachers had been gaining status and influence in relation to gender issues in the first round of data collection after the DramAide intervention. However, even they did not challenge existing gender notions. Instead, they tended to reinforce notions of women teachers as carers. In an interview in 2004, for example, Gugu told us: 'When the winter comes they have nothing to eat. So I organised to give them the blankets and everything. Then I went to the counsellor and organised some of the women to that community as a volunteer to help all those who have nothing to eat. Something that I'm trying to solve . . .'. Although this caring approach is admirable, nevertheless it raises questions about how much of the work of care devolved to the women teachers at Lilian Ngoyi, as we discuss in Chapter 10. The essentialised identities inscribed by so much of the school culture seemed extraordinarily difficult to shift. In these conditions the tasks of raising and maintaining concerns with gender equality were extremely difficult.

At Dingiswayo, the difficulties of 'doing gender equality' were openly acknowledged by teachers. The general feeling was that things had changed in terms of HIV and AIDS awareness, not because of the intervention, but mostly because the pandemic had begun to impact directly on the students themselves. As one teacher observed:

In terms of awareness and consciousness, yes, it has changed quite a bit, because now the learners are aware that there is AIDS and it has affected some of the learners now that we've had especially this year, we've had a girl who died of HIV and AIDS and she was pregnant when it happened . . . And we've got two learners in Grade 12 who are I should think it's fully blown AIDS because now they are beginning to be affected. They don't come often to school, but awareness is there. But as to whether we were able to make learners aware of the fact that if they are sexually active they have to go for testing, that we haven't really achieved. (Nomusa)

HIV and AIDS had come to be a central part of the school, but it had not brought a concomitant concern with gender equality with it, despite the introduction of these themes into the Life Orientation curriculum. Our follow-up study in 2004–05 revealed persistent sexist notions of leadership and continued subordination of women, suggesting that the DramAide intervention's attempts at cultivating principles of gender equality had fallen on stony ground and had never been taken up by the school's management team. Thus, the context of essentialised gender and cultural identities had not been shifted and the gender regime of the school had not changed. The persistent portrayal of men as mentally and physically strong and women as weak is demonstrated in the following interview transcript:

Thembeka: It's not easy here in school. This kind of school needs a male principal you see – because of the environment. Maybe we might have but this area does not allow to have a female principal . . . Because of these informal settlements and this place is dangerous.

Alex: So how is a male principal better?

Thembeka: A male principal – because of the build sometimes. You look at him as a person . . . because he has got dignity and he's not approachable. Females have got soft spots most of the time. They maybe not be respected by maybe the teachers. And also the female may easily be threatened. Because of gender. Because of being female. Anyone can say anything because she is female.

Alex: Can they not say the same things to males?

Thembeka: No. It's not easy to tell the male whatever you want to say. You have to . . . They're nature – they are born to be respected. It's part of our

culture. You can't talk anyway to a male person. Even at home you can't talk to your mother in the same way as your father it's, you say, 'Mummy can I have this?', it's a minor thing.

Alex: And is this the same with the students?

Thembeka: Yes. Especially boys. Boys don't respect females. You have to be harsh if you want the boys to respect you, to make them know that you are a parent, a teacher whatever. Because some are in adolescent stage – they tend to mind. Some with their grannies – they are spoilt.

Alex: How?

Thembeka: Maybe by giving them everything they ask. If they say they don't want to go to school today, the grannies they just let them. They are weak. A harsh person tells the child to go. The granny, no, they don't do that.

Observations in the school during this period also noted harsh regimes of punishment and teaching about HIV and AIDS, which stressed the deviance of children, particularly that of girls. For example, eighteen months after the DramAide intervention, an interview with a senior male teacher in the school revealed that:

[The students] don't engage in interactions, as they tend to remain in the stereotypes that prevail outside the school – that boys are dominant and girls submit to boys. So I mean you ask a girl to do a boy's thing, there's always the relation that they cannot do it because a girl is a girl and a boy is a boy. So the relationship situation does not help. I think it's to do with perhaps the teachers perceptions or assumptions in terms of arranging and organising their classes. (Sisa)

Predictably, in this authoritarian school environment, corporal punishment was commonly used, which, as at Lilian Ngoyi, was justified by reference to tradition:

Sisa: Discipline is quite a big problem in our schools, because we don't follow policies. And we don't even try to formulate policies based on what national policies are. I'll be honest with you that has been the practice.

Robert: You beat students?

Sisa: There are traditions, yes.

Robert: What other traditional methods are there?

Sisa: Um, when there is time sometime it's the punishment to ask them to work outside as a punishment, but the shortcut that teachers are trying to take . . .

Acknowledging that corporal punishment is contrary to the principles of gender equality as espoused by the DramAide intervention (and, indeed, illegal), Sisa went on to explain the difficulty of attempting to institutionalise the intervention:

> At a school level and within the context of a school place, like where we are, um we have all these novel ideas as to how to go about it, but children when they come to school with all the views that the local community . . . And the whole issue of gender equity which means that people have to look at a boy and a girl as a human being. It's still a very unclear concept. So for schools to be able to promote that with that view of novel ideas about gender equity . . . teachers themselves find themselves in a very awkward position. We are not taking this gender issue very seriously in class. And then make an observation whether it's really giving us the outcome that we desire or are looking for.

The teaching about HIV and AIDS that was being conducted in the school often stressed the deviance of children, seeing them as in need of correction and moulding. In this context, rather than challenge the existing unequal gender relations, guidance teachers' responses reinforced gendered notions of femininity and masculinities. Indeed, in our last interviews in the schools, it was apparent that the growth of 'virginity testing' of girls/young women was not only tolerated, but embraced and encouraged by some of the Life Orientation teachers and frequently accepted by learners of both sexes. This practice, seen and praised as traditional, laid the responsibility for abstinence on the young women, placing them in difficult positions in which agreeing to a virginity test was potentially both humiliating and dangerous to their health, but refusing it singled them out as, perhaps, sexually active.

Conclusion

In this chapter we have traced how our longitudinal involvement in the two schools, both through our own fieldwork and that of the researchers who worked with us,

enabled us to examine and re-examine the sustainability of gender equality in the wake of an intervention intended to promote both transformation of gender relations and education concerning HIV and AIDS. As we have shown, at Lilian Ngoyi the transformative language of equality was eagerly adopted in the early days during and after the DramAide intervention, but barely a year later discursive practices of gender were deeply unequal. It is thus important to recognise the fragility of transformative languages of equality in relation to the transformation of practice.

Indeed, in both schools, our later research revealed how the institutionalisation of teaching about HIV and AIDS in Life Skills lessons had become embedded in essentialised views about gender, with girls carrying much of the responsibility for their own and men's behaviours and with a focus on traditions that are generally patriarchal. In this way, messages about equality seem to have disappeared from view, perhaps even been driven underground. In this context, it is perhaps worth asking whether changing ideas about men's entitlement to sex (compare with Hunter 2004) and focus on abstinence have, perhaps, been accompanied by a growth in alternative displays of masculinity associated with corporal punishment, making sexist jokes and the acceptance of male leadership.

We need to ask why it was so difficult to institutionalise gender equality in the two schools, both of which welcomed the interventions by DramAide and the research project as a whole and which did seem to offer some promise of transformation. The gendered context in which the schools exist provides one of the reasons for the 'failure' of the intervention. The townships are places in which patriarchal values remain strongly entrenched, despite the progressive gendered policies and laws that have been put in place. There is widespread resistance, particularly among men, to the emancipation of women and many women equate the new gender laws with attempts to undermine traditional ways of doing things or, more specifically, they interpret state policy as an attack on 'Zulu culture'. In these circumstances, the success of an intervention is likely to be limited by community influences.

In the case of Lilian Ngoyi, we have noted how the rhetoric of gender equality, enthusiastically adopted in the immediate aftermath of the intervention by both learners and teachers, actually masked structural difficulties associated with embedded inequalities and difficulties in maintaining the stability of the school.

For Dingiswayo, the difficulties in institutionalising gender equality lay in the deep entrenchment of patriarchal gender structures, authoritarianism and what might be interpreted as a self-interested lack of interest in gender equality by a male management team.

CHAPTER 6

Struggling with Gender and Sexuality in Primary Schools
Teaching about HIV and AIDS

Introduction

Whereas the DramAide intervention in the two secondary schools discussed in the previous chapter was a one-off intervention (albeit over a period of several weeks), the interventions in primary schools discussed in this chapter took a different form. At one level, they could be seen as top-down initiatives: from province to school. At another level, they can be seen as a struggle to institutionalise the recommended approaches by teachers in their primary classrooms. The chapter begins with a brief discussion about the nature of these interventions. Here we show how, while policies and government (or provincial) strategies are important, there are dangers inherent in the top-down approach adopted. The next section of the chapter examines the transition from policy to practice through particular teachers who take on the role of implementing and, in some ways, attempting to institutionalise approaches to teaching about gender and sexuality in Life Skills lessons with young children. We then move to a detailed examination of two particular sex education lessons, exploring both the missed and avoided opportunities and those moments of potential transformation, as well as children and teachers' responses which, we argue, are related to the baggage they carry and their previous knowledge and understandings of gender relations. In this context we show how these are cross-cut by issues of gender, class and race/ethnicity. In the conclusion we draw attention to the difficulties of institutionalising interventions and some of the more hopeful indications of possibilities for change.

93

Intervention from above: From policy to practice

Policy interventions at national government level in relation to HIV and AIDS have been fraught with contradictions, resulting from the denialist stances by the then president, Thabo Mbeki, and his health minister, Manto Tshabala-Msimang, combined with efforts to introduce strategies and curriculum in schools that allow for teaching about HIV and AIDS. Simultaneously, attempts were being made to extend gender equality in schools. As far as interventions were concerned, many were in the form of peer education initiatives, often implemented by non-governmental organisations (NGOs). The introduction of Life Skills or Life Orientation into the curriculum in 2005 and the fact that it became examinable at matric level in 2008 facilitated teaching about gender and sexuality and initiatives to improve gender equality and to combat the spread of HIV and AIDS through the taught curriculum.

In this context, the messages from central government were confused, to say the least. Nevertheless, throughout the period of our research, the KwaZulu-Natal Department of Education offered special training courses for teachers on the Life Skills curriculum, teaching about HIV and AIDS and, in that context, sex education at both primary and secondary school levels. The provincial department also issued materials that could be used and suggested lessons. Teachers could adopt these with or without the training provided. These initiatives offered what can be seen as indirect interventions for both primary and secondary schools. In the secondary school context there were also more direct interventions. For example, older secondary school students could attend workshops to be trained as peer educators and also to do leadership courses of various kinds.

In the primary schools in our study, KwaDabeka and Bullwood, Life Orientation was taught by Mr Xaba, an African man, and Mrs Burke, a white woman, respectively. Both drew on materials disseminated through the KwaZulu-Natal Department of Education.

Children in Mrs Burke's class were presented with material from the departmental handbook (Department of Health 2002) for HIV and AIDS education in Grade 4. This presents a comic strip story about a child whose brother has HIV. In the original version, provided by the Department of Education, the story told is of a teacher talking to a class about a child, Vulani, whose brother is HIV-positive. It shows learners expressing their worries and the teacher explaining how HIV is,

and is not, spread. One of the frames is a picture of a heterosexual couple in bed, under the blankets, having sex with the caption: 'Adults can catch HIV if they are not careful when they have sex with someone who has HIV'.

Vulani's story was used by Mrs Burke as a prompt to raise questions about HIV and AIDS, but in the copy presented to the children she deleted the frame picturing the couple. Rather, Mrs Burke, in consultation with the principal, decided to substitute the following words and erase the picture: 'Normally blood has soldier cells that fight germs. If someone has HIV (human immunodeficiency virus), it attacks the soldier cells, so they find it more difficult to fight off germs and they get very sick.'

Mrs Burke had had little or no contact with the disease and was very reliant on the departmental handbook, but was unwilling to enter into the open discussion of sex to be found in it. She had not attended any of the courses on offer and was distinctly uncomfortable with the intervention; she was not sure that it was appropriate for the children she taught. Her well-regimented school showed little outward sign of the pandemic. Middle-class pupils arrived in cars and were collected in the afternoon. If any parents were HIV-positive, they were likely to be on antiretrovirals, remaining well, for the most part. Mrs Burke thus sustained the illusion that AIDS was 'out there', not in her school.

In contrast, Mr Xaba at KwaDabeka had been to HIV and AIDS and Life Skills training organised by the provincial Department of Education and had taught at the same primary school for ten years. He had close knowledge of the effects of AIDS. Some children at KwaDabeka came from households where a parent was ill or had passed away. He had colleagues whose families had already registered the death of a family member. He was thus very aware that HIV and AIDS were important and relevant to learners. While his physical size suggested a man of authority, he was very soft-spoken in the classroom. He told Deevia Bhana that at the training workshop he had learnt to call a spade a spade. He largely devised his own materials and approaches, drawing on what he had learnt on his courses, but was extremely uncomfortable using the word 'sex' – certainly in the context of a discussion with Deevia. Indeed, his discomfort in using the word was demonstrated by the fact that throughout her conversation with him after the lesson he kept referring to 'that word' by spelling it out, 's-e-x', rather than saying 'sex'. As will be seen later, he also avoided the term in his discussion with children in the class during her observation.

Irrespective of the race, class, gender and location of the teachers in this study, both found it difficult to engage with issues of sex and sexuality, but for different reasons. Mr Xaba was concerned about how parents might view such learning in his classroom. Mrs Burke felt that at Bullwood there was no need to mention HIV and AIDS and/or sex, not only because HIV and AIDS were not visibly present in the area and community, but also because she felt it was inappropriate for young children to engage with information about sex.

In this respect, it can be seen that the gap between policy and local classroom practice was made and entrenched through the teachers working to institutionalise the interventions.

Beyond the rhetoric in primary schools: Teachers, children and sexuality
Pedagogic relations
Within the outcomes-based education (OBE) approach advocated by the national Department of Education in the National Curriculum Statement of 2001, learners are considered to be co-constructors of knowledge in any pedagogical endeavour. This requires the arrangement of classrooms in ways that encourage participation and collaboration with each other, while teachers act as facilitators of the knowledge creation that is supposed to take place. However, more than twenty years ago, Valerie Walkerdine (1984) pointed to the problematic nature of 'child-centred' education. In this case, the child-centred and participatory arrangement of seating and classroom equipment was belied by the authority that both teachers exerted in their classrooms through their pedagogic styles. Both drew upon the general power of being an adult, monitoring and regulating children's activities, talking and discipline in the classroom (Thorne 1993). In this way, the lessons were examples of the adult power of teachers. They also drew on the power gained from having factual knowledge of HIV and AIDS, which both teachers tried to expand upon in their classes. Thus, while both teachers viewed their practice as learner-centred, their relative power to control the discussion and to shape what could and could not be said was evident.

While, as shown above, the Life Orientation curriculum and materials included overt discussion of sex and the fact that HIV is sexually transmitted ('Adults can catch HIV if they are not careful when they have sex with someone who has HIV'), the teachers in this study avoided discussions of sex altogether in their teaching

about HIV and AIDS. In other words, the seriousness of the HIV and AIDS pandemic and the efforts to contain and prevent its spread opened up the possibility of dealing with sexuality in the Life Skills curriculum and the provincial Department of Education's materials were an intervention designed, in part, to do this. However, by altering the frame in Vulani's story in which sex is mentioned and depicted visually and in her general approach, Mrs Burke rendered invisible both gender and sexual relationships in the context of a lesson in which these were supposed to be discussed. She was uncomfortable about linking HIV and AIDS to sex, gender and sexuality and remained silent on sex, trying, instead to substitute more 'innocent' ideas and words. The effect was to constrain efforts to teach more comprehensively about HIV and AIDS prevention and also to subvert the preferred reading of the original handbook.

In the course of her lesson, Mrs Burke used the word 'kissing' as an example of the many ways in which one could not become infected with HIV. When she did so, the controlled environment that she tried to produce was shattered by laughs and giggles. Anoop Nayak and Mary Jane Kehily (2001) claim that sexuality is an uncertain realm that gives rise to hyperbolic performance such as the laughter and giggles that erupt when children hear or read the word 'kissing'. Thus, in a moment, the silence surrounding sexuality was ruptured as the children themselves introduced the connection between sexuality and HIV and AIDS into the classroom. But laughter can also be a manifestation of shame associated with sex and Mrs Burke endorsed this by refusing to engage further. Instead of using the opportunity to expand the exploration of sex, gender and HIV and AIDS, Mrs Burke foreclosed discussion and continued a recitation of the ways in which one cannot become infected.

In Mr Xaba's class, in contrast, the silence surrounding sex was broken by the teacher himself, when he discussed the transmission routes of the virus. He referred to sex directly, adding a moral rider about parental wishes: 'We also get AIDS if we do unprotected sex (laughter, giggles). We should not do something our parents do not want us to do.'

As at Bullwood, laughter and giggles erupted in the classroom, making it very clear that such talk carried an emotional charge and was possibly embarrassing. This was Mr Xaba's only mention of sex and it was tied to morality and to parental objections to such behaviour. A connection was thus established between AIDS,

unprotected sex, disreputable sexualities and a moral position that mobilised the supposed views of parents. But having opened up the possibility of a conversation about sex, gender and sexuality, Mr Xaba moved swiftly away from this and returned to discussion of disease transmission. His embarrassment at talking openly with children about unprotected sex in an adult woman's presence was clear. He quickly walked to the back of the classroom to tell Deevia Bhana that he had had to say 'sex' because he could not let her down since she had come all the way to observe a sex education/HIV and AIDS lesson.

HIV and AIDS education has the potential to disrupt dominant discourse by destabilising a sanitised pedagogic approach to learning about sex and gender. However, this disruption was temporary in the two classrooms. Both Mr Xaba and Mrs Burke reframed the discourse, reinstituting closed views. They did so by deploying the authority vested in them as teachers. Any potential to expand the discussion beyond that which was privileged by the teacher was therefore not possible. Discussion of sexuality thus entered the classroom only to be suppressed and denied (Epstein and Johnson 1998; Epstein, O'Flynn and Telford 2003).

Pedagogies of dis/ease

Given that the two teachers were (differently) discomforted by talking openly about sex and sexuality, their approaches in the classroom were premised upon dissemination of information about HIV and AIDS. While this was a problematic interruption in the attempt to institutionalise a top-down intervention, nevertheless, the commitment of these two teachers to actually providing some education about HIV and AIDS must be recognised. The problem for them was that implementing what was required of them in relation to frank discussion of sexuality, gender and the infection was more than they could manage, given their personal and social histories and positions. Consequently, instead of using children as important resources in the articulation of knowledge of HIV and AIDS, the teachers' engagement with HIV and AIDS was based on calculated judgements about how to address the facts of HIV and AIDS, while avoiding too much discussion of sexuality.

In both classrooms children were grouped and both teachers attempted to involve children in their lessons. But the lessons on HIV consisted of prolonged explanations or repetitions about the facts of HIV and AIDS. Their use of question-

and-answer techniques restricted children's participation to brief, monosyllabic responses. Both teachers focused on 'facts'. Both used analogies to simplify and get through the content. Mr Xaba used the idea of 'road signs' to create discussion about the HIV and AIDS carrier. Mrs Burke talked of germs and soldier cells.

They thus paid little attention to the actual lives and voices of the children in responsive ways. To do so would have been to risk too much, perhaps. Much of Mr Xaba's approach was framed around caring for those infected with HIV and AIDS. Mrs Burke used learning material focused on AIDS-related discrimination. But in the two schools the children talked with different levels of knowledge. At KwaDabeka children made explicit reference to thrush, diarrhoea, weight loss and changing hair colour as markers of HIV and AIDS. Children at Bullwood had no such knowledge from direct experience. This is not surprising considering the fact that HIV rates are reported to be almost twice as high in informal settlements, such as that in which KwaDabeka was situated, as in rural and urban areas (Nelson Mandela Foundation/HSRC 2002).

Lengthy discussions took place in both contexts concerning the association between blood cells, germs and HIV and AIDS. Soldier cells (usually referred to as the white corpuscles, which defend and heal the body), said Mrs Burke, cannot fight off germs if someone has HIV:

Mrs Burke: HIV lives in the blood. It is very difficult to catch it from someone. HIV does not live in the air or in food, okay. Normally blood has soldier cells that fight germs. If someone has HIV, which is the Human Immunodeficiency Virus, it attacks the soldier cells so they find it more difficult to fight off germs and they get very sick. Okay, who can explain that to me? Who can explain what HIV is? What is, what does it mean, Tyler?

Tyler: It means like (little) germs that attack the, um, the other in the blood and you end up very sick.

At KwaDabeka, teaching also focused on biomedical facts:

Mr Xaba: Can you give me types of diseases that you know of? Does anyone have someone who is sick at home? There are many different diseases

like TB, cancer and today we're going to talk of HIV and AIDS. Do
you know someone who is sick at home?

(Children raise their hands)

HIV attacks the human cells. HIV is a long word but it is shortened.
H-Human, I-Immuno, V-Virus. Children, repeat HIV.

(Children say 'HIV' in chorus)

HIV attacks your cells and your immunity and your cells become
weak. What are the signs or symptoms? Do you all know road signs?

(Children nod their heads)

A sick person who has AIDS has some symptoms or signs. Have you
seen someone who has AIDS or any other kind of disease?

Bongumusa: Yes, TB.

Mr Xaba: How do you know that a person has TB?

Bongumusa: He coughs.

At Bullwood, Mrs Burke talked at length, explaining the nature of the virus:

Mrs Burke: Inside your body, you've got little, they, they call them cells, okay.
They're little um, um, little cells. Cells! We are made up of lots and
lots of little cells. Okay, we've got lots of cells in our body which do
different things, okay, and certain cells in our body we could call
soldier cells and when we get flu or gastro or something awful like
that, um, and we get the medicine, medicine helps those soldier cells
(Sit up, Bradley!) to fight and make sure that those that, that, um, that
the little virus that we have got that flu is taken away, and we are not
ill any more. Okay, does that make sense?

(Children say 'yes' in a chorus)

Now what happens with HIV . . . if someone has HIV which is the
Human Immunodeficiency Virus, it attacks those soldier cells so, all
the soldier cells in a person's body it starts attacking and those soldier
cells obviously can't fight the sicknesses any more. So if a person has
HIV, it means that when they get sick like, for example, like (Brittany!)
said earlier, um, about TB and, was it you, Jarryd? They get sick with
some other, um, with some other disease or some, some other illness,

right, their soldier cells don't work as well as someone without HIV, um, to fight off and make them get better. Do you understand that?
(Children say 'yes' in a chorus)
Okay, and so what happens is that um, illness they get really, really sick in the end, a lot quicker unfortunately than maybe someone who doesn't have HIV and, um, whose soldier cells are working properly and can fight it off. Do you understand that?
(Children say 'yes' in a chorus)
Right, do you understand what HIV is?
(Children say 'yes' in a chorus)
Okay, so nobody actually dies or, or gets really, really ill from, um, the AIDS or the HIV, HIV. What happens is, when somebody gets to a certain stage where their soldier cells aren't working that's when they have, have AIDS. Okay, does that make sense?
(Children say 'yes' in a chorus)

As can be seen from the above extracts, both lessons included a good deal of repetition, rhetorical questioning and intermittent teacher-learner interaction. But for long stretches the children were not directly drawn in, other than to chorus a response. The analogies of the soldier cell and the road signs were used to simplify and get through the facts of HIV and AIDS. The use of question and answer produced a semblance that pupils were engaged in the lessons.

Neither the apparent familiarity the children at KwaDabeka had with the course of HIV and AIDS nor the lack of familiarity of the children at Bullwood were put into social context for or with them. While Mr Xaba, in particular, was anxious to institutionalise approaches to teaching about HIV and AIDS, his nervousness about approaching sex and sexuality (and, consequently, gender relations) seemed to be a significant barrier to his success in doing so. Mrs Burke had no such qualms; her view of the infection as being of marginal importance to her class meant that she was able to teach about HIV and AIDS in a way that was almost dismissive (while charitably supportive of) those (like Vulani in the comic strip) who are infected.

Preventing HIV and AIDS: 'Put gloves on and you're going to be fine'

A second common pedagogical script operating in both classrooms was to establish a connection between HIV and AIDS and blood. Jonathan Silin (1995) argues that

the meaning of blood in discourses of HIV and AIDS is a potent symbol, summoning images of contamination. Someone infected with HIV and AIDS is signified not only as one with contaminated blood, but also as one who must be feared and as a threat to others. In both classrooms mention of blood was associated with fear.

Mr Xaba:	We want to prevent ourselves from getting AIDS and then we said that there are things to look at, signs and we talked of prevention. What should we not do in order to avoid AIDS?
Maria:	Do not touch the blood of an infected person.
S'the:	Do not touch a sore of an infected person.

Mrs Burke:	We can get food poisoning from food, isn't it? But HIV isn't as easy. It's not in food, only from blood . . .

In both classrooms, the teachers directed attention to information about blood and in so doing, minimised discussion of sex or sexuality.

Mrs Burke:	We need to be careful in situations maybe when we're faced with somebody who's cut themselves. How do you think they deal with a situation like that? If somebody has cut themselves, what do you think is the best way to deal with it, Amy?
Amy:	Maybe you can put gloves on.
Mrs Burke:	Maybe put gloves on, lovely, and, and work with them then, because no blood is going to come into contact and you're going to be fine.

Mr Xaba:	How do we get AIDS?
Bongani:	If someone is not wearing gloves and he has a cut he may infect us.
Gugu:	We should wear gloves before we touch a person's blood.
Thula:	We should be careful before touching a person's blood.
Siya:	We can use gloves when we bath the person.
Mr Xaba:	Why do we have to wear gloves?
Siya:	To prevent ourselves from getting AIDS.

By focusing on the powers of the glove, both teachers avoided important discussions about sex, condoms and protection. A similar silence with regard to

talking about sex was evident when the teachers dealt with the myths that prevail about 'catching HIV and AIDS' while, for the most part, not talking about sex:

Mr Xaba:	Can we get AIDS if we use the same cup from a person?
Bongani:	Yes.
Mr Xaba:	Can we get AIDS? No.
	(Children say 'no')
Mr Xaba:	Can we get AIDS if we use the same toilet?
S'the:	No.
Mr Xaba:	How do we get AIDS?
Silas:	When we touch blood of the infected person.
Bea:	When we use the same utensils . . .

Mrs Burke:	Right, you can catch HIV when you mix blood. People, who share needles when they inject themselves with drugs, risk catching HIV, why do you think, Timothy?
Timothy:	Because it touched theirs.
Mrs Burke:	Because yes, because the needle has gone (in). Unfortunately there are some people who do use drugs and they inject the drugs into their, into their arm. And because the needle goes into the arm and then they pass it onto somebody else and somebody else uses the needle and the blood is passed from one person to the next and they get HIV. Okay, so that is a way of catching it, isn't it? All right, hands down. I know you want to say something, but we are never going to get through everything otherwise. Right, but there are lots of ways that you cannot get HIV. For example, you cannot get HIV by sitting next to someone who has HIV. Okay, it's not like flu or a cold where lots of you get sniffy or you start going home and say 'Oh! my goodness I think I am going to get ill soon because I can just see all the germs and everything going around the classroom' and you know what, what that's like. As soon as one person gets sick, it tends to spread from one person to the next. Okay, so a cold is really easy to catch isn't it? Okay, but you can't get HIV from just sitting next to somebody, all right. Here are some other ways that you cannot get

> HIV, now the photocopier hasn't come out beautifully, but we can still read it. You cannot get HIV from sharing pencils and crayons, you cannot get HIV from swimming together or playing together, you cannot get HIV from kissing (children laugh), touching or hugging someone and you cannot get HIV from mosquito bites.
> (A few giggles from the class)

HIV transmission in sub-Saharan Africa takes place primarily through unprotected sexual intercourse (WHO 2002) and, as pointed out in Chapter 3, in heterosexual sex women are more vulnerable than men for both biological and social reasons. Nevertheless, both teachers in this study marginalised, ignored or overlooked the significance of sex and gender in HIV transmission and prevalence. Mrs Burke silenced the voices of the children, many of whom were stretching up their hands to be heard, on these issues.

In both classrooms, the lessons invoked moral responsibilities framed by discourses of care. These discourses have the potential to challenge, alter and disrupt gendered discourses, in particular those of masculinity, and to some extent the teachers gave information on this. By drawing upon Vulani's story, Mrs Burke was able to show how HIV and AIDS breed ostracism and stigma and to disrupt the idea that HIV and AIDS is only a biological condition. After reading the story out loud, she led a classroom discussion on discrimination and stigma:

Tarryn: Her brother's got AIDS and now they also think she's got AIDS, so they don't want to be next to her. They think they are going to get it.

Mrs Burke: They think they are going to get it. What do you think about that? What do you think about it? I think it's quite mean because you can only get it in the blood and also . . . how would you feel if someone, say Dean, wouldn't want to share his pencil with her and she really needed it? You would feel sorry for Vulani because maybe the children are being mean. Do you think that they are being mean on purpose?
(Children 'yes' and 'no' . . . mixed response)

Mrs Burke: Do you think they are sitting there being nasty and thinking, 'We're going to be mean and nasty to Vulani'?
(Children say 'no')

Mrs Burke: Why do you think the children are feeling this in the classroom, Jason?

Jason: 'Cause they don't, they don't know, they're thinking that it is like a cough, you cough someone else is going to get it. They think like that.

Mrs Burke: Okay, so they think you can catch it very easily, okay, um, and can you catch it very easily?

            (Children say 'no')

            Okay, all right, Jessica?

Jessica: Maybe they, just, like scared that they'll get it.

Mrs Burke: They are scared that they will get it, maybe through Vulani because her brother's got HIV. Yes, Tyler?

Tyler: Someone who has HIV is unable to give blood.

Mrs Burke: Okay! Okay, so what do you think? Do you think the class are reasonable? Do you think that it's okay that they are thinking these things or do you think, um, they shouldn't be thinking these things?

Adrian: Well, yes it's hard to say that. Anybody would worry about it, so it's either like yes or no . . .

Mrs Burke: You're right, it is. Anyone would worry about it, because if you think that there is a person, maybe, who is ill and you don't want to get that, do you? So how do you think these children, what's the best way if you were that teacher and you were sitting now and you were me and you were sitting now with a class of children and all these children are really, really scared because they don't want to get sick? What do you think is the best way to go about it to make sure that poor Vulani isn't unhappy at school? Because you know it is really awful to come to school and be so unhappy because other children don't want to be near you or play with you. How can the teacher solve this problem do you think, Mpumelelo?

Mpumelelo: The teacher can tell them to be nice to her.

Mrs Burke: She can tell them to be nice to her, but do you think that can change how they feel inside? Maybe it might, maybe some children will think shame we feel sorry for Vulani and maybe they might think, okay, um maybe I should just be nice to her. What, who's got another suggestion? Hayden?

Hayden: Um, she could tell them about AIDS and how you get it and there is
            no cure.

Often pernicious mythologies develop about the nature, cause and the transmission
of the disease which, according to the material used in Mrs Burke's classroom,
stigmatises Vulani. In this context, it can be seen that Mrs Burke proffers a position
which is anti-discriminatory and she draws out for the children the importance of
supporting Vulani and appeals to an ethic of care. But the development of a notion
of care at Bullwood was muted (see also Chapter 8). Mrs Burke did not, for example,
touch on the question of how or whether the children or the school should care for
Vulani if she, rather than her brother were infected. In this context, Adrian's
comment was significant: 'It's hard to say that, anybody would worry about it'.
While Mrs Burke tried to highlight a particular moral position and an ethic of care,
Adrian broke this position by questioning and suggesting his vulnerability in
accepting Vulani in the classroom. Mrs Burke's moral considerations come toppling
down as Adrian foregrounded issues of self-care. It can be seen that the concern
with care in the HIV and AIDS lesson is embedded within a host of competing and
intersecting discourses. Hayden however suggests if children learn more about
AIDS, perhaps it might be possible to adjust their behaviour to ensure non-
stigmatisation of people such as Vulani.

At KwaDabeka, the content of classroom discussion was framed around helping
someone infected with HIV and AIDS:

Mr Xaba: How can we help someone who has AIDS?
Mkhize: Take him to a doctor
Mr Xaba: Good. What else? We love this person and we want to help, but we
            don't want to infect ourselves.
Siya: By taking good care of that person.
Gugu: Wear gloves whenever we help.
Mr Xaba: What else?
Dlamini: We can help him use a spoon while eating.
John: Give him vegetables.
S'the: Give that person vegetables.
Gugu: Avoid letting him sleep on a dirty bed.

Mr Xaba:	What else can we do?
Bongani:	Help by giving medicine.

Here Mr Xaba encouraged an approach that depended on helping the infected person. His use of the word 'love' was particularly significant. Children in this school were living with poverty, HIV and AIDS and social fragility and many of the comments they made referred to caring for family members who were sick with HIV and AIDS. Male teachers are often viewed as not willing to engage with emotional issues in teaching (Mac an Ghaill 1994), but Mr Xaba enacted an alternative masculinity, which proffers love and care.

Mr Xaba brought to the classroom moral vocabularies and considerations which children responded to powerfully:

Mr Xaba:	What other things can help that person?
Bekhani:	Give that person fruit.
S'the:	Give that person vegetables.
Zondi:	Give that person treatment.
Mr Xaba:	These people who have AIDS there is something common. We love them and we want to help them. Let's revise the signs. What are they?
Behka:	They become skinny.
Zondi:	He becomes thin.
Sipho:	They have sores.
Siya:	He vomits.
Bekhani:	He has diarrhoea.
Busi:	The tongue has things on it . . .
Mr Xaba:	Thrush.
Siya:	The eye changes.
Busi:	The person becomes weak.
Nontho:	The colour of the hair changes.
Nokhutula:	The person can't walk properly.
	(Class breaks into laughter)
Busi:	The person sweats.
Nontho:	The person doesn't want to eat.
Sipho:	The person takes two spoons of food and vomits.

107

S'the: The person feels hot and cold at the same time.

Busi: The person becomes sick and thin.

Mr Xaba: You see these things at home? So you see you have to take precautions?
            Do you know that word?
            (Children say 'yes' in a chorus)

Mr Xaba: You see the person who has AIDS in not an animal. Let's revise.

HIV and AIDS was a living reality for young children at KwaDabeka. When Mr Xaba appealed to an ethic of care and love for those suffering from HIV and AIDS, the children readily embraced it in their suggestions for providing medicine, proper bedding and even suggesting the prophylactic powers of vegetables and fruit. The lesson, in effect, legitimated caring for the infirm, sick and dying as forms of relationship, which men, like Mr Xaba, could engage in equally with women.

Conclusion

Effecting gender change is slow, difficult work. In this chapter we have seen how two teachers eschewed the chance to raise issues of gender and sexuality directly while, at the same time, trying to implement and, in Mr Xaba's case at least, institutionalise the interventions developed at provincial government level. But their levels of discomfort came between them and the topic in hand. Choosing the comfort of a transmission approach to teaching, the teachers avoided engaging the learners in ways that would draw on their knowledge and life experiences, arouse their interest and raise issues directly concerning sex and gender. It would thus be easy to make the judgement that these two teachers were engaged with an interventionist approach, concentrating on knowledge transfer, while leaving gender relationships undisturbed. At the same time, it is important to note the very different approaches to the consequences of HIV of the two teachers. Mr Xaba, as a man who believed that loving and caring was not only right in the abstract, but also that it should be offered to those suffering from illness and disease, taught about this powerfully. In the context of widespread stigmatisation and very essentialised gender identities, he was making a different identity for himself and the children he taught. He was very concerned to establish teaching about HIV and AIDS at an institutional level in his school, but was nervous about doing so in a way that addressed gender and sexuality. Indeed, we could go so far as to say

that there were elements of an interactionist approach to gender equality in his pedagogy, but it was nascent and difficult to sustain.

However, it is also worth remembering that children learn and teachers teach in many different ways and the hidden curriculum is as (or more) powerful than that which is directly taught. These two lessons did not contribute overtly to gender equality, but there are many other lessons in each school day and many school days in a year. Other opportunities may present themselves. And for a teacher like Mr Xaba to address gender differently is a beginning. Gender equality is not achieved nor is patriarchy secured in one lesson or through intervention from on high. Nevertheless, these two lessons show how changes in curriculum do not automatically lead to gender transformation. Movement towards gender equality requires many different forces and influences to be working at the same time and over a period of time. Movement towards gender equality is seldom linear and is always tenuous. In the chapters that follow, we shall see more of how gender relationships respond to and are affected by gender equality interventions, attempts at institutionalisation and shifting forms of interaction.

'AmaBornFree'

Identities, Sexuality and the Negotiation of Gender Equality in Secondary Schools

And the other thing, people use this attitude thing, we are born free. Okay fine, you are born free, but do not be born free about your body. Your body is your body. Never sell out your body. Your virginity, when you lose your virginity, you lose it forever, it cannot be replaceable. And you will fall pregnant, and when you do that, think about the child support. And think about your family. What is your mother going to say about the baby? . . . And then think about what is going to happen to her when she finds out that you are pregnant? And then that's why people are never actually teenagers and never think before they do something, because they say we are born free. I can do whatever I want, it's my body. (Zandile, peer educator, Lilian Ngoyi)

Introduction

In the quote above, Zandile captures many of the themes addressed in this chapter, which focuses on the ways in which the relationships between learners and teachers in the secondary schools became a setting for negotiating identities in relation to gender equality, drawing on discussions of sex, sexuality and HIV and AIDS. The chapter, which draws on data collected at Lilian Ngoyi and Dingiswayo, identifies two distinct phases. The first was marked by the way that learners and teachers talking about sex and sexuality admitted to experiences that were sometimes good and sometimes bad. Generally in this phase, HIV and AIDS were seen as distant and affecting others; they were not associated with owned personal experience.

Gender equality was largely an abstract idea in the talk of some teachers. A handful of girls and boys aspired to relationships with each other that were more open and intimate and some girls expressed clear aspirations for work, independent income and 'to be someone'.

In the second phase of data collection, by contrast, there was an acknowledgement that HIV was close at hand, a matter of constant individual and collective concern in the school and in the community. This put a particular strain on relationships. This stress was marked, particularly at Lilian Ngoyi, by the emergence of the peer educators as a coherent group, with a defined identity and authority in the school. Nearly all the peer educators emphasised abstinence in terms that described sex as wrong, particularly for girls, who were expected to control themselves and, by implication, prevent the temptation of boys/men. Consequently, girls who became pregnant were seen as 'doomed', deserving harsh punishment. Calls for abstinence were associated with the encouragement of virginity testing, seen as an expression of 'traditional culture'. This practice, it was believed, would ensure that girls remained 'good' and 'pure' – that is, responsible for sexuality and sexual relations. There were only a few voices of resistance to compulsory virginity testing. While we noted in this second period attempts by a few young people to seek closer and more open relationships and interactions with each other and with adults (parents and some teachers) this tended to be expressed almost wholly through a language of care, which had the effect of occluding concerns with enhancing ideas about gender equality. The chapter traces this shift from a reluctance to acknowledge the seriousness and importance of HIV and gender equality to more active and proactive engagement with the epidemic and the many social factors, including gender and culture, which impact on it. It also explores how claims by girls for more equality come to be silenced in the process of this shift.

Distancing the epidemic and reinforcing dominant gender hierarchies

Our initial interactions with participants in both schools indicated that the general awareness of HIV was high among both teachers and learners. However, at this stage, the HIV pandemic was seen as being largely external to the schools. Participants spoke of the epidemic as occurring outside the school gates (and outside their families). They did not see it as affecting practices within the school,

community or family; neither did they see the epidemic as being affected by their practices. For example, in our early interviews at Dingiswayo, the principal spoke about a child who had left school and may have had AIDS in her family. His tone was impersonal and official. He implied that she was now somewhere both physically and socially very far from the life of the school. AIDS education, according to the guidance teachers at Dingiswayo in these early interviews, was a distant 'scientific' topic on which they requested technical assistance.

At Lilian Ngoyi, in this early phase, there was sometimes a frisson when we spoke about AIDS, but action, for a number of teachers, was part of a general commitment to community activism. Awareness of HIV within both schools was framed using narrow medical terms, and while, as we showed in Chapter 5, Lilian Ngoyi seemed to take up the terms associated with gender equality, HIV infection was not linked with any discussion of gender equality in either school. The role played by gender inequalities in the HIV pandemic remained unacknowledged.

In the first phase of the project, our data indicated that the teenage pupils and their teachers understood the explicit and implicit messages concerning gender, AIDS and sex in different ways. Girls viewed knowledge about HIV and AIDS as disabling, often associating this with their vulnerability to rape and other forms of violence. A common view was that it was better to know or say little and to hope for non-controversial sexual relationships and little questioning of gender equality, even if this entailed subordination. A few, however, refused this identity and instead viewed themselves as agents able to study, gain professional jobs and make a social contribution. But for both groups of girls, it was evident that HIV and AIDS were best kept at a distance, seldom acknowledged or talked about as having a potential impact in their own lives.

In both schools, the girls' knowledge about HIV and AIDS seemed set in the powerful context of their first-hand experience and fear of rape and other forms of sexual violence. All the girls who participated directly in the project spoke of a friend, a neighbour or a family member who had been raped or sexually assaulted and some disclosed to us that they had been raped themselves (see Chapters 1 and 4). Among these girls, there was also an understanding of the stigma attached to having been raped and the fear of telling anyone, including their mothers and teachers, if it had happened to them because, as one declared: 'I think if I told my mom, it [would] hurt me'. A second concurred: 'I had a problem when I was doing

Grade 2. After school the old man come to us with my friend, Philile and he take us to the . . . far away at school, at the bushes . . . Then he abuse us; he raped us . . . [He said:] "Don't tell anybody because I will kill you." So I was afraid to tell my mom until now.' (Thabi)

One of the letters submitted by a girl for the 'Sis Dolly' exercise[1] identified the reasons for this fear of telling as including not only the fear of the stigma, but also the fear of further violence and/or marginalisation and blame by significant others (parents, teachers, peers):

I am a girl of eighteen years. I was seriously abused in the past. I was in Grade 1 and I was six years old. This started like this. Some boy took my lunch money at school. If I had some lunch he ate it. He was in Grade 4. He looked much older. I was scared to tell them at home because my dad is very aggressive. When time went by he [the schoolboy] said I'm his girlfriend . . . One day when I was walking to school he took me into the woods. He asked me if [I knew] anything about sex. I said no. He said we'd do it anyway even if I don't know, he'll teach me. He took my panties off and raped me for several times. I was afraid to tell because I didn't know what he was doing to me. This went on for quite some time until my teacher noticed that there's something wrong . . . This still hurts because my mom says I'm a whore and I started [at] a young age. She says that boys are using me and I'm their ball . . . This boy has come back again and wants me to be his girl because he loved me a long time ago. He apologised for what he did and said that he is a grown man now. This is really troubling me because I got raped again last year. Now I don't know if I got the virus. I'm terrified. I tried killing myself but didn't succeed. (Girl, Grade 11)

We believed that the silencing of victims described in this letter was due not only to the stigma of having been raped, but also to the pervasive taboo against speaking about sex and sexuality generally, and against children speaking about these subjects to adults (teachers and parents included). This need for silence made the ways in which HIV and AIDS education was sometimes presented as merely 'technical' or 'scientific' particularly troubling. Without a link to gender equality concerns or an acknowledgement of the difficulties associated with sex and sexuality, an intervention might amplify the need for silence in the face of sexual violence.

Girls' fear of rape was powerfully linked to the fear of contracting HIV. Thus it was evident from the group discussions and class observations that they had a good understanding of how the virus was being transmitted primarily through unprotected sex. They knew that they were supposed to control sex, making sure to use condoms or to abstain and remain virgins. But the girls were in real, material fear of rape, over which they had no control, knowing that it brought with it the threat of contracting the virus. One girl declared: 'I'm scared of being raped because if they rape, you will get HIV' and a second added: 'I'm scared of AIDS. I'm scared of it because these days there are many people who get raped. No to sex! I am not ready for it. I am still a virgin.' (Gugu, translated from isiZulu)

This threat extended to intimate relationships between girls and boys, with girls afraid of being raped or being coerced into having sex with a boyfriend. Indeed, some saw having a boyfriend as automatically requiring a sexual relationship. Without 'giving' sex, a girl might risk losing him to another girl who would comply. This phenomenon is not unique to girls in South Africa. For example, Holland et al., writing about adolescents in the United Kingdom (1990; 1998) note that girls feel pressure to have unprotected sex with their boyfriends, disregarding their knowledge about the use of condoms to reduce the risk of pregnancy, HIV and other sexually transmitted diseases. Many of the South African girls explained their actions as linked to their needs to secure a permanent relationship. As Gugu explained: 'In order to keep the boy, you have to get pregnant.'

Holland et al. (1998) conceptualise the process by which young women take on identities defined by boyfriends, or men more generally, as *The Male in the Head* (the title of their book). This process, as experienced by the young women in our study, made them particularly vulnerable and rendered the issue of gender equality distant from their experience.

But a number of the young women we spoke with did not see themselves as victims of coercion. Some demanded unprotected sex, seeing it as an expression of love and trust. They saw insistence on a condom as an indication of distrust and/or lack of love. Becoming pregnant was a way of expressing love. Some also described having sex for money or to achieve status. One pondered the possibility of this route in the context of having an impoverished single mother: 'And my mother, she does not have any money. So I will think of something and that will make me to do something very bad, maybe I will sleep with someone to get money ... and it's very bad.' (Lungi)

Thus not all girls saw themselves simply as passive or vulnerable in relationships. However, they were generally silent on the links between HIV and AIDS, gender inequality and sexual violence. A few girls did not accept male sexual demands uncritically. For example, while they spoke of experiences of violence and domination, they also referred to instances where they rejected the essentialised identities ascribed to them by the gender and cultural hierarchies in the schools and communities. This was a strong theme in the 'Sis Dolly' letters where girls, answering the letters in the persona of Dolly exhorted the writers of the letters to finish their education and gain respect through their achievements: 'My girl, I will like to say to you EDUCATION IS THE KEY TO SUCCESS. Go to school and depend to yourself . . . education, your certificate always represents yourself.' (Girl, Lilian Ngoyi)

In another letter, schooling and self-belief were suggested as the answer to choosing between two boyfriends proposing different styles of love: 'Just now you better focus on your studies and afterwards see where you stand. You go girl! Do what's right for you.' (Girl, Dingiswayo)

While instances such as these were generally not associated with more detailed views concerning gender equality, they do show a level of negotiation of gender in terms of self and education, despite pressures towards identities of subordination. These processes of negotiating identities were evident when girls spoke about their decisions about sex. Dudu, for example, reported being 'afraid of being in a relationship' precisely because she wished to avoid a 'boy who says you must sleep with him to show him how much you love him'. One series of letters to and from Dolly showed a girl's dialogue with herself as to whether she should 'go with the crowd' and have boyfriends or develop a different stance of femininity. In her persona as the girl with a problem, she wrote:

> My mother is treating me like a baby. When I get back from school I have to hurry and I don't get a chance to walk back with my friends . . . [My mother doesn't] let me sing in the school choir because the practice ends late . . . I realise that my mother thinks that I spend all that time with boys but it's not like that. I know how to behave well like a woman.

In the persona of Dolly, the girl then replied to herself:

116

You must explain to your mother that you sometimes have to stay after school and study. You should do things that will make your mother happy so that she can let you do a lot of other things and let you study with your friends. You should be happy that your mother want to be respected in the community . . . so that you can have a bright future.

This girl is concerned both about what she experiences as infantilisation and, perhaps, about the terms on which she will spend 'time with boys'. As Dolly, she stresses the importance of study, community respect and good relations within the family. While she does not use the language of gender equality, she is evidently reflecting on what behaviour is most likely to be of value in her own future life.

In the first phase of data collection we found little concern with gender equality among the boys. Some appropriated knowledge about HIV and the AIDS epidemic and used this as a basis for claiming status in school. Others, often quite ostentatiously, ignored the messages from the DramAide intervention about HIV and gender equality and engaged in displays of hyper-heterosexuality, flaunting their experiences with many girlfriends and sometimes flirting with violence. A few resisted these identities and utilised more sensitive and emotional language to talk about their relationships with girlfriends, while one or two came out as gay in the 'Sis Dolly' letters.

Boys in both schools had a good knowledge of HIV and AIDS, which was not surprising given that, at the time, schools were saturated with AIDS messages coming from numerous sources (radio, television, billboards, church sermons and assemblies). For many boys, the mastery of this new knowledge became a source of power. Unlike their female counterparts, the boys did not construct their knowledge of HIV and AIDS in the context of their experiences of sex and sexuality. Instead they often linked them with existing gender hierarchies that privileged boys and men. Thus while many girls described sex as bad and/or wrong, often associating it with experiences of inequality or vulnerability, the boys spoke of sex as good and pleasurable and linked this with their entitlement to exert power over girls. In field notes we noticed boys' very overt flirtations with girls, their explicit references to the size of their breasts or the length of their skirts. In informal conversations and jokey asides, a boy might stress his large number of girlfriends or the significance of sitting with his legs astride so girls could admire him. These

comments confirmed dominant township gender hierarchies, associated with having multiple girlfriends and engaging in hyper-heterosex. As available literature has suggested (for example, NPPHCN/UNICEF 1997; Shefer and Ruiters 1998; Wood and Jewkes 2001) such constructions of masculinity are often accompanied by misogynistic attitudes, harassment, violence and hostility towards ideas of gender equality, particularly in the realm of relationships. The boys at both schools were often embedded in a version of what Hollway (1989) has identified in the United Kingdom as the 'male sex drive' discourse. They constructed men as the uncontrolled slaves of their sexual desire and as entitled to women's bodies, even by force if necessary. Some of these views were expressed in a whole-class discussion during one of the early data collection exercises. On the subject of rape, one boy said:

> Other girls who are wearing a short skirt, like this, and they are too keen, and they checks us, the way she walks, shaking herself, [and you say:] 'Ei, this girl is making me feel so nice' [laughter] and you imagine yourself over her rolling like a chicken in the oats [loud laughter] and you understand yourself and you say: 'I must get this girl' and having a sex with her and talk to her that you love her . . . and if you are asking to have sex with her, and maybe she don't say yes to you, then you think: 'I must rape this girl' [laughter].

During the DramAide workshops Mark Thorpe reported similar constructions of masculinity and femininity, which included

> the attitude among boys that it was desirable to have several sexual partners. Young men expected to be told in detail about their girlfriends' movements, but were very secretive about their own affairs. They regarded violence as an appropriate 'punishment' for the 'bad behaviour' of their girlfriends, and this attitude had been internalised by many girls too. There was a sense of risk inherent in male lifestyles, and an expectation that women could and should accept or tolerate this kind of behaviour. (Thorpe 2005: 201)

In the boys' talk there was little link between sexual violence, including rape, and HIV infection. But not all boys shared these views. Some openly rejected them and

did not see sex as an obligatory part of their relationships. Closeness was more important and these boys chose to have one girlfriend with whom they could be intimate friends. These statements were a far cry from the image of routine relational violence which much of the literature in this field reveals (see, for example, Dunkle et al. 2004; Human Rights Watch 2001; Jewkes et al. 2006; Sathiparsad 2005). One interview exemplifies this:

Siphiwe: The best thing that happened in school is that finally this year I met someone who can understand me in life.
Rob: Is that a girlfriend?
Siphiwe: Yes.
Rob: Yes.
Siphiwe: So it's quite good so . . . because she is in my class.
Rob: Aha.
Siphiwe: And we do some homeworks together and . . .
Rob: Is this your . . . is this the first serious girlfriend you've had?
Siphiwe: Yes.

Boys like Siphiwe chose to ignore peer prescription and essentialised views of gender. They tried to relate to girls in ways that were not marked by conquest and submission. Another boy described struggling against dominant ideas about masculinity: 'I go with my heart. What it tells me, and what I like. I don't just go as a friend. Seeing my friends are drinking alcohol I must drink it too. I don't do like that. Just tell my heart: I don't think I can do this and I don't do it.' (Nkosinathi)

Such struggles were also evident in the 'Sis Dolly' letters. One boy graphically set up the twin poles of hyper-masculinity and 'control', the need for education, and the search for more open relationships between men. In his letter to Dolly, the boy wrote:

I live with my father and my elder brother. My father do not have much time with me. He has not given me advice since I was born. I do not blame him though because he is working for me. My problem is that peer pressure is more than I can take. I have six girlfriends . . . I have slept with five of them . . . and my home boys praise me. They think I am one big charmer. However I am not sure whether I like what I am doing. Last month I told myself that I am going to abstain.

119

In the persona of Dolly, the boy then wrote back to himself:

> You are going through adolescence stage . . . There is no need to blame yourself. That will make it more worse than it is . . . I am glad that you do realise that being a womaniser is not good. It is even worse for you because you are a scholar and you are too young. My advice to you is to talk to your father. Try to close the gap between you and him. It's your life. Be in control of it . . .

In this imagined exchange, the writer is negotiating masculinity and family relationships, perhaps developing education or 'being a scholar' as an alternative way of being a man. What is notable, however, is that the boy is concerned with himself and his relationships with his father and friends, but is not able to express any concerns for the girls or for the quality of his relationship with them. This example shows clearly how gender equality was distant from the day-to-day life of the teenagers in the study.

The teenagers' difficulties with imagining their lives through ideas concerning gender equality were exacerbated by some teachers' ambivalent attitudes. One of the male teachers agreed volubly in discussions with a number of members of the research team that gender equality was a worthy course to take. However, his practice was not consistent with these views. Mark Thorpe's field notes recorded: '[In the] staffroom I watched a female teacher attempt to chastise [the same] male colleague. He had passed her class – whose students he knew on fairly familiar terms – and commented: 'Oh, I am sorry for you all . . . you will not know the pleasures of "flesh to flesh" like our generation!' (Thorpe 2005: 203)

The female teacher had attempted to explain to her male colleague that while such remarks were intended to be humorous, they served to undermine HIV education. However this was brushed aside. Alex Kent (2002) also recorded interactions where male teachers, who had made statements about their interest in gender equality, actively engaged in sexist behaviour towards other teachers. She concluded that gender inequalities in the school were reproduced through the hierarchical school structure, where positions of power were given to men, as well as through the hidden and the official curriculum. For example, she noted that while teachers expected girls to be more mature and responsible than the boys, discouraging them from taking risks, including being with boys, they encouraged and accepted such behaviours from the boys.

In the first few years of data collection it thus appeared that violence and HIV were fears for many of the girls. By contrast, for the boys, their sense of distance from HIV and AIDS allowed them to flirt with images of violent and dominating masculinity. For neither group did the language of gender equality have any resonance, and while there were girls who were determined to 'make something' of their lives and boys who wanted different kinds of relationships with girls, these ideas, which were necessary for giving substance to an equality that was not merely formal, were not yet sufficient to underpin wider efforts at instituting gender equality or to sustain different forms of interaction.

Bringing the epidemic close: Peer educators, pregnancy and abstinence
In the second phase of data collection, however, even these incipient moves towards equality appeared silenced as HIV and AIDS became a pressing reality. The pandemic was no longer a distant disease affecting other people 'out there', but affecting all people associated with the school – learners, teachers, friends, family members and neighbours.

By 2004 the disease was an emergency and both schools mobilised to meet it. At Lilian Ngoyi peer educators, a mixed group of young women and men, most of them in Grade 11, established themselves as a coherent group with a defined identity and authority in the school. They held key positions on the learner representative council (LRC) and were accorded significant status by the head teacher. They stressed abstinence, seeing and portraying sex as wrong, particularly for girls. In the discussions we had with them there was no concern with gender equality, only with appropriate behaviour.

The appointment of peer educators had been institutionalised through an initiative of the KwaZulu-Natal Department of Education and Culture. In 2003 learners from Grades 8–10 at Lilian Ngoyi had attended a residential course:

> So from there we were given certificates saying that we were, we have been trained and we having the capability of working with people infected and affected so the main reason for them taking us for training . . . [was] to get back into the school and make some of our colleagues to be aware of HIV and AIDS.

> The aim was to like educate other learners about HIV and AIDS, how do you get infected and what are the means and the ways of preventing HIV and AIDS.

At Dingiswayo, in 2004, young people had gone on a training course: 'I was chosen in school last year and we went to a camp in Glenwood. So we were trained and then we came back and organised sports festivals and told people about the principles about sport. And we were teaching them about HIV and AIDS and all that stuff. So, ja, I've been chosen this year to go to an international camp in London.'

The establishment of the peer education programme was an institutionalised response to the epidemic. But the peer educators saw their work as primarily concerned with disciplining fellow pupils. High rates of pregnancy were seen as a particular problem:

> Every year we found there is an increasing number of girls getting pregnant, and that worried us a lot. So, we tried going back to people over and over and speaking to them and then we even tried making drawing like you saw on the walls. And it's like each year there was an AIDS week where we invited people from Department of Health, various people from different high positions would come and even people who are HIV positive would come and talk to the learners and talk to the community about HIV and AIDS. (Thabo, peer educator, Lilian Ngoyi)

For Thabo the problem of girls becoming pregnant was connected with community mobilisation, awareness building about HIV and AIDS and building links with the Department of Health. Pregnancy was no longer to be seen just as a matter of excluding girls, as described in Chapter 5. Vuyo, one of the peer educators, insisted that the South African School Act No.84 of 1996, which allows pregnant girls to remain in school, was misdirected and should be abolished: 'I think the rule that says . . . teenagers who are pregnant should come to school, that rule should be terminated. Because we are losing concentrate in the classroom. We are falling asleep because of this person who's pregnant. We can't concentrate.' (Vuyo)

Pregnancy, in this context, was a disaster, but this could, perhaps, be resisted by the exhortations of the peer educators for abstinence. A handful of peer educators acknowledged that girls lacked the confidence and, by implication, the power to make decisions regarding their sexuality in relationships, but the general feeling among the group was that it was incumbent upon the girl to control herself, to 'just abstain' so as not to become pregnant. Lungile's assertion illustrates this contradiction:

The girls today just lack the self-confidence, they don't have the confidence. They don't believe to their selves, because if a boy promised a girl that if you don't want to do the sex with me, I will run away from you, and then the girl said: 'Oh, okay, let's do it'. Because I, I don't know what's happening today, because even if a boy promised you okay I'll run away, if you are yourself, if you are confident, you have to say okay, just go away, I don't care about you, I will get another one who will give me something – I don't know what. A man who can love me who can love me the way I want. Because it's not a matter of getting sex, you know it's about love. You can be in love but don't do sex. You can use everything, not even go about and think of the protection thing. Don't even use condoms. Just abstain. (Lungile)

Not one of the peer educators referred to boys' responsibility in sexual relationships. They seemed to accept, uncritically, the idea that a boy would be likely to leave a girl who did not agree to sex. At the same time, they believed that, while a girl could not change essential male nature, she could control her own urges and allow the boy to leave. One of the girls in the group declared:

I can say even boys, boys at this stage are not right. It's all about sex. You're too boring if you don't do sex, but girls in our age, I want to have a boyfriend. And that boyfriend will always say sex, sex, sex, sex. Sex will be the first thing that will come in their mind, and like us teenagers we want to have boyfriends, but we don't get the right one. There is no other one. That's the problem why teenagers are getting pregnant and like having sex. (Thando)

For the peer educators at Lilian Ngoyi abstinence, not equality, was the appropriate response to the epidemic.

At Dingiswayo teachers had taken up calls for abstinence: 'And there's this teacher . . . you know she's always talking about HIV and AIDS all the time and sometimes you get like, "Miss, you've been telling us this all the time." It's like, "I'm just trying to warn you and you know not to engage yourselves into sexual activity and to prevent yourself from [getting] this disease because it's really, really killing people."' (Linda) Girls at the school emphasised the importance of taking responsibility for their virginity:

123

I know as teenagers we all have the same mind, thinking the wrong things, but I wanted to change and make a difference and because I used to do wrong things last year, I will admit that. Go out with boys but include safety sex. And I've noticed one thing, boys plays around with us. During that time I didn't think about myself like okay I'm a girl I'm supposed to be proud. And like people at this school used to think, 'This girl! I hate this girl and this and that and that'. And I have noticed, I wanted to change and many people have seen, and I have changed. I've really changed. (Lungile)

In both schools virginity testing for girls was seen as a culturally appropriate strategy in the face of the epidemic. At Lilian Ngoyi, one of the teachers asserted a link between abstinence and virginity testing:

> I think to do the virginity test it's good, because it encouraged the girls, the teenagers to not doing sex. To protect them, 'cause if they don't go to the virginity test, anyone, if her boyfriend come and propose to her to do sex, she will say yes 'cause she know that my mother or anyone else doesn't check me whether I'm a virgin or not. Her mother maybe ask her are you still a virgin? She will say yes, 'cause she know that her mother doesn't take her to the virginity test. She will be lying and she will get pregnant or get HIV. (Jabu)

While Thabi, one of the learners, supported the idea of virginity testing, she was against its public staging and instead felt that it should be conducted within the boundaries and privacy of the family:

> I think it was the right decision [to have her virginity tested]. You know why? . . . You like putting yourself proud I'm a virgin, I'm a virgin, I'm a virgin. They are . . . boys out there looking at us, working our virginity unexpectedly. Like a mother, a mother should trust her child. Why shouldn't my mother or granny check whether I'm a virgin or not and expect somebody else to do it for her? It's her responsibility. It's her responsibility. It's her that should do that to me. Not just anybody.

Many of the peer educators we spoke to saw virginity testing as a cultural tradition and practice, which should be seen as a right and protected by law. Implicit was

the idea that it should be compulsory and practised by all those who belong and subscribe to Zulu culture:

> The government should not exactly [have] done this thing [making virginity testing illegal] because somehow if you check the Constitution of South Africa, it's about the rights of people, all the rights about equality of all the races of South Africa . . . The black people's culture is now being looked down upon because like they are not respecting, I mean, our positions as black. So I think the government has got to review this issue because it's one of the cultures that we're proud of as Africans. It's one of the cultures that we say that this belongs to an African person. It's belongs to an African girl . . . It's like bringing back the complete . . . among the races because this thing is a black people's issue and what are white people going to say? (Jacob)

Thus those girls who did not participate and who were not virgins (such as those who became pregnant) had to be harshly punished, shunned by peers and adults and excluded from school.

There were very few voices critical of virginity testing. One of the few was a girl at Lilian Ngoyi who argued that the government should protect girls and ensure that only those who wished to go for testing should do so: 'Because maybe the government is abolishing this [virginity testing] because some of the young people are being forced to go there for testing.' (Lungile)

Direct references to gender equality among the peer educators were very few. Discussions about issues of sexual violence, virginity testing and HIV and AIDS were devoid of any awareness of their links to unequal gender relations. For example, proclaiming her desire to be an advocate for all people's rights, the elected gender officer declared:

> I joined peer educators this year when I got a chance and when I was elected as a gender officer. That's where I wanted to be, just advocate gender equality of all the people . . . But it has been what I wanted all this time, because HIV and AIDS I always wanted to be part of it, to contribute in one way or the other. (Ziyanda)

Although she had a position as gender officer, her work with the peer educators did not see her taking any stance on gender equality and her views were

125

indistinguishable from the others with regard to the importance of disciplining girls.

The teachers, in recalling the DramAide intervention, saw it as a precursor to the work of the peer educators, apparently missing the crucial difference between the intervention's focus on gender equality and the lack of concern with this in the peer education programme. As one said: 'Their behaviour showed that they had been taught something [by the DramAide intervention]. They had something in their minds, because those who were sitting with their girls were speaking the language of condoms. It was funny, but it was like really working.'

There was a general feeling among the teachers that the need for the peer education programme had come about because the young people were mostly out of control, overly sexed and that the girls, in particular, needed stricter strategies to monitor their behaviour and control them. Among the strategies they identified for doing this were corporal punishment, abstinence by the girls and virginity testing to ensure that they comply. For example, some teachers expressed frustration at no longer being able to use corporal punishment to discipline the children. Others argued that the number of rights now held by children in the schools and their awareness of them were undermining what were seen as positive 'cultural' practices, including corporal punishment and virginity testing: 'Some of the kids say virginity testing, it infringes their rights. They talk about their rights of not wanting to be tested, and yet that belief was a good one.' Others spoke with great concern about the high levels of pregnancy among schoolgirls. While some demonstrated great sympathy for girls who become pregnant, others were less sympathetic and felt that such girls should not be allowed to stay in school. Again, these views were silent on the responsibility/culpability of teenage fathers.

The mural that the peer educators arranged to be painted at Lilian Ngoyi as part of AIDS awareness week expressed all these ideas in a compressed form. A large red ribbon symbol was painted on the central block of the school in 2006 and pupils and teachers, as well as well-known media personalities were invited to write messages. These included exhortations to abstinence: 'Guys, don't play games with your body! Please look after it! Just wait!!' and 'Abstain from sex' and 'No sex. No regrets.' There were no overt messages about gender equality on the wall. The messages from the guidance teachers expressed care and support, an ethic which, while not involving equality, did express qualities that were not about

aggressive heterosexuality. These included 'Take care' and 'We are here to help and support all of you'.

There were only two messages on the wall that gave a faint indication that a language concerned with equal rights or gender equality had any resonance. One read: 'Our former president Nelson Mandela said AIDS in the world is no longer a disease but a human rights issue'. This appears to resonate with the community location of Lilian Ngoyi, with the strong African National Congress (ANC) political links in the neighbourhood. The message stresses human rights generally and does not highlight women's rights. In the opposite corner of the mural, one girl has written in a small space: 'I might have an attitude, but I am no one's prostitute. You wanna force me to have sex, see what happened to my ex.' Here the idea that a young woman could freely choose with whom, and on what terms, to have sex is boldly expressed, but this is by no means the dominant view on the wall.

Lindile, one of the most thoughtful of the Lilian Ngoyi peer educators, described her association with the group as emerging from her concerns with a friend who was an outsider:

> The other reason why I started joining this peer educator thing is that I had a friend which is from Tanzania and she had to come here in this school, because other schools they have been, they were full, they were full of students and the things was our class and then they couldn't communicate with her. Because she was speaking Swahili language and then they were like, they'd just laugh at her. They'd find it amusing when she speaks. And then she was forced to speak English. Then I thought maybe if I joined this thing with her, the peer educating thing, maybe it would be very helpful for her, to communicate with other students who try to understand her language and her background, where she comes from, and why is she here in South Africa. So that is the reason why I've joined the peer educating thing, to stop the xenophobia thing, because it's killing us.

For Lindile, peer education and activism about AIDS was partly about trying to live out relationships of equality, even though gender equality was not pre-eminent here. Similarly one of the Life Orientation (LO) teachers described how she struggled to develop more tolerant attitudes to girls who were pregnant, trying to explore issues about empowerment for girls and:

127

talking about maybe the advantages and the disadvantages of the early pregnancy. Making the child to understand the feelings, the problems the girl will encounter after having a baby. And telling them that it is not about sex that makes you to be more recognised or to be more beautiful to, to belong. It's not about having an affair. You can belong to a group without having an affair. And that does empower them, it does help them to have this self-esteem to believe in themselves . . . in the LO classes in Grade 8, when they were talking about this pregnancy, the early pregnancy, the child in Grade 8, now I'm talking about the background, the family upbringing, the child from Grade 8 said that it's not the brain that is pregnant, I don't think with the stomach . . . If a child can tell an educator that I don't think with my stomach, I only think with my brain. It's not the brain, which is pregnant; it's the stomach, so there's nothing wrong. They should be allowed. But when they are empowered now, when they are informed because some of them were denying the facts. They were saying, 'No, no it's not right to be pregnant at school.' Now these ones are more empowered. They do have the knowledge and the different backgrounds situation which they can oppose this other one.

Such comments indicate some willingness to engage with the issues a little, to challenge accepted norms of behaviour and move some way towards a deeper and more equitable understanding of HIV, gender and relationships. However, the trauma of the epidemic made these views difficult to sustain against the general emphasis on abstinence and fear.

Conclusion

This chapter has shown how fragile the language of gender equality has been, how difficult it was to institutionalise after the DramAide intervention and the extent to which the interactions, as initiated by the peer educators, became much more concerned with messages about abstinence and cultural propriety than with gender equality. However, there were voices that went against the mainstream: girls and boys who have refused particular identities, taken actions based on care and begun negotiations linked to gender equality. These have not been fully supported and the despair and anxiety associated with the HIV and AIDS epidemic

have made these hard to bring to the forefront. But that these views exist at all is testament to a potential to develop alternative discussions and to make rights and freedom not only words on the mural, but something substantive.

Note
1. This exercise is discussed in Chapter 4. In quoting from the letters here, we have not altered the use of English but reproduced them verbatim. It should be noted that the young people writing the letters were not native speakers of English – for the most part, their mother tongue was isiZulu.

Sexuality, Race and Class
The Making of Identities through Young Children's Accounts of HIV and AIDS

It's quite a tricky one. I don't do anything except health issues. It's a tricky one. I believe that parents are not comfortable with us doing their job. Actually I don't think that the kids know one person who has it. Also at the senior primary level there's lots of experts that talk about explicit stuff, but I think they're too young to be told issues. I do say "don't touch others, don't take suckers from others" but these are basic hygiene stuff. But no, I belong to a mothers' group and, no, parents will not be comfortable with this. (Mrs Hobbs, Bullwood)

Children, all our children, should know about sex and the AIDS, because this is the thing which the children live [with], in an environment which is affected in their homes by this AIDS and sex. Children should know about AIDS because the parents, the people they are living with, they've got these diseases so the schools are supposed to teach about this and the teacher should freely speak about this disease and also about sex, because our media, especially the TV, they talk about condoms, the uses of using the condom and then also even the movies, they include sex. So our children and the teachers are encouraged to talk about it, even the parents at home. (Ms Ngcobo, KwaDabeka)

Introduction
In Chapter 6, we explored the conducting of lessons on HIV and AIDS at two primary schools. We showed how there were both similarities and differences in the ways

that the teachers discussed sex, gender and the consequences of the epidemic. In this chapter, we draw on ethnographic and interview data from the two primary schools to unpick what children aged between six and eight understand about sexuality and HIV, how they construct their identities in relation to these issues and the discourses they deploy in doing so. Throughout the chapter, we draw attention to the contrast between the differing experiences and cultural worlds of the children in the two schools. We argue that not only is the South African HIV and AIDS epidemic profoundly gendered, classed and raced, but that young children inscribe themselves within particular gendered, raced and classed discourses with regard to sexuality and HIV. We explore both the continuities and differences between the discursive frames of sexuality and HIV and AIDS in the two schools, drawing attention to the way that children inscribe themselves simultaneously in discourses of pleasure and danger, fatalism and hope, distancing and an ethic of care.

Young children's constructions of sexuality

Notwithstanding Mrs Hobbs's claim, quoted at the beginning of this chapter, that the children she teaches are too young to talk about or understand sexuality, there is a growing body of literature that shows that children of this age can and do engage in discussions of sexuality and related topics. Jon Swain's (2003) work in primary schools demonstrates that the body is a highly sexualised site and Emma Renold (2005) explores both how young children police the borders of heterosexuality and how they invest themselves in discourses of pleasure and desire. Debbie's Epstein's work with a number of colleagues (Epstein and Johnson 1998; Epstein, O'Flynn and Telford 2003) and others working in this field (for example, Jackson 1982; Kitzinger 1988; 1990) have argued that discourses of childhood innocence do more than disguise or make invisible children's sexualities; they also create dangerous spaces for children for two main reasons. First, they render the child who has been sexually abused as no longer sexually 'innocent', in ways that can lead such children to be cast as sexual predators, at fault for their own abuse. Second, they have the effect of rendering innocence as sexually alluring, as can be seen in pornographic images of the 'gymslip girls'.[1] Valerie Walkerdine's (1997) work on gender, class and young children shows not only that young

working-class girls in the United Kingdom invest in sexualised versions of popular culture, in part as a way of escape, but also that they are active agents in the construction of meanings for popular culture, sexuality and class. Paul Connolly (1998) argues that such constructions are highly racialised. There has been very little work in South Africa specifically on young children and sexuality, other than that described in this book and the work done previously by Deevia Bhana (2002). Such other work as there is tends to be primarily statistical, enumerating, for example, the number of children under the age of sixteen who are engaged in sexual activity or who have become HIV-positive through sexual transmission, rather than vertical transmission during childbirth (for example, Nelson Mandela Foundation/HSRC 2002). Or it is concerned with dangers to children, rather than their own agency (for example, Richter, Dawes and Higson Smith 2004).

In this chapter, we explore children's understandings and constructions of gender, sexuality and HIV. We show that children are neither innocent nor ignorant about these matters, but that the ways they produce themselves are contextual and specific. The material and cultural differences between the two schools provide significant constraints and resources for the making of identity and production of discursive strategies in the context of HIV and AIDS. We show how the children are invested in both the pleasures and dangers of sexuality in ways that are gendered and classed and raced. In conclusion, we draw out some of the key theoretical and practical implications for the future of HIV and AIDS education in the context of the South African epidemic.

KwaDabeka and Bullwood: Giving meaning to sex

At Bullwood meanings of sex and sexuality were articulated in a variety of forms, including talk about boyfriends, girlfriends and heterosexual games. In the extracts below, children aged between six and eight responded to Deevia Bhana's question about what they understand HIV and AIDS to mean by seeking immediately to define sex:

Rachel: I don't know what's it [sex] called, but it's like ... you know like when they, when they [are] like in love and then they like stick together.

133

Steven: I know what it is. Something to do with sexual?
Deevia: What is that?
Steven: Stuff.
Deevia: What stuff?
Steven: Sex stuff.

In spite of this grappling with difficulties of definition, the young children at Bullwood used the informal discussions to negotiate ideas about sexual identity, often stressing heterosexuality:

Deevia: And you Cassy?
Cassy: I had a boyfriend in Grade 1 and his name's Grant [giggling] and he's in this class.
Nicole: I had Brandon, Kyle and Carl.
Dana: Who's Brandon? Does he have orange hair?
Cassy: I kissed Grant in Grade 1. He kissed me on my lips [Other children: 'Aahhhhhh'). And he did a piggyback.
Nicole: We played catches. Bianca also used to like Grant. Bianca always used to sit on the grass next to Grant.
Cassy: Do you know who Grant likes now? Ruby, do you know Ruby? Grant loves her. She has so many boyfriends. She has Travis, Daniel and I dunno the rest, but she told me that she has them . . .
Deevia: Did you kiss Grant, Cassy?
Cassy: Yes. Please keep that secret and don't tell anyone else.
Deevia: I won't tell anyone. I wouldn't.[2]
Cassy: Especially the boys.
Deevia: I thought you can get AIDS if you kiss.
Nicole: No you won't. Only if you kiss someone who has AIDS you'd probably get it . . . I met Kyle and Carl in preschool and I met Brandon in Grade 1. Kyle and Carl were fighting over who was going to be my boyfriend. So I said, 'Both of you'. So then when I was in Grade 1 when I saw Brandon, he loves me so much, he was holding my hand in assembly [giggling].
Amy: When we went to Grade 1, this boy Brian, I liked him because I don't know . . . [giggling]

Cassy: I got this neighbour called Brandon. He's got red curly hair, short curly hair and . . . and we've kissed on the lips a thousand times before.
Deevia: Oh my gosh.
Cassy: 'Cause this is what we did. We played hide and seek with his sister and his sister's name's Sara and she's three years old and we played hide-and-seek with her and then we hid behind the flower bushes. While we waiting for her to find us, it took so long so we picked these little white flowers and we played, 'I love you, I love you not', and kept on landing on 'I love you' and we kept on kissing all the time.

The young girls in this discussion sought and found a great deal of pleasure in performing heterosexual femininity. For them, learning to be a girl involved fantasising about boyfriends, engaging in heterosexual games and kissing. Collectively, here, the girls were constructing their sexuality and their friendships in relation to heterosexual actions. The children in this discussion 'knew' that talk about kissing and love was forbidden in the classroom, but this did not prevent them from exploring heterosexual pleasure in their talk with Deevia Bhana. Sexuality was a subject for exploration, pleasure and agency among these young children but also a domain of danger:

Deevia: What's sex?
Cassy: My mum told me about it and if somebody has AIDS and you go and have sex with them, you can get the AIDS too . . .
Deevia: What's sex?
Rachel: When some people get together and they have a secret time together.
Deevia: A secret time together?
Rachel: Yes . . . Everybody is sleeping 'cause it's like a secret thing to do and they don't want anybody to see them and that's how they get pregnant and they sleep naked. [giggling]

In the conversation above pleasure (the 'secret time together') and danger (sex and AIDS) were juxtaposed. The children held the dyad of pleasure/danger in tension, enjoying the idea of the secret time and even sleeping naked, but aware, too, of the connection between sex and AIDS.

At KwaDabeka, with children from very different backgrounds, but the same age range, there were similarities and significant differences in the children's understandings of sex. Their talk about sex and sexuality included discussion about boyfriends, girlfriends and play. Hide-and-seek was one such game:

Mandla: They [girls] always say: 'Let's play hide-and-seek; let's play hide-and-seek'. And we play fathers and mothers, then we have sex.

Deevia: What do you do?

Mandla: The boy sleeps on top of girl and then they touch each other's private parts.

But 'playing sex' for the children at KwaDabeka was not only about pleasure. In their talk the broader context of rape, violence and HIV and AIDS was present. For a group of boys, sex was both the same and different from rape:

Sanele: Sex is ukubhebha [sex] but only adults are allowed to do that.

Sipho: Sex is when a man sleeps on top of a woman and she gets pregnant.

Thabo: Sex is rape, when a man rapes a woman sometimes the girl likes it, sometimes she doesn't.

Bongani: Sex is when people touch each other's private parts.

In another discussion, one of the other boys took the equation further with AIDS being fused with sex and 'rape':

Sandile: AIDS is rape, ukubhebha.

Deevia: What is rape?

Sandile: An old man inserts his penis in a small girl's vagina, by force.

Thus sexuality was constructed both as a resource for pleasurable play (in hide-and-seek and pretend games of mothers and fathers) and also as a source of significant danger, with a threat of male violence. The conditions of life in informal settlements are such that there is little opportunity for privacy and it is, therefore, commonplace for children to witness sexual activity – which, while not necessarily always difficult or traumatic, may well be.

KwaDabeka and Bullwood: Giving meaning to HIV and AIDS

Young children's meanings of HIV and AIDS are not simple. They are produced through social and cultural differences that constrain and create meanings in ambiguous ways. At Bullwood, while the children were clearly conscious of the relationship between sex, sexuality and HIV and AIDS, they also made discursive moves through which they distanced themselves from the disease. As Jonathan Silin points out: 'It is easier to blame the victims of an epidemic when they are conceived of as different, [as] the space between self and other increases' (1995: 13). A similar tendency was found among these children at Bullwood:

Deevia: Have you heard of a disease in our country?

Lance: Yes, yes, yes. What's it called? HIV. I heard it on the news. It's mainly on the news. It's babies who get it and they [tell] us on the news but I don't know how they get it.

Veronica: Once on TV, it's like . . . H-I-V. I don't know. I can't read it.

Amy: I heard from the radio.

Veronica: They look very, very sick.

Carol: No. They look like normal people. They don't walk properly and they don't do stuff properly. I haven't seen anyone with AIDS.

The children were aware of HIV and AIDS through television and radio reports. While there was disagreement as to whether infection is visible or not, there was no dissent from the notion that people infected are 'they'. Lance went on to explain some of the differences and this was taken up in racialised ways:

Lance: They can't read, write and speak properly. Sometimes they can die from getting AIDS.

Carol: Mainly Africans have AIDS.

Veronica: Lots of other people have it too.

Carol: Mainly Africans, some whites and some Indians.

Lance: Babies also get it.

Carol: I've only seen big people.

Lance: They look like normal babies, but they look sick as well. Not that sick. They don't look good.

Carol: Maybe they've been in dirty places.

Veronica: They get infected by their moms and dads.

Deevia: How do they get it?

Lance: That's a hard question.

Carol: Maybe they get it from HIV. Maybe they got born in dirty places.

The children at Bullwood associated HIV with the poor and, primarily, with black people, despite Veronica's interjection that 'lots of other people have it too' and Carol's acknowledgement that some whites and some Indians might get it. At an epidemiological level, the information they had was correct. Most of the people infected with HIV in South Africa are, indeed, African, both in terms of absolute numbers and in terms of proportions of the population group. They may have obtained this information through news reports on television and radio or from their parents. What is important to note, however, is that this knowledge was supplemented by racialised assumptions of dirt and pollution (see Douglas 1966), associating it with illiteracy, disease and, implicitly, poverty. The children were not confident that they knew how the disease is caught. Lance's statement, 'That's a hard question', was one which any of them could have made. It was in their attempts to answer this 'hard question' that the racialisation primarily occurred. As the conversation continued, the children turned to questions of their own vulnerability:

Lance: *I don't want to catch it*. If you catch it, you will die (original emphasis).

Carol: On Blue Couch they talk about people who are sick. You can't get affected if you touch a person.

Deevia: What's Blue Couch?

Carol: It's on every morning, but it's not on now. It's like K-TV, but they talk about AIDS and people and stuff.

Lance: Maybe they're wrong and if you cough and you can get it.

Amy: I don't want to get it if they cough in my face.

Here the children began to move away from their distancing strategies, exposing their own vulnerabilities. The fear inherent in their knowledge came through clearly in this context. Lance began by making an explicit statement of fear – 'If you catch

it, you will die' – and Carol offered a degree of comfort, explaining that 'You can't get affected if you touch a person'. However Lance did not accept her reassurance, pointing out that the presenters on *Blue Couch* might be wrong, and Amy backed him up, entering the conversation with her own fear. The fears were contained, however, by recourse to a discourse of care of the self:

Lance: You might not even get it. How do you know that they have it? I always eat my vegetables. My parents say I must eat healthily. If you don't know and if you go to church, people are sick and if they cough on you and you get it, but if you keep healthy, you won't catch it.

Amy: And we won't catch it by eating vegetables.

Lance: We all know that.

Amy: Keep ourselves clean and we won't get germs.

Lance: We all know that.

Carol: Ja. If they come near me, I'll walk away.

Veronica: They will feel lonely.

Carol: I'll walk away and keep far from them. Not far from my parents. I'll walk further away from others. If they come near me, I'll walk away. I won't say anything to them in case they cough and I'll catch it.

Lance: AIDS, you can hardly stop it. My gardener died of AIDS. We drove him to the hospital and when he went to the hospital, he died. My mom didn't tell me, but the doctor knew. I wasn't anywhere near him.

Amy: Tomatoes keep you away from flu and from bad diseases.

Carol: My cousin almost got blind, but she didn't eat carrots and now she does.

Amy: My dog died of AIDS. We had to put her down. She died of AIDS. So my mom's going to get me a Jack Russell.

In this extract from the conversation with the children, they deployed knowledge about healthy eating and living to defend themselves against the possibility of becoming infected. In their view, if eating vegetables and keeping clean can keep you healthy, the punishment for living in dirty conditions and not eating vegetables is illness, whether HIV or almost going blind. The children take gendered subject positions. Lance constantly adopts the role of a superior knower of the answers – or, at least, when to doubt public authority. He was generally the one who moved

the discussion forward, bringing up new angles. He was the first to respond to the query that began this part of the conversation (he had heard about AIDS on the news), the first to identify the main carriers of AIDS as illiterate, the first to express his own fears and to introduce fresh vegetables as a prophylactic. Towards the end of the conversation, he repeatedly told the girls 'We all know that' when they tried to raise a point and his final contribution established him as the one who knew someone (his family gardener) who had died of AIDS. The only exception to Lance's role as protagonist in this conversation was when, in a concluding move, when the whole conversation became too much for her, Amy shifted attention via her ill dog (which, she claimed, had died of AIDS) to her new Jack Russell, an altogether more comfortable topic of conversation. From this point on, the conversation was primarily about pets.

As can be seen, the children here construct HIV and AIDS as a disease of 'other people'. At Bullwood, AIDS was about danger and about people 'we do not know and in whose lives we are not implicated' (Silin 1995: 24). In this way the distancing from HIV and AIDS worked in ways to reproduce AIDS as a problem for the 'mainly African' Other. As these young children struggled to form meanings about HIV and AIDS, they appropriated and reproduced a widely circulating logic in which HIV and AIDS, race, disease, dirt and poverty were linked and seen as mutually productive of each other.

For the children at KwaDabeka, in contrast, HIV and AIDS were real and something they lived with on a daily basis. The children made clear connections between HIV and AIDS, disease, sex, gender and death. This is readily apparent in the following extract of a conversation between Fana (a girl aged seven or eight), Mandla (a boy of a similar age) and Deevia:

Deevia: What is AIDS?

Fana: It's a disease, an infection. And you get very sick.

Mandla: The lover or the girlfriend, you sleep with them and you get AIDS.

Fana: We get AIDS from boys if we sleep with them. *U hambe ulala nabantu abaningi* [If you sleep around you get it].

Mandla: In Hammarsdale, I saw a funeral of a boy and he was so dark. He had AIDS.

Fana: From rape you get AIDS.

These children knew not only that AIDS is an infection that makes you 'very sick', but they also had a very precise understanding of the sexual transmission of the virus and were personally acquainted with AIDS-related death. It is relevant to notice that the girl, Fana, also made the clear connection between sexual violence and the transmission of HIV. Her comments echo the fears expressed by the older girls at the two secondary schools discussed in Chapters 5 and 7. These children were able to articulate clearly their knowledge of HIV and AIDS, as well as their understandings of the use of condoms to provide protection against both disease and pregnancy:

Mandla: I know that a condom is an elastic balloon that adults use when they have sex.
Deevia: Why do they use a condom?
Mandla: It protects them against AIDS or when the woman does not want to get pregnant. But I only use it to put water inside and play with it.

In contrast to the Bullwood children, who had little or no familiarity with the use of condoms, Mandla not only knew exactly what a condom was, but he also knew how to subvert its sexual association by using it to put water inside and playing with it. The object was not strange to him – he knew both its official and playful potentials and uses. Similarly, the two boys in the following conversation understood the relationship between sex, condoms and HIV and AIDS prevention:

Deevia: What causes people to get AIDS?
Kano: They don't use condoms. They are told to use condoms and they don't use it.
Deevia: What's a condom?
Kano: I saw the condom when I visited neighbours.
Michael: On the radio, they say that we must use a condom. Even if you buy *Ilanga* [Zulu daily newspaper], you can see pictures of condoms. My mother said not to sleep with girls. I'm too young, only when I'm grown up and I must use condoms.

Kano revealed not only that he knew what condoms were for, but also that he was aware of the perennial problem of health promotion professionals, that people

'are told to use condoms and they don't use it'. These children thus saw the use of condoms (in the future) as their key to safety when they are 'grown up' and old enough to 'sleep with girls'. In this context, Michael was very clear that sex should be delayed and that being grown up did not exonerate one from the responsibility for avoiding infection. In contrast, the children at Bullwood never mentioned condoms, but believed that they could and should take care of themselves and their health by eating properly (including vegetables) and keeping clean.

As with the older girls, discussed in Chapters 5 and 7, the younger girls were conscious of particularly female vulnerability. One of the girls said: 'From boys we get AIDS.' Fana was not the only girl who linked HIV infection to sexual violence. Girls' vulnerability was also pointed out by Fezile and Gugu:

Fezile: From rape you get AIDS.
Gugu: AIDS is rape.

Here we see an unequivocal association of HIV and AIDS with rape. Not only do girls and women get AIDS from rape, AIDS *is* rape – a point made in no uncertain terms. In this way, the children's talk at KwaDabeka both brought the genders together in their knowledge of sex, condoms and HIV, and simultaneously threw them apart. Both boys and girls associated each other with AIDS. Boys accused girls of spreading AIDS and vice versa. A number of boys and girls emphasised men's power and branded women – especially infected women – as prostitutes. In the following conversation two girls, Thabi and Pindile, and two boys, Sithole and Sihle, discussed this issue:

Thabi: Boys and men spread AIDS because they sleep around with different women.
Sithole: Women spread AIDS because they sleep around with different men and sell their bodies too and they do not use condoms.
Pindile: Girls spread AIDS because they sleep around. They sleep with different men.
Sihle: Boys spread AIDS because they sleep with different girls, even the ones that sell their bodies, *izifebe* [prostitute or girls who sleep around].

These young girls and boys knew that dangerous sexual behaviour – sleeping around and not using condoms – is associated with HIV risk. The *izifebe* provided Sihle with an evaluative discourse, one which disapproves of girls who sell their bodies. The sexual objectification of women, the degradation of girls as *izifebe*, was a means through which Sihle's asserted his power as girls/women were stigmatised – not only as prostitutes, but also as AIDS carriers.

Yet we do not wish to suggest that young girls are always or inevitably victimised. Epstein et al. argue elsewhere that the changing circumstances in South Africa are producing new narratives 'through which people can tell themselves who they are, and how they can live in the context of HIV' (Epstein et al. 2004: 12). The children at KwaDabeka were willing and able to engage with issues of reckless sexual behaviour in the context of HIV and AIDS:

Pindile: Now, I will not play around with boys I don't know during the day or at night. I will not go to strangers, especially men, when they call me.

Nami: When I'm older, I will use a condom when I have sex with a man because you get protected when you use a condom.

Sandile: No, I will not sleep with girls until I am old enough to take care of myself.

Nkosi: I will not sleep around with girls. I will only sleep with my girlfriend, but will take her to the clinic to check if we both don't have AIDS. I don't want to get AIDS. It kills.

Gugu: When you have AIDS, you get very, very sick and you make your family cry when you are sick and when you die.

As can be seen in this extract, the children at KwaDabeka had developed a discourse of a projected care of the self, based in the everyday presence of HIV and AIDS, sex, rape and death their lives. They 'knew' what not to do and, indeed, what they will do, in their fantasies of their adolescent and adult lives. While we cannot predict their behaviours in the future (any more than they can) what we can say is that in their talk it was evident that they were developing a range of discursive strategies with a vocabulary of care, respect and responsibility, which may enable them to perform gender and sexuality differently in future.

Conclusion

In this chapter we have demonstrated that young children's discursive frameworks for understanding HIV and AIDS, gender and sexuality are marked by class, race and gender. While all the children interviewed were aware of the HIV and AIDS epidemic and connected it to sex and sexuality, the affluent children from Bullwood adopted a distancing strategy. Their fears and concerns were contained and controlled through recourse to invoking HIV and AIDS as a disease of the Other, usually an African (black) person living in poverty. Their most typical resource for making meaning was the television and the only direct contact any of them talked about was when Lance said that his family's gardener had been taken to hospital by his mother and had died and Amy said that her dog had also died of AIDS. Significantly, Lance was the most fearful of the children in this context.

In contrast, the very much poorer children at KwaDabeka, all of them black, drew on much more personal knowledge of the disease. Not only did they know how it was transmitted through sexual activity, but they also had a very clear idea of its impact. While not deploying a distancing strategy in relation to the Other, they did distance themselves from illness by projecting the possibility of becoming infected into the future and inscribing themselves in discourses of safer sex through the use of condoms.

In both schools discussions of HIV and AIDS and sexuality were associated with gendered identities, but the forms were very different. At Bullwood one of the boys took on the role of the knowledgeable authority, whereas at KwaDabeka the children's constructions of gender emerged through discourses of male violence, including rape. For these children living in an informal settlement with the effects of HIV and AIDS in their everyday lives, sexuality could be pleasurable – a source and resource for play – but it was also strongly overlaid with narratives of danger and distress. In contrast, the children from Bullwood may have been aware of sexual dangers, but these took second place to a pleasurable exploration of heterosexual possibilities, particularly among the girls who positioned themselves as heterosexually desirable and desiring in entirely feminised ways.

This raises a series of theoretical, practical and political questions. While this chapter is in line with the 'new childhood studies' literature (for example, Alanen and Mayall 2001; James, Jenks and Prout 1998; James and Prout 1995; 1997; Renold 2005) in emphasising children's capacities to be actors in their own lives, it also

raises questions about the limits of such agency. If, as we have shown, children make themselves in conditions not of their own choosing, how far are the affluent children, distant and distancing themselves from the epidemic, able to constitute their identities in ways that might help them protect themselves as they become sexually active? On the other hand, the children from KwaDabeka are unable to escape the daily grind of poverty simply by choosing to do so. Their paths and possibilities are much more complex than that. These impoverished children show themselves to be resourceful, knowledgeable and confident navigators of what many would regard as unchildlike contexts. Nevertheless, they constantly reinscribe themselves within the very dangers that make them vulnerable. The theoretical and practical question arising from this is: What resources would enable them to make different choices and produce themselves differently, as less vulnerable and endangered sexually and in terms of their health?

In practical and political terms, our evidence has serious implications for HIV and AIDS education in the South African context. As can be seen from the quotations at the beginning of this chapter, Mrs Hobbs, echoing her colleague, Mrs Burke, saw no necessity for engaging with the children on matters of sex, sexuality or even HIV and AIDS and displaced the responsibility for this on to parents. Given that rates of infection are generally high and are increasing among young people in the affluent middle classes, the urgency of changing this perception cannot be underestimated. It is clear that, as in other countries, parents do not, in practice, take on the sex education role that Mrs Hobbs expects them to. The children at Bullwood had no clear knowledge about HIV, the vectors of transmission or sexual practices. This must be a matter of concern for those responsible for health promotion and the development of policy and practice in the area of sexuality, gender and HIV and AIDS education.

Our findings also have a bearing on what kinds of actions to promote equality and address the spread of HIV and AIDS might be appropriate in working with children at schools such as Bullwood. The assumptions of the teachers would seem to indicate that they may not even go so far as to make any interventions in this regard. However, the understandings of the children reveal the need for actions that take into account context and recognise the knowledge and understandings of the children themselves. The children, especially at KwaDabeka, are skilled and active knowers. They reveal themselves as able to reflect in impressive ways,

not only on the disease, but also on relations of sexuality and, by implication, gender. They are, then, ready to participate in interactions as active agents. In the next chapters we consider whether conditions formed by the institutions of the schools and the identities of teachers provided opportunities for children to expand interactions.

Notes
1. In many cultures and countries the notions of youth and virginity are seen as very alluring. In the United Kingdom, for example, soft pornographic imagery often shows young women dressed as schoolgirls in the now rather old-fashioned but still iconic school uniform, the gymslip, thus offering connotations about schoolgirl sexual attractiveness.
2. Quoting from this interview may appear to be an ethical breach as Deevia promised not to 'tell'. However, the context was that Cassy didn't want anyone in her school (teachers, other children in the school and parents) to know. We have chosen to present the passage because we have already provided the disguise of pseudonyms to both the school and the children and this means that Cassy's concern about having her disclosures revealed will not be betrayed.

Teacher Masculinities and Gendered Pedagogy

Introduction

As school communities in South Africa responded to the coming of HIV and AIDS, the gendered attitudes and practices of male teachers changed, with some significant consequences for their response to gender equality policy. This chapter focuses on the views of male secondary school teachers in three Durban schools and presents brief life histories of eight teachers, paying attention to the gendered arrangements of their lives. We then analyse their attitudes towards gender equality, locating these within the gendered regimes of three quite different schools: a township school, a school formerly catering only for Coloured learners and a single-sex, sports-focused, Model C (elite) boys' school. In the final section, their views on HIV/AIDS, sexuality and caring are explored.

The ways in which teachers experience gender, respond to the policy imperative of gender equality and relate to the AIDS pandemic are not uniform. The different school contexts, as well as their life histories, exercise a profound influence on their views. There is no neat or snug fit between race, class, school type and the teachers' views about gender equality and their own gender practices. A wide range of opinions was expressed by the teachers so that, for example, among the three African teachers, we found strong support for gender equality, as well as antipathy.

Not surprisingly the nature of each school was very important in shaping teacher views. The middle-class, boys' school, like Bullwood primary school, was not faced with the pressure to accommodate or advance boys academically, nor did it have much exposure (through its learners and parents) to the AIDS pandemic. The coeducational township school, on the other hand, had sexual harassment,

high rates of pregnancy, HIV infections and AIDS deaths, resource shortages and high levels of poverty to contend with, all of which have implications for gender relations. The third formerly Coloured school lies between these poles. It has an increasing number of African learners, but is relatively well resourced and poverty and AIDS encroach in only a limited way into the affairs of the school.

There are contradictory assumptions in the literature about the role that male teachers have in the achievement of gender equality. On the one hand, they are often regarded as an obstacle to gender equality and the reason for the existence of gender inequities. This idea is strongest in the literature that identifies high levels of teacher abuse and sexual violence, particularly towards students (for example, Leach and Mitchell 2006). One of the most alarming features of teacher attitudes to gender equality is the apparently widespread nature of sexual harassment by teachers of female students, especially in coeducational township schools. Many teachers in township and rural schools historically have had 'love affairs' with students. While these are now prohibited, they appear still to be commonplace. On the other hand, there is a recognition born of the literature on changing masculinities and the involvement of men in work for gender equality that male teachers are critical of the process of gender transformation (Morrell 2007a; Morrell and Epstein 2008; Walker 2005).

The achievement of gender equality in South African schools depends a great deal on the values and practices of teachers, male and female. The secondary literature on teachers identifies a number of important trends, including the division of the teacher corps along a number of lines. Racial differences remain, although the different terms and conditions of employment that characterised apartheid education have now been removed. Racial divisions have been expressed in forms of political organisation and approaches to teaching. Teacher organisations that historically organised along political lines and aligned themselves to South Africa's anti-apartheid and trade union movement, for example, the South African Democratic Teachers' Union (SADTU), emphasised the role of teachers as workers. Historically, SADTU drew on African teachers for the bulk of their support. Rival organisations stressing the professional and apolitical nature of teaching, for example, the National Professional Teachers' Organisation of South Africa (NAPTOSA) drew historically from staff in white, Coloured and Indian schools where the issues of poor resources and highly politicised student bodies were not pressing.

Since the 1990s all teacher organisations have thrown their weight behind official commitments to gender equality. In reality, however, the organised teacher bodies have made only a limited contribution to gender equality. They have supported equal pay and conditions of service for teachers and the advancement of women into management posts in which they were under-represented. Their voice has been weakest in relation to curriculum change and gender equality within schools.

Using the material from Robert Morrell's life history interviews,[1] we begin by presenting pen portraits of eight teachers. Although each of the teachers has a life trajectory quite specific and indeed unique to him, we present a selection of biographies in order to suggest similarities between their testimonies. These similarities reflect the importance of race and class in shaping, but not determining, gender identities and relationships and allow us also to show how specific school contexts influence gender relations. The eight teachers we discuss, therefore include two white teachers (Matthew and Peter) from Oak High, two Coloured teachers (Johnnie and Fulton) and an Indian teacher (Ramesh) from Gladstone Secondary and three African teachers (Bongani, Sandile and Muzi) from Dingiswayo Secondary.

The teachers

Peter is the first-team rugby coach at Oak High, in his mid-thirties, an athletic white man. He is married to a teacher and they have two young sons. He was born in Portugal and came to South Africa when he was ten years old and spoke no English. His father lost his business when Frelimo took power in Mozambique in 1974 and then established himself in Durban, working with shipping electronic systems. These experiences made his extended family suspicious of African governments and this is probably why most of them returned to Portugal after 1994. He attended all-white schools in Durban, matriculating at a local technical, single-sex school. The school was strict and Peter was often caned for trivial offences. He then completed a diploma at a local teachers' training college. Like most white South Africans of his generation, he did a year of national service in the army before beginning his teaching career. Peter played rugby and cricket at school, but he excelled at chess. After school he played club rugby, squash and did cross-country and marathon running. His mother was the dominant force in his

life, although his father was also supportive. Family provides Peter with stability in his life. He has a patriarchal idea of the family, although his wife also contributes to the family finances and they make decisions together.

He explains that his position at Oak High, where he is head of the boarding establishment, is because he is used to hard work and fits in, adding, 'but I'm also a very competitive person, so I think I got into the swing of things quite quickly'. Peter favours single-sex schools for boys because they are competitive. His view is that coeducational schools make boys 'more relaxed'. He is aware that boys' schools can exclude on grounds of race, class and ability, but says Oak High promotes 'tolerance for each other, whether it be cultural or racial or whatever'. One of the major mechanisms used by Oak High to integrate boys is sport: 'It certainly brings chaps together. Team sport teaches you to rely on your partner and rely on the person in your team. It teaches you to lift each other up when you know you've had a loss and so on. Basically a reflection of life.'

Matthew, in his early sixties, is beyond the official retirement age. He teaches at Oak High, where he is a legend, having taught there for 30 years. Matthew grew up in Durban with his parents and four sisters. On the surface, they were a study of middle-class white respectability. His father was a lawyer and the main breadwinner – a veritable patriarch, admired in white middle-class circles, but a bully at home towards both Matthew and his mother. His behaviour 'verged on the autocratic or almost despotic'.

Matthew had trouble dealing with his father. His father saw virtue in achievement and public recognition and Matthew seems to have tried to satisfy his father in this regard. He excelled at sport, particularly rugby, which made the most demands of a physical masculinity and gave the greatest recognition, captaining rugby teams all the way through school and university and playing representative provincial rugby. Here he brought glory to the family name, but he also rebelled.

> I revolted against, I mean, I got into the university thing and I became a little bit of a rebel, you know the normal . . . through those years of the hippies and that sort of stuff. And I was probably the first in fact I think when they described me as I ran onto the rugger field at Kings Park as lacking sartorial elegance. I was the first rugger player from varsity to run onto that field with long hair.

Despite his talents, Matthew never neatly fitted the standards of the day. He was left-leaning politically and was never promoted up the teacher hierarchy. Although never actively involved 'in politics', he has a robust anti-racist view and a desire to promote friendship and communication across the racial divide. He was often at odds with headmasters at the school whom he described as 'terrible bullies', finding them unfeeling, cruel and competitive. Despite his investment in sport, he was highly critical of physical cruelty (corporal punishment) and authoritarianism. He critiqued the masculinity of these white male teachers: 'The male, you had to be strong and you had to, you know, bump people out of the way and assert yourself and drink your beer fast and this sort of rubbish. And here they were in these positions of authority . . . that I couldn't handle.'

In some respects, Matthew is an archetypal white, middle-class man. Sport is important in his life. He occupies space in a confident way, comfortable in his body, sure of his views. He is heterosexual. He has children, although he is no longer married. However, it is not easy to read a hegemonic masculinity into or off Matthew's own accounts of his manhood. The contradictions are strong and conformity with conventional white, English-speaking, middle-class values is marginal. In fact, Matthew's biography is another example of what Nikki Wedgwood (2005) identifies in an Australian context as both contributing to and deconstructing hegemonic versions of masculinity and his life has generated tensions that may be generative of or contributory to shifts in gender relationships and identities.

Fulton has worked at Gladstone Secondary for nearly twenty years. He is 40 years old and is unmarried. He is Durban born and bred. He attended high school in a rural boarding school, went to the local university and graduated with a degree and a teacher's diploma. He teaches history to the senior students of the school, but also has counselling duties. He lives in a tight-knit family and sees his parents and siblings daily. His father was a painter, but no longer works. His mother was a self-employed dressmaker. The family all support one another financially. He describes his family of origin as lower middle class (because the family owned its own property) but there was never any excess cash and his horizons as a Coloured person were limited by apartheid: 'I could only go up to a certain level and could never progress beyond that'. The Group Areas Act meant that he associated largely with Coloureds and had a few Indian friends, but no black friends. He was drawn

to Marxist thinking at university, but apart from a few United Democratic Front (UDF) marches, he did not get involved in politics. Fulton's school experiences were of authoritarian practices, little racial mixing and sport – soccer in his school, although he was also a competitive gymnast.

Johnnie is in his early thirties. He teaches history and maths to the senior classes at Gladstone Secondary. As a young child he and his family, who were classified as Coloured, were forced by the Group Areas Act to move to an area reserved exclusively for this racial group. After finishing school he attended the 'Coloured' teachers' training college, but finished his studies at the 'white' college as apartheid began to crumble and his college was closed. He has taught for nine years at Gladstone. His father was a welder, earning enough to permit his wife to stay at home. Johnnie thinks of emigrating and some of his siblings are already in Australia. He thinks South Africa has deteriorated since the end of apartheid. Two of his three brothers also trained to be teachers, but both now work for the petroleum industry where the opportunities and pay are much better. His family was tight-knit and ties were strengthened by membership of the Roman Catholic Church. He and his cousins spent a lot of time together. His father had a permanent job and there were never money problems. His parents paid for his studies and supported him until he got a job. His parents were strict and moralistic. Talking about having girlfriends in his early teenage years, Johnnie says his parents prohibited it: 'They would probably have given me a good hiding. [They] . . . said this is not the time for girlfriends, wait until you are older.' Johnnie played sport, although he describes himself as 'not very sporty'. Baseball was his favourite and, during his school years, he played for a club, but not with great intensity or success. Johnnie doesn't speak isiZulu and has no African or white friends. His world is ethnically bounded and held by a strong sense of family.

Ramesh was born in Durban in 1959, the fifth of seven children. His mother was from Johannesburg and his father from Durban and his parents moved around a lot because of his father's job as a truck/bus driver. His mother was a housewife, who, presumably, kept on making a new home in each new place. Ramesh had numerous relatives on both sides, but seldom sees those on his mother's side because they mostly live in Johannesburg. He says that most of his ten uncles on his father's side have died ('most of them are late now').

Although Ramesh grew up in a large family, this has not been replicated in his generation. He and his siblings have two children each, but they still keep in close

touch, moving between the families every couple of weeks and sharing meals to which all contribute. Family events are also marked by religious (Hindu) observance. They also have extended family outings to the beach. He is very close to his own children, a daughter aged 21 and a son aged 17, who both live at home.

Ramesh played professional soccer until he had a leg injury, representing his school and province in 1974. As a child, he played every day before and after school and at break and would then go to the 'association ground', where he and other talented youngsters were coached. All Ramesh says about football is redolent of enjoyment and he has no regrets that his soccer impacted adversely on his examination results, as: 'I really didn't want to take a job that took me away from my family. I had to do something that I would be always with my kids. You know, be there with them especially academically. That's why I stayed in education. That relationship that I missed in my life, you know, make sure my kids get it.'

Ramesh has career ambitions, but would miss being with the pupils: 'I would like to [get promotion], but my real love is with kids now. When you go higher up, you'll find you have fewer classes to teach. I don't really want to get to a managerial position, you know. I like to be more interactive with students.'

In the past it has been difficult to survive on a teacher's salary, but now his wife has paid employment and things are financially better. Of more concern to him is his perception that teachers have lost status: 'When I was at school the reality is, if you were a teacher you were somebody', but now there is little respect from young people for teachers.

Ramesh is not particularly politically aware. Although his life as a child was affected by apartheid – 'you know we weren't allowed to own our own homes' – he did not feel particularly disadvantaged. He is not interested in teacher or national politics.

Sandile was born in an African area of Pietermaritzburg in 1962. His mother was a trained teacher but his father, a factory worker, preferred her to stay at home and look after the children. His father was a worker who worked for a big manufacturing company in Durban. He was 45 years old when Sandile was born. Family was important from a very early stage in Sandile's life. When he was four, the family moved to live with his maternal grandmother in a Durban township and when she died, his mother's brother told them to leave the home (he had inherited it) and so Sandile's father built his own house in Umlazi.

Sandile's family was not wealthy, but nor were they poor. His father paid the education costs of all his children and 'was a great inspiration to me'. Sandile matriculated and attended a teaching training college where he obtained a teaching certificate. This was the major route for Africans wanting to enter teaching, but generally involved only rudimentary training. He only recently upgraded this qualification to a diploma through part-time university study. Sandile's father was deeply involved in trade union activity and was also, illegally, an ANC member. He was detained in 1976 in the wake of the Soweto student rebellion.

Sandile followed in his father's footsteps and was involved in student politics at school, especially in the school boycotts of 1980. He was a member of the UDF and SADTU. He was an activist, but managed to avoid arrest. Although Sandile has a respectable job, he is not wealthy. His wife is a nurse (a poorly paid profession in South Africa). He does not own a car. He has two children and has adopted his deceased sister-in-law's child. He likes living in the new South Africa. He identifies with its values, but also appreciates the benefits it now provides, such as subsidising his recent university studies. He hopes that his eldest daughter will go to the local university and train to become a psychologist.

Muzi teaches at Dingiswayo. He was born in 1957, the middle child of seven children, growing up in a township area outside Johannesburg. His father was a migrant, born in Mozambique who managed to establish himself on the Reef by securing a job in the manufacturing industry. Here his father prospered and, unusually, was taken by his white employer on a trip to Britain. Muzi's father earned a reasonable salary and the family had a car when he was a child and was the first in the community to have a television.

Muzi attended a college of education and qualified in 1976, just as the school boycott, which he did not support, was beginning. He says his parents were 'very strict' and that the thing he liked about his father was that 'first and foremost, he was a disciplinarian', who, 'whenever he had a problem with you, he would call you to the bedroom to give you some beatings in the bedroom'. Muzi had many domestic responsibilities as a child and was, therefore, 'too busy in the house', but managed to find time to work hard and play softball for the school and, later, college teams. Simultaneously he developed an active heterosexual social life, becoming interested in girls in the first form of secondary school, which he describes as usual in his peer group, though: 'I was scared of my parents, that once they discovered I was having a date with a girl, they'd kill me. That's why I think it

took me such a long time.' He learnt about sex and sexuality by listening to his older brother and friends and some adults before trying out various sexualised activities: 'We had some adults, like people whom you used to stay and talk with. Then you would go and try and exercise what you have heard from them . . . what's kissing? How do you kiss? Then they'd tell you, and then you would go and try it.'

Despite always having several girlfriends, Muzi felt that the girls were 'more advanced than us' and knew what to do in sexual relationships. He felt himself to be in heterosexual competition with other boys since the boys who were successful heterosexually became a 'sort of hero' in the peer group. The outcome of this approach to heterosexual relationships is that Muzi has fourteen children, none of them with his wife, whom he met when he started teaching at Dingiswayo in 1992. Her first husband died in 1998, but it is not clear how long after that they were married.

Of his fourteen children, Muzi has a relationship with three. His oldest was born in 1975, when he was eighteen, the second one a year later and the third nine years after that. He sees these (adult) children and pays maintenance for his youngest child, now two years old. His father supported the older children while Muzi was studying.

Muzi says he loves teaching and wants to stay in the classroom. Initially, he was motivated by money but, 'became so interested in teaching'. Muzi is not interested in politics or active in the teacher unions, holding a generational and traditionalist commitment to the existing order and he expresses distaste for the student rebellion of 1976 when 'children were ruling adults'. He is not particularly enthusiastic about the new South Africa and says that getting the vote has made little difference.

Bongani was born in 1962, one of six children, and grew up in Clermont, one of Durban's older townships (created in terms of the 1923 Native Urban Areas Act). Unlike Umlazi and KwaMashu, Clermont was close-knit, small and relatively close to the city. His mother was a nurse and his father 'was a manager with the now Durban Transport'. The family moved to Umlazi when Bongani was still a child. His mother was a strong presence in his life. 'She used to tell us that my father did not want her to go to work. He wanted her to be with us to look after us and now and again she would go back to work.'

Bongani says his paternal grandfather worked on the gold mines and accumulated enough to buy 31 acres of land in Inanda in 1908 (before the Natives

Land Act of 1913 made African land purchase illegal).[2] Bongani's father worked the land (sugarcane) and became a businessman. He was a polygamist (he had two wives) and lost a lot of money after being attacked and nearly killed in what might have been a robbery or attempted contract killing. Bongani has inherited the land, but is confronted by landless people's attempts to occupy and seize the land. He expresses acute frustration about this. Bongani's mother's father was a chief from Hammarsdale. Bongani is proud of his family, noting their achievements, status and wealth.

One of Bongani's brothers has died: 'I can say one of the tragedies in this country [is] to die from HIV/AIDS. We have the highest death rates. He's one of the statistics.' Bongani laments the behaviour of his brother – he had many girlfriends, two children out of wedlock and mistreated his wife: 'He was treating her, his wife, not in the way he was brought up to treat our women'.

After Bongani matriculated, he worked as a labourer because his mother and father were in a state of dispute and his father was giving his mother no money. With his savings he paid for a teacher's training course at Mpumalanga College in the mid-1980s. He was then placed in various schools and was promoted to head of department and moved to Dingiswayo. He continued to further his studies, attending courses at Leeds University and obtaining an Honour's degree in science at the local University of Natal. At the time of the interview he was registered for a Master's degree.

Bongani remembers apartheid as a bad system that caused his father to be arrested for all sorts of reasons, including not paying tax and not having his papers. He was aware of the violence in the area throughout the 1980s, but says: 'I think I tried to keep my head down, because one thing that my father told us was don't get into politics'. He describes himself as: 'politically aware, not someone involved in politics'.

Gender in the teachers' lives, in schools and in the context of HIV and AIDS
Descriptions of South African men frequently feature oppressive, violent or misogynistic images (Morrell 2001a) and teachers are no exception. Rape studies in South Africa identify them as a significant perpetrating group (Jewkes and Abrahams 2000; Jewkes et al. 2002). Furthermore, various studies around the world testify to the involvement of teachers in sexual harassment and corporal punishment

of learners (Hallam 1994; Holdstock 1990; Morrell 2001a; Niehaus 2000). However, there is not much work specifically on male teachers and their gendered practices. As we saw above, the domestic living arrangements of the teachers in this chapter include the full range from living alone, to living with a woman, to being divorced (though none living with another man). There are significant racial and class differences. African men are much more likely to have had children before they were married, although divorce is more common among white men. Indian and Coloured men are more likely to be meshed into residential extended families. However, formal living arrangements can only tell us so much about gendered life, so what follows is an analysis of the gendered content of the lives of our teachers.

Matthew, the oldest, is divorced. He talks with insight and affection about his ex-wife and children:

> So James [his son] has stood up like my wife – very strong woman, far brighter than me, which is not difficult, but far brighter than I am. My father bullied my mother into submission. Marvellous lady, who ended up sort of depressed and unhappy and just existing. Gloria, my wife, was far too strong for me, and I am so glad that I met her and married her, because she has said, hey none of that. And has really sorted me in that way and made me properly balanced I think. And my daughters are the same.

Peter is married to a teacher and they have two young children, a boy and girl. His own father was in charge of the family, the only breadwinner and made the decisions. Peter's understanding of gender dynamics in the family echoes his upbringing, but his wife is also a breadwinner and there is much more equality in his domestic environment. His immediate family is close-knit, possibly because most of the extended family now lives in Portugal.

Fulton and Johnnie, the Gladstone teachers, are both very close to their families. Many of Johnnie's family have emigrated, but he describes life very much through the lens of extended family membership. Fulton is the same. Membership bestows obligations of care for the parents, but it also serves as a security blanket if things go wrong. And for both of these men, the family is very much a patriarchal institution in which men make decisions, although women are recognised as important and are expected to participate in deliberation.

157

Ramesh grew up in a big, extended family and he still talks as if this is the case, even though many of the family have died and his wife has only one sibling. Ramesh shares domestic work with his wife and makes decisions together with her: 'We normally discuss them. I discuss it with my wife.' Ramesh is not sympathetic to feminism or the goal of gender equality: 'Women feel they get a raw deal, which I don't see how they get a raw deal'. He continues: 'The only women I have had relations with is my wife, there's my mother and my daughter. And they get a fair deal even when it comes from commands and demands and instructions.' Ramesh believes that when men take decisions in the interests of family, they do so fairly and there is nothing wrong with this. He cannot see the logic of introducing gender equality. He doubts that sexual harassment is a problem and says that females 'often use it as a form of getting back at the males'.

Muzi is married, but it is not his wife or his family that sustains him. This support is provided by his fellow male teachers. He drinks with them each day after school, returning home in the evening. His interview shed little light on the quality of his relationship with his wife. He described their relationship in terms of the division of labour:

> Almost every day then my wife will go to the toilet and put water for us to bath. We bath together. Then I would do the bedding. That is putting back everything – the pillows and things. Then I wash, I do wash my underwears and my socks/ stockings together with my handkerchief. I wash those almost every day. And my wife washes hers too. Then go back, dress and iron. I do iron for both of us because she can't iron. She's a bad ironer. She puts many creases in it when she irons. Then she goes. I like porridge in the morning: white mielie-meal[3] porridge or sour porridge or cornflakes. When she's busy in the kitchen, I'm busy ironing. After that ironing, we put on our clothes. Then we go and eat. That is our breakfast, and we go to the car.

Muzi's involvement in domestic chores is unusual for his generation, in which many African men expect to be waited upon. But these routines are not rooted in an egalitarian view of gender roles. Muzi's sexual politics are conservative. Women, he believes, should not be 'the same than a man or equal' and he is uncomfortable with the idea that 'our wives are trying to be the heads of our homes.' He says that, while the boys he teaches are growing up in a new egalitarian environment,

nevertheless they 'are not that enlightened', think that they are superior to girls and act 'bullish together'. Muzi generally has patriarchal, traditionalist views. He regards it as a traditional practice for (Zulu) men to have sexual relations with many women. He muses about marital fidelity: 'I am not sure. I'm not sure. But even being faithful, even to us, because we are from this culture our grandparents had two wives, three wives you see. It's hard to accept it.'

Sandile is married and lives with his wife, their children and his wife's sister's daughter whom they took over when she died. Sandile is a democrat and believes in equality. He shares the household chores with his wife, but bemoans the fact that

> before we got married, she really looked at herself as inferior to me. I don't know if it was a government influence or it was a male-dominated one, which made her to look like that. But I tried to be on her side. With the legislation now in place, the women are very much empowered to address their concerns with regards to . . . being ignored when it comes to implementing certain things. So, my relationship with my wife is that of a wife and a husband, but if there are things that need to be discussed, we both are equal in those matters.

Sandile does not believe in corporal punishment and has never beaten his children. He believes that respect is the key to human relations.

Teachers and gender equality

The nature of gender relations in schools differs markedly – some schools are strongly committed to gender equality, others are not; some schools still have an authoritarian, even gender-toxic (Kenway and Fitzclarence 1997) teaching style, while in all schools there are varying degrees of support for gender equality and different gendered personal styles.

As we showed in Chapters 5 and 7, in some schools, such as Dingiswayo, the gender regime is still emphatically patriarchal. Sandile believes that this is because of teacher attitudes and behaviours:

> It is the teachers at school (that) need to be empowered. I don't know whether by the government or what, to address this issue of gender equity. There should be workshops. Because they are the ones who should teach the children about gender

equity and so on, *because that thing is still with them – that boys are superior than the girls*. I don't think it has been erased so far. I think we should inculcate in them that gender equity. [emphasis added]

However, the current state of affairs is being addressed. Sandile provides an example of what has happened at Dingiswayo:

There was a teacher who was holding prayer, so-called prayer. It was Women's Day on Saturday, so the whole of the school, she has been trying to, she has been inviting very successful women, and calling on children who are a bit [more] intelligent than others – women, girl children – to address us about the issue of race and the importance of women in our country. So, I think gradually, I think by the year 2006, 2007, we won't be having so much problems when it comes to gender issues among the boys and girls at this school.

'Why has gender equity not been embraced?' Robert Morrell asked Sandile, 'Why is there still so much male violence in the school?'

Well, in fact maybe those are men, maybe of my age and even a bit younger. During our growing times here, it was only the survival of the fittest. We couldn't go from one section to the other without being harassed in any way or the other. So the only way we thought was to be a man was to fight for yourself. That has to be. I don't know how it can be addressed, but we usually knew that if you are a man, you are a good fighter, an aggressive soccer player and you were not afraid of anybody. And we used to even to ridicule each other if you had a girlfriend, and you haven't hit her even once. So it was that bringing up – that wrong upbringing and it is shared amongst many of my peers even now.

Despite this situation, gender equality is official department policy and enshrined as a core value in the Constitution. In schools, learners are well aware of this, although girls favour gender equality more strongly than do boys (Morrell 2007a). Among the teachers interviewed, not one spoke against the principle of gender equality as enshrined in the Constitution, although some doubted its relevance in the school context.

In the sports-focused, boys-only Oak High that does not have to deal with the most obvious challenges of gender equality (promoting the interests of girl learners), Peter was cautiously in favour of gender equality. He spoke about the promotion of women into senior school posts: 'If they are as good as a male, they deserve to have the same post. I don't think a person should get a job or whatever just because they are a woman or whatever, you know, they've got the same right as any person to any job, to anything. Really that's not an issue.'

At Gladstone Secondary, Johnnie, who is conservative in his political attitudes, argued: 'It's only fair. And at this school we allow for equal opportunities. We have females that do Physics and males that do Physics. We don't offer subjects specifically for males or females.'

At Dingiswayo, Sandile, the teacher with a past involving anti-apartheid activism, was strongly in favour of promoting gender equity:

It was long overdue, because you find that you are with the women in the same classes, and you are clever in the same way. You are capable, not clever, in the same manner. But you will find that when it comes to senior positions at the industries, at the hospitals, at schools, you find that it was male dominated. Not because males were far more intelligent than . . . the female counterparts, but it was because of this male chauvinism that men are greater that women. I find it very interesting that women have been elevated to senior positions in every sphere of life and it doesn't affect me at all. In fact, I think it more needs to be done. It hasn't been fulfilled yet, but more needs to be done to ensure that, because women are great leaders, they are great leaders; I agree in total that they are great leaders as long as they have the potential to do so. I do not mean all of them, but though there are those and if a woman is given a chance to be a leader – a leadership chance – then she accepts. So I think it is very good. It was a positive step by the government to implement.

Bongani was not quite so effusive in his support, although he spoke strongly about gender equality in sport. Girls, he said, 'should be involved in all sporting codes on an equal footing or on an equal basis. Whether it is rugby, whether it's soccer, whether it's cricket, because sport now is open to all. There is no sport which is meant for ladies only, or meant for boys or girls and so on.'

Similarly, he supports gender equality as it pertains to his female colleagues, but only if the women are competent: 'As long as they are able to do what they are supposed to do'. However, as we show below, while supporting equal participation, he believes that women are abusing gender equality and pushing forward their interests too aggressively.

While teachers like Sandile support gender equality wholeheartedly, some believe that it has contributed to problems in schools and has generated new tensions. At Gladstone Secondary, which has had an influx of African learners since the early 1990s, Fulton and Johnnie believe that gender equality and the general emphasis on human rights in the school has led to discipline problems. Fulton commented on a lack of control and a loosening of power, but made his observation carefully:

> It makes me feel old when I think about it. It's like, there's almost a generation gap that exists already. The kids that I have now tend to be less well behaved, I don't know what you'd call well behaved though, less well behaved in the sense that you know you really have to . . . don't expect them to greet you for one, [laughs]. You don't expect them to stand up for you, whereas all of this was expected in the past . . . but sometimes that's nothing to do with the way the school is run, or the way the kids behave, it has something to do with the teacher themselves. But there just seems to be, you know the kids have a lot more freedom, they can express themselves a lot more. They have access to things like learner representative councils, which they're now beginning to realise can work for them and they're making use of it.

Another factor that Fulton believes has contributed to the new situation is the female principal. She is less authoritarian than previous (male) principals: 'There seems to be less control over the student value, I'm talking about pupils with female principals. I don't know if it has some sort of racial implication attached to it, but from what I've gathered African kids tend in my experience to have less sort of respect for a female who's in charge.'

Johnnie is blunter, commenting both on the situation in the school and the country: 'There is too much freedom amongst the blacks. They feel they can do anything that they want to do and nothing will happen to them. But when the

white man was ruling, they were a bit scared. There was more discipline in our country, now there is no discipline in our country at all.'

At Dingiswayo, Bongani believes that women are using gender equality unfairly, saying that 'women [are] becoming more aware of their power in terms of holding men in check'. Bongani did not say that he himself has been a 'victim of gender equity'. Rather he had 'heard stories from men who have been abused by women on the basis of this gender equity'.

> The thing is there is this connotation among women to get even with men . . . There are instances where men, even here among my colleagues, they experience where the women will do something unthinkable. Like for example, say we are doing this and that, and you assume to be a very gentle person, they can even smack you and get physical with you. Not all men are men in a sense. There are women who physically abuse their men.

This sounds far-fetched, a rather fanciful attempt to discredit the goals of gender equality, which in Bongani's own life is not a goal or an underpinning value.

Sandile made a more telling point about male backlash, one of the effects of and responses to gender equality in schools: 'Some men used to be very powerful, and when they see now that that power which they had over women in the past is slowly declining, they decide [to act]'. He goes on to link this to increasing rates of suicide and family murder among Africans: 'Most men decide that I should kill this person [the women who is no longer heeding male authority], and thereafter commit suicide, or maybe kill the children, and thereafter commit suicide.'

Teachers and AIDS

One respect in which the AIDS epidemic has impacted on teachers is that they have, perforce, become very engaged in caring work, particularly in the township schools most affected (see Bhana et al. 2006). The caring work of female teachers and how it is involved in femininity is discussed in the next chapter. What we want to note here is that it is not only those in township schools who participate in pastoral care and behave in empathetic ways towards learners and neither is it only women who do so. There have always been both male and female teachers who care deeply for their students and put a great deal of effort into pastoral care.

Equally there have been those who do not. What we found in focus groups with Life Orientation teachers, however, is that there is more space opening up for men to care in the context of HIV and AIDS.

Matthew is an example of a very empathetic teacher who works in a school largely unaffected by AIDS. He tries to develop relationships with the boys that are respectful, but which allow them to express their concerns and their personalities. He is opposed to authoritarian approaches and the resulting distance that develops between teacher and pupil. He explains his attitude to teaching by describing much high-school teaching as 'bullying', 'or lack of concern about how pupils feel or young men or young boys feel'. He adds: 'Intimidation and dominating a person through insults and that sort of thing' was the norm when he was at school and he committed himself to 'never be a teacher like that or a parent like that, whether or not I have I don't know'. He criticises teachers and pupils for being 'tight . . . because they are too scared to be, let the feminine side express itself. I keep saying to them: people, you have a feminine side. My rugger players, you have a feminine side, express it, nothing wrong with that. "Sir, they'll think we're odd, queer, homosexuals." So, what's wrong . . . express yourself.'

By sharp contrast, Muzi and Ramesh, both in schools where AIDS is more common, are comfortable with formal distance between themselves and pupils. They are not particularly interested in the lives of the boys and focus more on the transmission of knowledge. They talk of pupils in abstract terms, 'othering' them. Muzi believes that pupils lack direction:

And that thing is caused by . . . their parents are illiterate; most of the parents are illiterate. And these kids come from these squatter camps. They don't have something like what can I call it [pause] . . . they don't have a model of some kind whatever . . . they don't have something to look upon . . . I want to be this, you see. Because what they see here is not something you know.

And for Ramesh, what is important is discipline and obedience: 'They must be obedient, follow instructions'. If there is a transgression, Ramesh is inclined to 'reprimand' pupils. After this, he will 'talk to them': 'You know the role of drugs and alcohol. I'd speak to them about it . . . use my scientific knowledge, explain to them the nature of these things, what it does to you. It can damage your cells.'

His view is that the relationship between teachers and pupils is 'too casual. Kids don't have the real respect for the teachers now. They just talk to them anyhow. You know, shout your name outside school.'

For Fulton, the epidemic has led to some limited involvement in an AIDS support group. He comments that he has 'been involved with this type of stuff, although not recently' and was more involved when he did more counselling and taught 'guidance lessons'. Although he believes that some of the learners at his school are HIV-positive, he is not sure who they are. Yet he is very conscious that 'a lot of them are affected by their parents having AIDS and having lost a parent, you know, to AIDS recently' and this is on his radar as far as pastoral care and counselling are concerned.

All these teachers were aware of HIV and AIDS as a general issue, although like Fulton, they did not necessarily know how many or which (if any) learners at their schools were sexually active and/or HIV-positive. Peter commented that at his school 'it's difficult to know . . . because people [hide] that kind of information'. However, he was aware that some boys in his school were sexually active and that there were two whose girlfriends had become pregnant in the past year. But for him this was a minor issue in comparison to his previous (coeducational) school where 'there were more than two people falling pregnant'.

The Dingiswayo teachers were very aware of the sexual activity of their students. Sandile contrasts his own youth, in which having sex and having a girlfriend were

> different because I grew up observing a lot of poverty. And one thing that came to my mind, and the worse thing one can do is to have too many kids outside of wedlock . . . And also my upbringing also came through. Because my mother, she bought us up to become gentle with women, and in a way made me not to have children before marriage, if you know what I mean. And to have girlfriends for me was to have a pretty face next to me. And sex wasn't really an issue so much.

He contrasts this with the practices of the learners in his school, saying that 'sexuality among the kids is not like what it was when I was the same age', identifying poverty as one of the reasons why young people, particular young women, might be sexually active: 'Some of the children we teach at this school, they come to you

and say: teacher, I have not had food . . . there is always an element of poverty. To answer the problem, we as teachers are faced with an uphill battle with regard to how do you say abstain from sex? For women it's a way to earn money.'

For some of the teachers, the place of HIV and AIDS education was in the Biology curriculum, while for others it took place in guidance or Life Orientation lessons – although Fulton explained that cuts in education expenditure had borne down particularly hard on the teaching of these subjects. As a Biology teacher, Fulton explained that he talked in these lessons

> about other aspects of poverty, that raises things, we talk about those sorts of things. And we discuss them with the kids. Now there is AIDS and there is no ways that we can avoid talking about safe sex because if children cannot really control their hormones we have to talk about it. *It is difficult to talk about it among kids, but there is no other way. It's the way to survive, you talk about it with an aside or in passing, but we stress the point to overcome the social problems that we have.* [emphasis added]

There is, then, a sense in which the need for teaching about safer sex in the context of the epidemic is shifted, when guidance or Life Orientation lessons are lost, disappearing into other curriculum areas like Biology, which are adjusted to take account of the social aspects, as well as the medico-biological aspects of the spread of the disease.

Fulton worries that parents might object to schools handing out condoms, despite official policy:

> I think the parents would give us a bit of upheaval if we did that. You know it's always been a battle for me that . . . you know parents, view it in a different light . . . and then you're dishing out condoms . . . it's like you're almost encouraging your children to have sex. You know they don't look at it from the other point of view . . . that it's preventing a spread of you know a killer disease and stuff like that.

Sandile, however, talks from the perspective of a father wanting his children to know about the sexual transmission of the virus: 'We should be open enough to

our children. Tell them what the causes are. Don't beat about the bush. I don't beat about the bush . . . I tell them about the dangers of HIV / AIDS.'

Conclusion

We have shown in this chapter that issues of gender equality feature in the thinking of male teachers, although not necessarily in the context of their views, teaching or thinking about HIV and AIDS and not necessarily always in a positive light. The different ways they think about it reflect a combination of their school contexts, their own life histories and the classed and raced history of South Africa.

The teachers from middle-class families, whether white, Coloured or Indian, are often mindful of, for example, sharing domestic duties in their everyday lives. Nonetheless, they often, but not invariably, take quite orthodox and conservative social and political stances, stressing the importance of 'discipline', rather than tackling entrenched disadvantage. In terms of our grid in Chapter 3, we could see them as approaching the issues of gender equality, HIV and AIDS through an interventionist emphasis in which dissemination of knowledge is the key.

The black teachers have more diverse views, ranging from the very conservative, with little sympathy for gender equality in any part of their lives or in relation to policy, to a much more radical view in which they locate gender equality as part of a broader worldview associated with the ending of apartheid and the need to press forward with major social change that addresses inequality. In some instances these are teachers who have already identified themselves, in practice, with social movements committed to social justice. For these latter teachers institutional transformation takes a central place in their thinking.

It is among black teachers in schools that have been affected most directly by HIV and AIDS and among learners who have had to deal with HIV infections and AIDS deaths in their families that the impact of the disease is most obvious, albeit not uncomplicated. Where HIV and AIDS are understood as being among the challenges (along with the legacy of apartheid) facing education by teachers who are committed to responding to these challenges, it is most likely that their practices and attitudes will reflect a greater commitment to gender equality. There is a place, here, for the interactionist approach.

It is clear that masculinities among our teachers can shift and are shifting and that HIV is playing a part in these shifts. Nonetheless, old forms continue to be

powerful and conservative, patriarchal attitudes still form an important part of the picture. Our findings, then, offer grounds for both hope and despair. In some cases, one could argue from them that nothing ever changes. In others, we can see the spaces for potential transformation. In the next chapter we turn to a study of female teachers to see how femininities and gender relationships have been affected by the HIV and AIDS context.

Notes
1. See Chapter 4 for a discussion of the methodological approach.
2. In all likelihood, this was his great-grandfather.
3. 'Mielie' is a South African word for maize.

Women Teachers, Women's Work and the Challenge of Care

Introduction

Women make up the majority of South Africa's teachers (see also Chapter 2). They should have been among the major beneficiaries of the removal of apartheid and government commitment to gender equality and in some senses they have been. The many grievous inequalities suffered by African female teachers during the years of the discriminatory administration of the apartheid period have been removed. Women now have the same salary scales as men, are eligible for the same medical and pension benefits, are entitled to maternity leave and have been identified as a group that should be moved into management positions, where they were formerly under-represented (Moorosi 2006b).

Yet all is not well for women teachers. The Human Sciences Research Council (HSRC) reports low morale among a majority of teachers and women, particularly, stress heavy work loads and lack of career development opportunities (see Chapter 2 and HSRC 2005a; 2005b). At the time this research was done, higher education institutions battled to attract sufficient recruits to education degrees to fill the gaps left by AIDS deaths and resignations (Badcock Walters et al. 2003; HEARD 2005). Gender analysis of teachers by seniority shows that while women make up 74 per cent of basic grade teachers and 66 per cent of heads of department, they comprise only 41 per cent of deputy-principals and principals (Moorosi 2006b). Why do women teachers appear to fare so ill in their jobs? Why have the commitments to eradication of gender and race inequalities not fostered a new dynamism and commitment among the hundreds of thousands of black female teachers in South Africa?

Perhaps an initial answer is to be found by looking at the form of the gender- and race-segregated labour market. For many years teaching and nursing provided the only two professional opportunities for African women. As long as these professions were regarded as avenues of upward social mobility and revered by families and communities, teaching and nursing bestowed respectability on African women and gave them some financial independence (Marks 1994; Unterhalter 1999; 2002).

The advent of Bantu Education (first implemented in 1955) created an unprecedented demand for teachers as the apartheid government expanded education provision to Africans (Hyslop 1999). Numerous teacher training colleges were opened across the country for African teachers, as the logic of segregation entailed that they alone were to teach African children. These colleges by and large provided hasty and superficial training to the thousands of trainees that passed through their doors. Education was delivered through transmission, with little critical thought encouraged and authoritarian practices (including corporal punishment) entrenched. African girls entering these institutions were not encouraged to express opinions and were often subject to the rough sexual overtures of their male colleagues (Marks 1987).

Once teachers had graduated, they generally found themselves in under-resourced schools, which in the 1970s became the centre of militant anti-apartheid resistance by students. While some teachers associated themselves with these struggles, many did not (Gultig and Hart 1990) and they became alienated from students. Many schools became educationally arid. Teaching as a vocation lost its gloss and in many schools, teacher professionalism waned (Christie 1998; De Villiers 1990). Female teachers were subjected to high levels of sexual harassment by male colleagues and were overlooked for promotion and marginalised in school-level decision-making (Sebakwane 1993/4). A number of autobiographical accounts by women teachers in the 1980s highlighted the ambiguous ways in which they understood their work and viewed their students' aspirations (Unterhalter 2002).

The 1990s and the transition to discussion and debate seemed to offer opportunities for women teachers to re-engage with ideas about the direction for education policy and practice. Some former teachers found themselves playing key roles in negotiating new national strategies for education and training (Lugg 2007), teacher unions took up gender discrimination in the areas of pay and working

conditions, with considerable success (Govender 2004) and changes in the curriculum provided opportunities for advancing teaching on gender equality, human rights and sexuality (Chisholm 2003; Carrim 2008; Mitchell 2005). However, despite the emergence of a distinctive and authoritative voice on gender equality by a number of women teacher leaders (Mannah 2005), many women teachers working in township schools struggled to find a language in which they could contribute to the development of new practice and, even for women who took principals' posts, there were considerable difficulties in translating the headlines of government policy on gender equality into the institutionalised practices of work (Moorosi 2006a; 2006b).

The ups and downs of trying to live out new gender-equitable relationships in schools were apparent in the reflections of the teachers at both Lilian Ngoyi and Dingiswayo over our six years of working together. The DramAide intervention offered women teachers at these schools a chance to challenge unequal gender relations, to assert some agency and to develop new femininities. In the weeks of planning and reflecting on the intervention, as discussed in Chapter 5, some found a different space for thinking about their work and a language to legitimate their concerns with gender equality. But as we have described, despite the intervention, its message, its resources, the law and Constitution, the schools' female teachers found it very difficult to shift the gendered positions of their male colleagues, to make much impression on the gender regime of the schools or, for that matter, to embark on any major project of presenting a new assertive and independent femininity within the school context. At both schools some actions of male teachers not only undermined messages regarding HIV risk and safe behaviour, but also had damaging implications for gender equality within the school. Alex Kent's analysis of space at Lilian Ngoyi explored the ways in which sexualities were performed by learners and teachers, exposing the ways in which masculinities and femininities were maintained and policed within the school and, as a result, how gender inequalities were reinforced (Kent 2002: 78). Places where only men congregated were spaces where socially useful knowledge was exchanged and these served 'to police subordinate females' as girls and female teachers were excluded (51).

Kent's data showed how gender inequalities in both schools were reproduced through the hierarchical school structure, where positions of power were given to

men. Over the whole period of the project, women remained the minority in management positions and were thus largely excluded from the structures of power and decision-making in both schools. While there were some indications by the end of the project that this might be changing, albeit slowly, many teachers seemed to accept the situation as normal, with one teacher describing the lack of women in management as something 'indigenous'.

Men and women teachers worked very separately in many day-to-day activities linked to lesson planning, pastoral support and extra-curricular activities. We were struck by the repetition of gender stereotypes in women's talk about men and vice versa. However, there were exceptions. At Lilian Ngoyi, a number of male teachers kept up an interest in discussing gender over a number of years. At both schools a number of women teachers asked us for support on how to incorporate gender into plans to implement the new curriculum. But in the context of some of the hyper displays of masculinity, sexual harassment and the subordination and exclusion of women and girls from male spaces of power within the schools, the extent to which these interests could grow and bring about wider-reaching and more profound changes in both discourse and behaviour was not clear. However, our understanding of what gender equality might look like in practice came to change as we documented the ways in which women teachers were giving expression to a new dynamic and political engagement with the onslaught of AIDS.

Into the epidemic: An ethos of care

After the DramAide intervention the schools settled back into the daily and weekly demands of delivering the curriculum, but the gathering pace of the HIV and AIDS pandemic meant that teaching could not be 'business as usual'. Our repeated visits over the years alerted us to high rates of mortality and morbidity. In conversations with teachers we became aware that they knew many people who were becoming sick and dying. Their conversations were full of references to attending funerals, colleagues and learners who were absent and learners who suddenly found themselves looking after their siblings or their parents. In some of our early conversations with women teachers, young members of staff had been able to talk to younger members of the research team about sexuality and begin to engage with issues of gender inequality in their own lives. However, the maturing pandemic presented new challenges associated with the imperative of care and this generated a reassessment of what gender equality might mean.

In 2005 we saw how a number of men and women teachers had adapted and changed aspects of their attitudes and behaviour. Most striking was the way in which in both schools, an ethos of care, generally associated with the work of women teachers, was being increasingly adopted and valued.

Care work is stereotypically the work of women and it is through women's association with care that some of the justifications for unequal pay and lack of access to decision-making are supported (Robeyns 2007). The work and the skills associated with care for the young, the old and the sick have historically been understood as women's work. Women teachers are often constructed as caring and nurturing, and as moral heroines of school learners. Mairtin Mac an Ghaill (1994: 37), for example, suggests that teaching has been viewed as a soft job involved with caring and nurturing, constructed as 'women's work'. Worldwide, care work has generally been undervalued, unrecognised and invisible. Thus the work of women teachers is often seen as a natural trait, requiring no special skill and oriented to no particular political goal.

As the AIDS pandemic has swept through eastern and southern Africa, the need for care work has been made much more visible (Epstein 2007). At Lilian Ngoyi and Dingiswayo we saw how illness and death created conditions in which women teachers gained the opportunity to reposition themselves as indispensable to the humane and successful operation of both schools. It was they who led concern for children who were orphaned, who were facing fears about their parents' illness and their own. Generally it was women teachers who organised to feed children who were hungry and contact relatives to make arrangements for those who were homeless. Part of the work of these teachers was forcing an acknowledgement among their colleagues that HIV and AIDS were not distant, but central to day-to-day work. Agnes, a teacher at Lilian Ngoyi, said in 2005 that it was important to acknowledge 'there are a lot of kids who are affected by HIV in this school'. She explained: 'So I could see it now that in the class there are cases, because you will find that certain individuals are absent, and when you ask them, "Why are you not coming to school", they say, "My mother is sick; my father is sick". "How sick is he/she?" "Sir, she couldn't stand up on his own; she couldn't wash; she couldn't walk".'

Sometimes students themselves were infected with the virus. Despite continued stigma, which made it difficult to be open about their status, a number of women

teachers made a concerted effort to establish relationships so that students could confide in their teachers. One teacher explained how she helped a Grade 9 boy. He had told her of his HIV-positive status and she had taken a great deal of time outside formal lessons to talk to him and help him:

> And I stayed with him; I talked to him; I tried to counsel him . . . And I remember it was before June last year when he said he doesn't feel that he would be able to write the exams. And I told him, I called him and I even told him, you know who will die first between the two of us? Because it was just the two of us. And I said who will die first between the two of us? Is it you or me? And he said he doesn't know . . . And I became very much more happy . . . And when I encouraged that boy, he did very well, because we talked about the food, we talked about the behaviour, we talked about everything. (Lungi)

Another teacher described the satisfaction she felt in helping a girl find the confidence to tell her mother about her status: 'And that girl was happy about it. She came back to me and she has told her mother. So, the battle of trying to make her accept the situation and to try and break it out she told her mother. She came to me and she hugged me and she told me thank you.' (Bono)

Ensuring that they provide emotional and material support to those who need it has required teachers in the schools to develop particular skills to read the signs in children's actions and identify those who were having problems, even if they don't come forward themselves. One teacher described identifying a girl in need of support:

> So there was one student there, a female student with a very you know pale shirt. It's like you can see you know that she is having a problem. So I had to take her out of the classroom and have a talk with her. So I ask her and it eventually came up that she doesn't have parents and she is living with an auntie who is not working. So I talked to her. Then there were some other problems, which I identified . . . in my mind I have to do something about this child. I'll have to get a shirt for her . . . So some problems . . . you just identify children, because they don't come up themselves and tell you even if they do have problems. So in most cases we are the people who see that there is a problem here. (Nancy)

Providing material support is another important dimension of care work for the teachers in the two schools. As the HIV and AIDS epidemic has worsened, there has been a deepening of the poverty experienced by the poorest learners. In both schools teachers recognised this poverty and its harmful effects. They tried to help learners with uniforms or 'old clothing' and tried to raise funds for learners' most basic needs. At Lilian Ngoyi some women teachers started a feeding scheme. They raised money among their colleagues and neighbours and brought food to school to ensure that children had bread and tea at the beginning of the day or a snack during break. Although only 50 children out of a school of 1 100 came for the food and others were too ashamed, the women were determined to continue. Although they were beginning to feel the pressures of continuing with the fundraising and the extra work entailed in preparing the food every day, the extreme poverty of the children was a matter of major concern. One drew our attention to the crusts of bread, cut off in a domestic science lesson, which were later taken out of the rubbish bins by hungry children. Another spoke about how seeing the children at the feeding scheme was an opportunity to note who might not be washing or dressing properly because of illness or poverty.

Teachers explained that in situations of great need, they could not but respond. They saw their caring and support work as something they simply had to carry out, given the needs around them. As one explained: 'When the child has been hurt, when the child has been abused physically by whoever, when the child has just found out that she is HIV or he is HIV whatever, you've got to – he's crying, he's desperate, thinks the whole world is falling and you've got to be there.' (Agnes)

Providing care came with challenges and sacrifices. Teachers expressed their frustration at the obstacles they encountered when trying to help pupils. Some felt exasperated by the difficulty of obtaining outside assistance and treatment for those infected by HIV. Others described the stress of having to deal with caring for learners on top of worrying about their own families: 'You end up working so hard, but having to attend to so many things including your own family and your relatives and also going under the same thing.' (Lungi)

However, the sense of carrying on and caring as much as you could, because that was the only thing to do, was a theme we heard again and again at both schools. Some stressed the pain of coming up against illness, dispersed families

and children in need of emotional support: 'Painful. Very painful . . . Referring them to social workers. Referring them to hospitals . . .' (Mabel)

In our field notes in March 2005 we continued to notice the energy and drive of the teachers, despite what they were going through:

> The discussion with Mrs Khumalo and other teachers at [Lilian Ngoyi] this morning leaves me stunned. I think they are drawing on something I can only call religious in the deepest sense. Mrs Khumalo was wearing a bright red dress. She seemed alight. Every day they talk to children with the most terrible experiences of loss and hunger. Yet they go on giving out. They don't just theorise about living with AIDS, they have to talk about it face-to-face every day. I don't know how they do it. There is just this sense of giving.

> I asked Debbie about what had happened when the buses from [Dingiswayo] came back from the funeral. She said everyone was just so sad. They were singing and singing. Teachers and pupils. Just together a lot. [Teacher] was quiet. I think she knew we would find it too much.

For some teachers, responding with care came with a sense of frustration at the lack of support and training to help them deal with the range of situations they encountered. Zodwa at Lilian Ngoyi explained: 'Well, it's a bit difficult because we as you put it we have not been trained to confront this, but we do our ultimate best.'

Dudu and Mabel felt there was a lack of training and a lack of resources at the school so that when there was a crisis, a Life Orientation (LO) teacher was called to deal with it and everything else had to be dropped:

> [Another difficulty is] the question of not attending the problem at that point in time [when the child raises it] because we have to go to class, and teach geography, something totally different from LO.

> And sometimes there's a problem that you can't ignore. So you attend to the problem and sometimes it takes you the whole day away from your classes. So for the whole day we have not taught the other classes.

However, for many of the teachers, formal training was not so much the issue. Rather, they were anxious to provide informal support and show concern. Relating to the children's needs was not always a simple matter of extending love and care. For many, the epidemic presented them with difficulties where they had to examine their own beliefs about how they expected teenagers to behave:

> We have gone even an extra mile of bringing the condoms. Because bringing the condoms, other people can think they are they are saying they must have sex, but we start with abstinence. But the rate of pregnancy at school tells us that there's no abstinence. To try and protect them we rather put the condoms here and tell them that here are the condoms, protect yourself . . . We are trying to protect them against all this situations, which they encounter. (Lena)

Some disapproved of the frank sex education in the schools, but accepted it was important to give it as a means to support the children:

> 'Cause the way we talk about sex here at school, what is mainly good about [Dingiswayo] even the principal of the institution, allows this HIV, this sex education. You don't have a problem of saying that maybe the principal is going to say this. No, it's free. We teach them and they know. They can come to anyone of us to ask if they have a problem. They can come to any one of us to ask for help if they need it. (Tina)

> We inform them as much as we can. It's one of the schools in which we are proud that we try to empower them as much as we can, both of them, boys and girls, because it is not only about the girls. Even the boys, they are affected by HIV and AIDS, therefore we also empower them to use the condoms to protect themselves against different diseases. We are proud that no child can tell you that they are not taught about HIV and AIDS. They are informed. There's nothing that they do and can then say it is because I didn't know. (Lungi)

While some did not approve of teenagers being sexually active, they tried to understand the reasons for this:

And the problem with our kids, some of them when they come to us they complain about this love from their parents. They lack this love . . . now they end up saying because they are not love[d], they don't get the love from the parents, they fall to the boyfriends. When they fall to the boyfriend, they think they are getting much more better than from their parents, and she has mentioned they don't have this confidence to themselves; they aren't confident to themselves. They rely when the boy say the boy loves her, they think that oh now I'm belonging to this group, now people, the peers, will think now I also have this status of being loved, of which is wrong. (Lena)

In the schools, the crisis precipitated by HIV, as well as the poverty and violence experienced by the students, has resulted in increased value being placed by the school management, not only on care work itself, but also on the 'feminine' skills and characteristics associated with it. The head teachers were proud of the work that the team of women teachers was doing and emphasised this to us at our meetings at the end of the project. This was partly because, over the course of the project, LO came to be an examinable subject. Although a number of women teachers commented that this did not mean LO secured the respect it deserved, given the seriousness of the issues it dealt with, our assessment of the value that the head teachers accorded the teachers involved most directly with providing care was partly because they were giving content to the delivery of LO, but also because they were helping the schools to connect with their communities: 'Mr Shabalala [at Lilian Ngoyi] was really friendly. He greeted us warmly. He said straightaway we must go and see Mrs Mbali [who worked with the peer educators]. He walked us over to their block. Until last year I don't think we'd even been aware of the work the LO teachers do. Now they are number-one for visitors to see.'

The way in which some of the teachers were using care as a response to the epidemic did not go unnoticed or unappreciated by students. At Lilian Ngoyi, the peer educators spoke warmly about their teachers and the way in which they have tried to understand, respect and support them. They confirmed that both female and male teachers have provided support in the school, mentioning in particular two teachers who were particularly supportive and who they felt really made an effort to understand teenagers. One of these teachers is a woman, and the other is

a man. Referring to the male teacher, one of the girls explains: 'It's like he's a man ... And in the meantime you see him as a teenager, girl. I don't understand this, but like he can talk; he can advise us as like girls.'

At Lilian Ngoyi, the students clearly articulated the same ethic of care seen in the interviews with teachers. They spoke with pride of the way they could deliver care:

> I'm proud of what we've achieved. Firstly we gained the respect and the trust of the learners. Ja. They like changed, they like trusted us, they were able like to tell us problems as I have said and we were not only looking at an aspect of HIV and AIDS, we also were like we looked at the aspect of poverty. Like we said that one of the things that I'm proud of ... we were able, us learners, to sit down with the governing body and tell them, look at the problem we have got. Learners coming from poor backgrounds, they come to school with empty stomachs and then go back home with empty stomachs. So we need you guys to help us. We need to start a feeding scheme or something. It has been successful and it has helped quite a number of people. And another thing it's like I'm proud of, it's like we've made a lot, although some of them do not come back and tell us like say, 'hey you've changed my life' and everything, but I see that it has made a lot of people aware of the fact of HIV and AIDS.

It can be seen that a number of women teachers at both schools were taking gender identities associated with care and remoulding them. Some were challenging the stigma and stereotypes attributed to those with AIDS. Their actions enabled their fellow teachers to be frank about HIV and learners to talk about their needs and fears. Even though the support given by the school, social workers and health officials was not always adequate, a significant group of women teachers were pushing the demands of care and registering its urgency. These women appeared to have set up an almost parallel stream of esteem in the school. While academic and sporting success carried great kudos, the glamour associated with sexual conquest was frequently mentioned by teachers and pupils, as was the esteem accorded to what was seen as appropriate cultural behaviour, the women teachers who enacted care were also evidently valued and esteemed by senior management and their most articulate pupils. The work of care was enacted with a form of

robustness, resilience and engagement with difficulty and difference that was striking. It seemed to be neither dictated by government training schemes nor community policing. It seemed wholly born out of the problems of the epidemic and was attempting to be responsive to them. This appeared to be a form of gender equality as interaction, not easily recognisable, because so innovative, but powerful nonetheless.

But such a reframing of the meaning of gender equality had two dangers: first, it might confirm conventional readings of femininity and thus confound gender equality initiatives that are explicitly designed to challenge patriarchal orthodoxy; second, it might prevent men from sharing the care work and contributing to new, gender-equitable school regimes. Despite the emerging concern with care and the increased recognition of the need to engage with the gendered issues arising from the HIV pandemic and its ramifications within the schools, gender hierarchies in both schools persist.

Conclusion

The HIV and AIDS pandemic has provided women teachers with an opportunity to pursue the goal of gender equality. In their case the goal is not an abstract set of political objectives, but involves a number of highly personal interests. Among these are valorising their own identities, obtaining space and recognition within the decision-making processes of the schools for issues they value, particularly those relating to the welfare of learners, and relating more equitably to learners and colleagues.

As the epidemic has progressed, an initial reluctance to fully engage with the disease, particularly its gendered dimensions, was replaced by a recognition of the need to become creatively and proactively involved. In getting close to the epidemic, women teachers developed a strong ethos of care and support for those affected. As the importance of care came to be recognised and valued within the school, the femininities associated with care were increasingly validated and demanded by both men and women. Such changes opened space for the development of alternative visions of masculinity and femininity and a different positioning and understanding of care.

These changes represent a real and important step forward in understanding and responding to the epidemic and engaging with approaches to gender equality.

However, at the same time, despite this articulation of alternative views concerning HIV and gender, more traditional visions continued to play a role in the formation of identities and social relations in both schools. Gender inequalities continue to be accepted and normalised in power structures within the schools. Examining such persistent inequalities further and finding ways to address them will be critical to ensuring that the changes witnessed so far in the two schools can be built on and sustained and their transformative potential realised.

Progress towards Equality

A Conclusion

This book has set out to examine whether gender relations in South African schools have been structured differently as a result of the changes wrought in the post-1994 period. How effectively have the hopes and aspirations of those living under apartheid been translated into a reality associated with gender equality? How have particular features of contemporary times associated with violence and the HIV and AIDS epidemic affected this process? In this chapter we review our findings, pointing out, on the basis of data from the schools studied, what has been achieved, what has not, and what the reasons for this have been. We also consider the theoretical approaches we have used in order to assess whether they provide a strong enough explanatory framework or whether they point to further conceptual clarifications that will be needed to help frame future work in this area.

We framed the study with a discussion of different ways of understanding gender, equality and the emphasis of policy. We linked these different understandings with particular ways of approaching the problem of violence and the gender dynamics associated with the HIV and AIDS epidemic. As we argue below, the data indicate that at different moments and among different actors we can trace aspects of each of these frameworks. The aspects articulate with one another, in some cases complementing one another, in other cases confounding the goal of gender equality. The articulation of the different forms of gender equality seldom, however, produced the empowerment we named as the deepest expression of gender equality (the fourth approach discussed in Chapter 2). Similarly it has been difficult to talk about the HIV and AIDS epidemic outside essentialised language and to think about violence as anything other than a set of aberrant acts.

Our key argument is that the social conditions within schools and the interaction with those conditions by gendered actors (teachers, learners or researchers) have made it difficult to realise the multidimensional aspects of equality in the process of engaging with violence and HIV and AIDS. However, we have seen some movements towards realising some elements of equality.

Our understanding of gender equality involves changes in gender structures, relations and identities. Investigating the processes involved has demanded detailed qualitative research of various kinds (see Chapter 4 for details) for, as Wendy Hollway and Tony Jefferson (2000) point out, quantitative research can answer only those questions requiring measurement. Since we have been concerned to understand the complex processes involved in the changing nature of gender equality, we drew on a range of qualitative approaches. Drawing on the work of Amartya Sen, we have argued that it is important to see how rights are linked with expanded capabilities. For the capability sets of different girls and boys to be expanded and for their range of opportunities to have some chance of being realised, there has to be widespread and deep structural, relational and identity changes. While improvement in access and academic performance may well constitute a contribution to this, on their own they are not enough to offer expanded opportunities, provide for the conversion of resources into valued achievements and transform life experience.

We have argued that there are three basic approaches to achieving gender equality: interventions, institutions and interactions. Interventions are generally limited in their ambition and scope. They frequently have a limited timeframe and treat gender as a 'noun' (a thing that can be measured). Institutional approaches utilise active understandings of gender, regarding it as a constitutive part of processes and therefore requiring more than one-off efforts in order to sustain change towards gender equality. Institutional approaches are often associated with creating structures or instituting processes specifically designed to impact on gender relations and to promote gender equality. The commitment of resources, human and economic, is often substantial in such endeavours. Interactionist approaches grasp the complexity of gender and seek to effect equality by working intensely at the individual and organisational level. These efforts are intended to impact on gender relations and identities, which is precisely where capacity – a way in which to see capability in the Sen sense – is realised. However, interactionist work is

arduous, progress is never obvious and advance is never guaranteed. For interactionist work to succeed, it needs the buy-in and support of not only political leaders and policy-makers, but also of learners and teachers. This cannot be assumed. Gender identities are always fractured and contradictory. They reflect histories, life trajectories, social structures and patterns and these often may mesh in unpredictable ways to undermine the commitment to gender equality. Interactionist work is also thus always incomplete and partial. A fourth approach, which we named empowerment, links interventions, institutions and interactions together and suggests that it is in their articulation that a more multidimensional form of gender equality can be seen.

In a recent critique of Sen's ideas about the positive effects of education, Craig Jeffrey, Patricia Jeffery and Roger Jeffery (2008) argue that educational progress, even the attainment of higher education qualifications, does not necessarily translate into 'freedom' in a context where the labour market is shrinking and factors such as caste and social class are at play. In a study conducted in the Indian province of Utar Pradesh, the authors found that many well-educated men were unable to translate their higher educations into stable jobs. The widespread expectation of the payment of bribes and the importance of family connections prevented many young men from securing jobs and this in turn prevented them from starting a family. Jeffrey, Jeffery and Jeffery show that young men did not generally respond to this crushing situation by rejecting education, but rather held on to the belief of the power of education by weaving their educational achievement into a form of distinction that separated them from those less educated. The authors are thus sanguine about the possibility of improving individual capability in contexts which still reproduce class, caste, gender and other inequalities.

Political economy plays a significant role in limiting or enabling the achievement of gender equality. In India's case the achievement of independence in 1948, the gradual erosion of the caste system, a history of economic growth and the recent adoption of some neoliberal policies have come together to shape the capacities of the young men who are the subject of the Jeffrey, Jeffery and Jeffery study. A significant feature of this study is that the number of men who resorted to crime and violence (as a sign of disillusionment) was very limited. However, the South African situation is different. Liberation and the formal ending of apartheid are in the very recent past. The gap between rich and poor has not shrunk and poverty

affects a large minority of the population who are overwhelmingly African. Education achievement has not translated into 'freedom' for many young African men and women and an important sociological question is whether or not this can be linked with increases in violence. Constructions of masculinity, as we have argued, impact on the capacity of girls to enjoy equality, as well as expressing the gender limitations (for example, dropping out of school early) that boys experience in the schooling system. The links between these processes and some of the everyday patterns of violence we have documented need further research.

So what progress has been made towards gender equality in schools? Our first answer is provided by the most recent batch of matriculation results. When the results were released in December 2007, a major Sunday newspaper ran the bold headline: 'Girls chalk up victory over boys'. It reported that only 167 513 boys had passed compared to 200 404 girls (*Sunday Times*, 30 December 2007: 5). Moreover, while 3 529 boys passed with distinction, so did some 5 381 girls. These results are striking, and particularly noteworthy for Africa where boys routinely outperform girls (UNESCO 2008).

The improved performance of girls in the matriculation examination is a significant but limited measure of gender equality and reflects gendered understandings that are embedded in interventionist approaches to gender equality. If the improvement of girls' performance is taken as evidence of progress, we nevertheless need to ask the question posed, in a United Kingdom context, by Epstein et al. (1998: 9–12), 'which girls?' and 'which boys?' In South Africa the answers to this question are both racial and gendered. At the lowest end of the performance spectrum (among African girls in poorly resourced areas and schools), boys still outperform girls. In other words, progress towards gender equality is not experienced universally or uniformly and the major beneficiaries are likely to come from families that already provide gendered advantage.

The matriculation results do not and cannot say anything about the experiences that girls and boys have in school. For many boys and girls, but particularly for girls in township schools, violence, whether personally experienced, observed or anticipated, clouds their encounter with schooling. Schools are often not safe places and although suburban schools are safer, they are not immune from the violence. There have been interventions at school and state level to make schools safe (from violence occurring between learners and from the threat of violence posed by people

external to school populations). These have included conflict-resolution courses and lessons on respect, but mostly the focus has been on erecting high walls, installing security gates and employing guards. As important as these interventions are, they have limited impact on securing enduring and sustainable progress towards gender equality.

As discussed in Chapter 3, the most influential and visible initiatives on South Africa's gender terrain have been taken by the state, whose laws and policies have had wide-ranging effects. They have created bodies to monitor gender (from gender focal points to organisations like the Office on the Status of Women) and infused the bureaucracy with administrative processes that ensure the visibility of gender. In schools, principals and governing bodies are required to show sensitivity to gender when teachers are employed. The state's laws and policies have also strengthened and legitimated a human-rights framework for the country that has, in turn, fed into new public discourses. New gendered identities (most notably gay identities) have emerged (Gevisser and Cameron 1994; Carrim 2009).

However, the state's mainstreaming approach has the limitations, identified in Chapter 3, of all institutionalist approaches. Mainstreaming is most effective at the policy level where injustices can be corrected via interventions from above. While the approach correctly identifies the causes of gender inequality in material inequalities and patriarchal power, it has little capacity to reshape gender relations or to impact on gendered experiences.

Aware of its lack of capacity to effect change, the state and individual schools have enlisted the support of a variety of organisations to mount gender interventions. This book assesses the impact of and associated responses to one such intervention, that of DramAide.

In Chapter 5 we examined how the DramAide intervention was experienced in two township schools and how it impacted on the learners and teachers of these schools. The particular gender regimes of the schools, Lilian Ngoyi and Dingiswayo (described in Chapter 4), played a large part in determining how the intervention was received. At Lilian Ngoyi, a school with a democratic history and ethos, gender transformations were more evident, although they coexisted uneasily with practices such as sexual harassment, which continue to bar the way to gender equality. In Dingiswayo, where authoritarian governance and pedagogy were the norm, gender transformation took a more modest form, with little of the critical enquiry about

existing gender practices and norms occurring that were a feature in Lilian Ngoyi, where the emergence of feminist ideas amongst the girls was particularly striking. The impact of an intervention often is limited by its short duration and this was no different in these two schools. After DramAide ended its work, the enthusiasm for gender equality gradually waned. Teachers did not teach about gender in HIV-prevention lessons and while gender equality discourses still circulated, they neither reflected the existence of gender equality nor contributed to gender equality within the schools.

One way of entrenching gender equality is to mainstream or institutionalise it. In Chapter 6 we discussed how Life Orientation lessons in two primary schools were conducted. Life Orientation provides the opportunity to talk to young children about gender and sexuality, yet in these two lessons both teachers, one African man and one white woman, shied away from dealing with gender and sexuality and thus lost the opportunity to develop and strengthen gender-equality practices and discourses.

In ongoing interactions between learners and teachers, new ideas about gender equality find expression and enactment. In Chapter 7 we showed how learners grappled with the presence of HIV and the importance of gender equality and how they came proactively to engage with issues of sexuality. In this engagement, however, the HIV discourses promoted by government that endorse sexual abstinence were highly influential. Gender equality as a goal was marginalised and this made it very difficult for alternative voices among learners to be heard. There are boys and girls who are drawing on equality discourses to fashion new identities, but the discursive and institutional environment makes it difficult for these models to become more popular and even hegemonic.

In Chapter 8 the understandings of sexuality and HIV among children aged six to eight in two schools were explored. While teachers may shy away from issues of sexuality, the children approach and understand the body as highly sexualised. Their curiosity, however, is tempered and shaped by the very different material and cultural contexts of their schools. At Bullwood, middle-class security allows children to discuss issues of sexuality by drawing on media and television images in a way that distances them from the AIDS pandemic. At KwaDabeka, young children draw on personal knowledge of HIV and AIDS and of sexual violence. Their experience and expectation of sex is shaped by fear, danger and distress.

The two school environments are thus critical for understanding how gender identities are formed and in grasping the limits of children's agency.

In the final two chapters we examined how teachers are making sense of post-apartheid schooling. In Chapter 9, eight male teachers talked about themselves, their own schooling and their current location in the education system. The position of men as agents for gender change in schools is often questioned. They dominate school management hierarchies; some are involved in sexual harassment and many believe in naturalised ideas about the superiority of men. Can they promote gender equality? In this chapter we show that some men explicitly support the goals of gender equality and many are aware of the need to break with stereotypical attitudes and practices that perpetuate sexism. In the context of HIV and AIDS, there is a willingness on the part of some of the teachers to engage in care work and this constitutes a significant shift in the sexual division of labour and a challenge to gender stereotypes in schools. Yet while masculinities are changing, there is also evidence of unyielding support for men's rights and a defensive or hostile response towards gender equality. The responses of the male teachers were by no means uniform, yet each strongly reflected their school environment. Schools have different resources and locations and face different challenges. In making sense of and responding to these challenges, the teachers drew on their own experiences, which bore the powerful imprint of apartheid. In some cases, for example, involvement in the anti-apartheid struggle facilitated support for gender equality. In other cases, strong ethnic belonging led men to yearn for the old status quo, particularly the use of corporal punishment and respect from learners.

In Chapter 10 we showed that women have responded positively to the HIV and AIDS epidemic by stressing the care component of their work. This work has been practical, for example providing food for children whose parents are afflicted, as well as emotional and affective, by providing reassurance and support for learners affected by the pandemic. It has also involved the strengthening of personal agency and the development of more assertive and confident models of femininity. Yet this response has not always challenged unequal gender values and relations, even though it has created space for the development of alternative visions of masculinity and femininity.

Both male and female teachers are key players in gender equality work. Their co-operation with one another in respectful and equal ways and their commitment

to the goal of gender equality is crucial to gender transformation. Their contributions, however, are substantially contingent on their collective and individual abilities to work gender equality into their own identities and practices. Only in this way will gender interactions become more equal and empowerment be experienced broadly.

Has the goal of gender equality been helped by conditions accompanying the AIDS pandemic? This was the question we posed early in this book, prompted by the contrasting views of Edward Kirumira (2004) and Cindy Patton (1994). Kirumira identifies three areas in which the pandemic has fostered progress towards gender equality. First, it has dented the taboo related to talking about sex and sexuality, and AIDS interventions have tested and expanded the permissible sphere for communicating about sex and sexuality. Second, the use of an empowerment discourse in HIV prevention campaigns has strengthened women's agency. Third, the pandemic has 'moved the realm of sex education from the family – and therefore domestic/private sphere – to the school environment, which is associated more with the public sphere. Such education has challenged sex education for social reproduction and by so doing has interrogated gender relations, especially among the young generation.' (Kirumira 2004: 157)

Patton, on the other hand, observes that the 'most striking feature of the HIV epidemic' was its ability to 'reinforce socio-culturally constructed inequalities of gender, social status, race and sexuality' (1994: 2).

In some ways the pandemic has indeed created conditions that have fostered gender equality. It has provided the imperative for interventions, such as DramAide, to be introduced into schools. The introduction of Life Orientation into the curriculum has provided a place in school timetables to discuss gender issues, though the ability and willingness of teachers to yoke gender to HIV and AIDS has limited the impact. Morbidity and mortality, the human face of the pandemic, have created new needs for care work which have, in conjunction with gender equality messages, given women more recognition (though not necessarily more power within schools) and this has encouraged some male teachers to embrace care as a new element of masculinity.

On the other hand, the AIDS pandemic has not been noticeably associated with or causally linked to a decline in gender-based violence. Young African women remain the most vulnerable to HIV infection, often as a consequence of such

violence. Sexual violence, male dominance in sexual situations and sexual harassment in schools all persist.

In the last period of data collection, we were, as explained in previous chapters, very struck by the enthusiasm and commitment of the peer educators at Dingiswayo and Lilian Ngoyi. In some of their talk, these young people reflected awareness of the need for gender equality and an end to sexual violence, while simultaneously enacting the continued dominance of boys even within their immediate group:

> What I always wanted was to balance the gender; to stop sexual abuse; to stop sexual harassment, which I could not do. I don't have the power over them to stop them. Oh earlier this year when I met [Siphiwe] he introduced something that we could have done long ago. It's called *School Mag*, it's a project, which we never succeed on doing, because the problem was sponsors, the problem was how are we going to get it? How we going to make the mag being done? . . . The idea of a school magazine is to introduce all these topics of sexual harassment, sexual abuse; it's the lacking of women to overpower. It's to empower women. We wanted to empower women to take part in leadership; to take part, to try to participate in subjects like science and commercial subjects. And we wanted to introduce this thing of HIV and AIDS. It's not easy to talk to them, like face-to-face and say that HIV this and HIV that because they are ignorant and we wanted to write something. This mag will be for all the school members, which will contain the career guidance, and HIV and AIDS, early pregnancy. Consequences of early pregnancy, which never succeeded, but we haven't given up. We still trying and hoping it will work out. (Zakhele)

> We've got the Representative Council of Learners (RCL), so it's like he's a member, he's part of that council; he's part of the leadership. I've been encouraging her [Sibongile] and other girls here at school to take part on leadership, because for the last four years that we've been into the school, Londi is now the president of the RCL and last year I was the president of the RCL, and the ladies are just like taking the back seat and I was trying to encourage all the ladies to take part in leadership, so, we came up with this idea of a magazine and everything, but at the moment it hasn't succeeded, but we haven't actually given up hope. (Siphiwe)

The desire on the part of these boys to 'empower women' can be seen as evidence both of changing masculinities and of the difficult structural issues that need to be overcome. 'The ladies' are 'taking the back seat' despite the 'encouragement' of the boys in the leadership of the RCL to become more active. The 'mysterious gap between hope and happening' (Kenway et al. 1997: 1) is evident here. But what is hopeful is that a desire for change that could perhaps be mobilised through interaction is clearly present. These same boys, in other parts of the discussion, cling to what they see as 'traditional practices', such as virginity testing. Yet they are also facing the epidemic with a commitment to ending gender abuse and the empowerment of women through their roles as peer educators. That both discourses exist side by side does not negate the power of either, but shows where the spaces are for dialogue and interaction in the development of strategies for gender equality and for the prevention of HIV and AIDS.

In this book we have identified the obstacles to and constraints in establishing and consolidating gender equality. These include the prevalence of poverty and unemployment, the difficulties for the state in keeping its promises of a just post-apartheid order and the way in which patriarchal discourses continue to legitimate gender inequalities. In the context of schooling, the difficulties vary from school to school and are more insurmountable in schools which historically serve African working-class communities. Following Jeffrey, Jeffery and Jeffery (2008) we recognise that these contextual issues limit efforts to achieve gender equality. On the other hand, we have shown that gender changes are occurring. Institutional approaches to mainstream gender have ensured that at the policy level, gender is a fixed agenda item. In schools the curriculum has ensured that there is a place for gender to be taught. Numerous state and non-state gender interventions have reached learners and teachers. While some of these have been undermined by the dispositions and pedagogical styles of teachers (as shown in Chapter 6) there has been a gendered impact on learners. Girls, in particular, are more aware of their rights and their agency and boys are increasingly accepting that girls have rights. Gender shifts are not linear and invariably there are contradictions. Boys hold on to the idea of male superiority and girls accept patriarchal limitations on their agency. And both boys and girls still seem to accept that violence is an everyday reality.

Gender relations in South African schools are changing but transformation, particularly understood in terms of the empowerment approach, is very slow.

Constitutional protection against gender discrimination is the bedrock of change upon which rests the state's mainstreaming initiative and non-governmental organisation (NGO) interventions. The AIDS pandemic has given impetus to gender equality work even as it has contributed to tragic hardship. The task for the future is to build on these beginnings, knowing that equality will continue to be elusive.

References

Abdool Karim, S.S. and Q. Abdool Karim (eds.). 2005. *HIV/AIDS in South Africa*. Cape Town: Cambridge University Press.

Adonis, A.N. 2008. 'An analysis of change in the management practices of school principals in the context of an external intervention from 1997 to 2000: A case study of the Imbewu Project in the Eastern Cape province'. Unpublished Ph.D. thesis. Edgewood: University of KwaZulu-Natal.

Aikman, S., E. Unterhalter and T. Boler (eds.). 2008. *Gender Equality, HIV and AIDS: A Challenge for the Education Sector*. Oxford: Oxfam.

Alanen, L. and B. Mayall (eds.). 2001. *Conceptualizing Child-Adult Relations*. London: Routledge/Falmer.

Alegi, P. 2004. *Laduma! Soccer, Politics and Society in South Africa*. Pietermaritzburg: University of Natal Press.

Altbeker, A. 2007. *A Country at War with Itself: South Africa's Crisis of Crime*. Johannesburg: Jonathan Ball.

Ames, P. 2005. 'When access is not enough: Educational exclusion of rural girls in Peru'. In S. Aikman and E. Unterhalter (eds.). *Beyond Access: Transforming Policy and Practice for Gender Equality in Education*, 149–65. Oxford: Oxfam.

Amone, J. and P. Bukuluki. 2004. *The Impact of HIV/AIDS on the Education Sector in Uganda: Examining the Impact of HIV/AIDS on Governance in the Education Sector*. Paris: IIEP.

Anderson, K.G., A. Case and D. Lam. 2001. 'Causes and consequences of schooling outcomes in SA: Evidence from survey data'. *Social Dynamics* 27(1): 27–59.

Attwell, P.A. 2002. 'Real boys: Concepts of masculinity among school teachers'. Unpublished Master's dissertation. Pietermaritzburg: University of Natal.

Badcock Walters, P., C. Desmond, D. Wilson and W. Heard. 2003. 'Educator mortality in-service in KwaZulu-Natal'. Research paper. Durban: HEARD, University of Natal.

Ball, S.J. 1994. *Education Reform: A Critical and Post-Structural Approach*. Buckingham: Open University Press.

————. 2008. *The Education Debate: Policy and Politics in the Twenty-First Century*. Bristol: Policy Press.

Barbarin, O. and L. Richter. 2001. *Mandela's Children: Child Development in Post-Apartheid South Africa*. New York: Routledge.

Baxen, J. and A. Breidlid. 2004. 'Researching HIV/AIDS and education in sub-Saharan Africa: Examining the gaps and challenges'. *Journal of Education* 34: 9–29.

Baxen, M.J. 2006. 'An analysis of the factors shaping teachers' understanding of HIV and AIDS.' Unpublished Ph.D. thesis. Cape Town: University of Cape Town.

Baylies, C. and J. Bujra. 2000. *AIDS, Sexuality and Gender in Africa: Collective Strategies and Struggles in Tanzania and Zambia*. London: Routledge.

Bhana, D. 2002. 'Making gender in early schooling: A multi-sited ethnographic study of power and discourse from Grade One to Two in Durban'. Unpublished Ph.D. thesis. Durban: University of Natal.

————. 2005. 'Violence and the gendered negotiation of masculinity among young black school boys in South Africa'. In L. Ouzgane and R. Morrell (eds.). *African Masculinities*, 205–20. New York: Palgrave.

————. 2008. '"Six packs and big muscles, and stuff like that": Primary-aged South African boys, black and white, on sport'. *British Journal of Sociology of Education* 29(1): 3–14.

Bhana, D., R. Morrell, D. Epstein and R. Moletsane. 2006. 'The hidden work of caring: Teachers and the maturing AIDS epidemic in diverse secondary schools in Durban'. *Journal of Education* 28: 5–24.

Blackbeard, D. and G. Lindegger. 2007. '"Building a wall around themselves": Exploring adolescent masculinity and abjection with photo-biographical research'. *South African Journal of Psychology* 37(1): 25–46.

Blignaut, S. 1979. *Statistics on Education in SA, 1968–79*. Johannesburg: South African Institute of Race Relations.

Boler, T. 2003. 'The sound of silence: Difficulties in communicating on HIV and AIDS in schools'. http://www.actionaid.org.uk/doc_lib/146_1_sound_silence.pdf, accessed 28 October 2008.

Boler, T. and D. Archer. 2008. *The Politics of Prevention: A Global Crisis in AIDS and Education*. London: Pluto; Johannesburg: Jacana.

Bonnin, D., G. Hamilton, R. Morrell and A. Sitas. 1996. 'The struggle for Natal and KwaZulu: Workers, township dwellers and Inkatha, 1972–1985'. In R. Morrell (ed.), *Political Economy and Identities in KwaZulu-Natal: Historical and Social Perspectives*, 141–78. Durban: Indicator Press.

Booth, D. 1987. 'An interpretation of political violence in Lamont and KwaMashu'. Unpublished Master's dissertation. Durban: University of Natal.

Budlender, D. 2000. 'The political economy of women's budgets in the South'. *World Development* 28(7): 1 365–378.

———. 2005. *Expectations versus Realities in Gender-Responsive Budget Initiatives*. Geneva: United Nations Research Institute for Social Development.

Bullen, E., J. Kenway and V. Hey. 2000. 'New labour, social exclusion and educational risk management: The case of "gymslip mums"'. *British Educational Research Journal* 28(4): 441–56.

Campbell, C. 2003. *Letting Them Die: Why HIV/AIDS Intervention Programmes Fail*. Oxford: James Currey; Bloomington: Indiana University Press.

Campbell, C. and C. MacPhail. 2002. 'Peer education, gender and the development of critical consciousness: Participatory HIV prevention by South African youth'. *Social Science and Medicine* 55(2): 331–45.

Carrim, N. 2009 (In press). 'Human rights and the limitations of releasing subaltern voices in a post-apartheid South Africa'. In R. Cowen and A.M. Kazamias (eds.), *International Handbook of Comparative Education*. Dordrecht: Springer.

Casale, D. and D. Posel. 2005. 'Women and the economy: How far have we come?' *Agenda* 64: 21–29.

Chisholm, L. 2003. 'Gender equality and curriculum 2005'. Paper presented at the University of London Institute of Education seminar, 'Beyond access: Curriculum for gender equality and quality basic education in schools'. London, 16 September.

Chisholm, L. (ed.). 2004. *Changing Class: Education and Social Change in Post-Apartheid South Africa*. Cape Town: HSRC Press.

Chisholm, L., S. Vally and S. Motala. 1998. *Review of South African Education 1994–1997*. Johannesburg: Education Policy Unit, University of the Witwatersrand.

Christie, P. 1998. 'Schools as (dis)organisations: The "breakdown of the culture of learning and teaching" in South African schools'. *Cambridge Journal of Education* 28(3): 283–300.

Clarke, D. 2008. 'The road less well traveled: Gender-based interventions in the education sector response to HIV'. In S. Aikman, E. Unterhalter and T. Boler (eds.). *Gender Equality, HIV and AIDS: A Challenge for the Education Sector*, 105–28. Oxford: Oxfam.

Connell, R.W. 1987. *Gender and Power: Society, the Person and Sexual Politics*. Cambridge: Polity Press.

———. 1989. 'Cool guys, swots and wimps: The interplay of masculinity and education'. *Oxford Review of Education* 15(3): 291–303.

———. 2002. *Gender*. Cambridge: Polity Press.

Connell, R.W. and J.W. Messerschmidt. 2005. 'Hegemonic masculinity: Rethinking the concept'. *Gender and Society* 19(6): 829–59.

Connolly, P. 1998. *Racism, Gender Identities and Young Children: Social Relations in a Multi-Ethnic, Inner-City Primary School*. London: Routledge.

197

Coombe, C. and J. Godden (eds.). 1996. *Local/District Governance in Education: Lessons for South Africa*. Johannesburg: Centre for Education Policy Development.

Corden, A., R. Sainsbury, P. Sloper and B. Ward. 2005. 'Using a model of group psycho-therapy to support social research on sensitive topics'. *International Journal of Social Research Methodology* 8: 151–60.

De Bruyn, J. and N. Seidman-Makgetla. 1997. 'Engendering the budget process'. In D. Budlender (ed.). *The Second Women's Budget*, 60–77. Cape Town: IDASA.

De Villiers, E. 1990. *Walking the Tightrope: Recollections of a Schoolteacher in Soweto*. Johannesburg: Jonathan Ball.

Deacon, R., R. Morrell and J. Prinsloo. 1999. 'Discipline and homophobia in South African schools: The limits of legislated transformation'. In D. Epstein and J. Sears (eds.), *A Dangerous Knowing: Sexuality, Pedagogy and Popular Culture*, 164–81. London: Cassell.

Department of Education. 2003. *Revised National Curriculum Statement Grades R-9 Life Orientation*. Pretoria: Department of Education.

Department of Health. 2002. *HIV/AIDS Lifeskills and Sexuality Education Primary School Programme Grade Four Teacher's Guide*. Cape Town: Department of Health.

Dobash, R.E., P.R. Dobash, K. Cavanagh and R. Lewis. 2000. *Changing Violent Men*.Thousand Oaks, CA: Sage.

Douglas, M. 1966. *Purity and Danger: An Analysis of Concepts of Pollution and Taboo*. London: Routledge Kegan and Paul.

DramAide. 2000. 'Mobilising young men to care: Addressing gender issues in relation to HIV and AIDS'. Draft discussion proposal for a partnership of the University of Natal, Durban, School of Education and DramAide project. Durban: University of Natal.

Duffett, A. 2006. 'Contesting masculinities: Sport as a medium for development in three HIV/AIDS initiatives in South Africa'. Unpublished Master's dissertation. London: University of London.

Dunkle, K.L., R.K. Jewkes, H.C. Brown, G.E. Gray, J.A. McIntryre and S.D. Harlow. 2004. 'Gender-based violence, relationship power, and risk of HIV infection in women attending antenatal clinics in South Africa'. *The Lancet* 363: 1 415–421.

Dunne, M. (ed.). 2008. *Gender, Sexuality and Development: Education and Society in Sub-Saharan Africa*. Rotterdam: Sense Publishers.

Eliasov, N. and C. Frank. 2000. 'Crime and violence in schools in transition: A survey of crime and violence in twenty schools in the Cape Metropole and beyond'. Report for Social Justice Research Project, University of Cape Town.

Epstein, D., J. Elwood, V. Hey and J. Maw. 1998. 'Schoolboy frictions: Feminism and "failing" boys'. In D. Epstein, J. Elwood, V. Hey and J. Maw (eds.), *Failing Boys? Issues in Gender and Achievement*, 3–18. Buckingham: Open University Press.

Epstein, D. and R. Johnson. 1998. *Schooling Sexualities*. Buckingham: Open University Press.

Epstein, D., R. Morrell, R. Moletsane and E. Unterhalter. 2004. 'Identities, activism, identities: Gender and HIV and AIDS in Africa'. *Transformation (Special Issue, Identities, Activism, Interventions: Gender and HIV and AIDS in Africa)* 54: 1–16.

Epstein, D., S. O'Flynn and D. Telford. 2003. *Silenced Sexualities in Schools and Universities*. Stoke-on-Trent: Trentham Books.

Epstein, H. 2007. *The Invisible Cure: Africa, the West and the Fight against AIDS*. London: Viking Adult.

Freund, B. 1996. 'Confrontation and social change: Natal and the forging ahead of apartheid 1949–72'. In R. Morrell (ed.), *Political Economy and Identities in KwaZulu-Natal: Historical and Social Perspectives*, 179–95. Durban: Indicator Press.

Gaillard-Thurston, C. 2008. 'Raw girls: A gender study at an urban KwaZulu-Natal co-educational high school'. Ongoing Ph.D. research. Edgewood: University of KwaZulu-Natal.

Gevisser, M. and E. Cameron (eds.). 1994. *Defiant Desire: Gay and Lesbian Lives in South Africa*. Johannesburg: Ravan Press.

Gibson, D. and A. Hardon (eds.). 2005. *Rethinking Masculinities, Violence and AIDS*. Amsterdam: Het Spinhuis.

Gibson, D. and M.R. Lindegaard. 2007. 'South African boys with plans for the future: Why a focus on dominant discourses tells us only a part of the story'. In T. Shefer, K. Ratele, A. Strebel, N. Shabalala and R. Buikema (eds.), *From Boys to Men: Social Constructions of Masculinity in Contemporary Society*, 128–44. Cape Town: University of Cape Town Press.

Gouws, A. 2006. 'The state of the national gender machinery: Structural problems and personalized politics'. In S. Buhlungu, J. Daniel, R. Southall and J. Lutchman (eds.), *State of the Nation South Africa 2005–2006*, 143–66. Cape Town: HSRC Press.

Gouws, A. (ed.) 2005. *(Un)Thinking Citizenship: Feminist Debates in Contemporary South Africa*. Aldershot: Ashgate; Cape Town: University of Cape Town Press.

Gouws, E. and Q. Abdool Karim. 2005. 'HIV infection in South Africa: The evolving epidemic'. In S.S. Abdool Karim and Q. Abdool Karim (eds.), *HIV/AIDS in South Africa*, 48–66. Cape Town: Cambridge University Press.

Govender, L. 2004. 'Teacher unions, policy struggles and educational change, 1994 to 2004'. In L. Chisholm (ed.), *Changing Class: Education and Social Change in Post-Apartheid South Africa*, 267–91. Cape Town: HSRC Press.

Grant, M. and K. Hallman. 2006. 'Pregnancy-related school dropout and prior school performance in South Africa'. Population Council Working Paper No.212. http://www.popcouncil.org/publications/wp/prd/rdwplist.html, accessed 23 October 2008.

Grundlingh, A., A. Odendaal and B. Spies. 1995. *Beyond the Tryline: Rugby and South African Society*. Johannesburg: Ravan Press.

Gultig, J. and M. Hart. 1990. ' "The world is full of blood": Youth, schooling and conflict in Pietermaritzburg, 1987–1989'. *Perspectives in Education* 11(2): 1–19.

Gustafsson, M. and F. Patel. 2006. 'Undoing the apartheid legacy: Pro-poor spending shifts in the South African public school system'. *Perspectives in Education* 24(2): 65–77.

Hallam, R. 1994. 'Crimes without punishment: Sexual harassment and violence against female students in schools and universities in Africa'. African Rights Discussion Paper No.4. London: African Rights.

Hames, M. and K. Koen. 2006. *Beyond Inequalities 2005: Women in South Africa*. Harare: Southern African Research and Documentation Centre.

Hamlall, V. 2003. 'Boys' narratives of violence in a technical high school in Chatsworth, Durban'. Unpublished M.Ed dissertation. Durban: University of Natal.

Hammersley, M. 2006. 'Are ethics committees ethical?' *Qualitative Researcher* 2. http://www.cf.ac.uk/socsi/qualiti/qualitative_researcher.html, accessed 29 March 2008.

Harber, C. 2002. 'Schooling as violence: An exploratory overview'. *Educational Review* 54(1): 7–16.

Hargreaves, J. and T. Boler. 2006. *The Impact of Girls' Education on HIV and Sexual Behaviour*. Johannesburg: ActionAid International.

Harrison, A. 2005. 'Young people and HIV/AIDS in South Africa: Prevalence of infection, risk factors and social context'. In S.S. Abdool Karim and Q. Abdool Karim (eds.), *HIV/AIDS in South Africa*, 262–84. Cape Town: Cambridge University Press.

Hartshorne, K. 1992. *Crisis and Challenge: Black Education 1910–1990*. Cape Town: Oxford University Press.

Hassim, S. 2006. *Women's Organisations and Democracy in South Africa: Contesting Authority*. Pietermaritzburg: University of KwaZulu-Natal Press.

Health Systems Trust. 2007. 'Health statistics: Teenage pregnancy'. http://www.hst.org.za/healthstats/2/data/eth, accessed 2 July 2007.

HEARD (Health Economics and HIV/AIDS Research Division). 2005. 'Teacher attrition and mortality in South Africa: A study into gross educator attrition rates and trends, including analysis of the causes of these by age and gender, in the public schools system in South Africa, 1997/8–2003/4'. Unpublished report, Education Labour Relations Council, March.

Heward, C. 1988. *Making a Man of Him: Parents and Their Sons' Careers at an English Public School 1929–1950*. London: Routledge.

Holdstock, T.L. 1990. 'Violence in schools: Discipline'. In B. McKendrick and W. Hoffman (eds.), *People and Violence in South Africa*, 341–72. Cape Town: Oxford University Press.

Holland, J., C. Ramazanoglu, S. Scott, S. Sharpe and R. Thomson. 1990. *'Don't Die of Ignorance': I Nearly Died of Embarrassment, Condoms in Context*. London: Tufnell Press.

Holland, J., C. Ramazanoglu, S. Sharpe and R. Thomson. 1998. *The Male in the Head: Young People, Heterosexuality and Power*. London: Tufnell Press.

Hollway, W. 1989. *Subjectivity and Method in Psychology: Gender, Meaning and Science*. London: Sage.

Hollway, W. and T. Jefferson. 2000. *Doing Qualitative Research Differently: Free Association, Narrative and the Interview Method*. London: Sage.

HSRC (Human Sciences Research Council). 2005a. 'Absenteeism among educators in South African public schools'. Fact sheet 2. Pretoria: Human Sciences Research Council.

———. 2005b. 'HIV prevalence among South African educators in public schools'. Fact sheet 6. Pretoria: Human Sciences Research Council.

Human Rights Watch. 2001. *Scared at School: Sexual Violence against Girls in South African Schools*. New York: Human Rights Watch. http://www.hrw.org/reports/2001/safrica/ZA-FINAL-04.htm#P60, 20 January 2009.

Hunter, M. 2002. 'The materiality of everyday sex: Thinking beyond "prostitution"'. *African Studies* 61(1): 99–120.

———. 2004. 'Masculinities and multiple-sexual-partners in KwaZulu-Natal: The making and unmaking of *isoka*'. *Transformation* 54: 123–53.

Hyslop, J. 1999. *The Classroom Struggle: Policy and Resistance in South Africa 1940–1990*. Pietermaritzburg: University of Natal Press.

Ibrahim, S. and S. Alkire. 2007. 'Agency and empowerment: A proposal for internationally comparable indicators'. OPHI Working Paper. http://www.ophi.org.uk/pubs/Ibrahim_Alkire_Empowerment_FINAL.pdf, accessed 22 October 2008.

Jackson, S. 1982. *Childhood and Sexuality*. Oxford: Basil Blackwell.

Jacob, F. 1996. 'Empowerment: A critique'. *British Journal of Community Nursing* 1(8): 449–53.

James, A., C. Jenks and A. Prout. 1998. *Theorizing Childhood*. Cambridge: Polity Press.

James, A. and A. Prout. 1995. 'Hierarchy, boundary and agency: Toward a theoretical perspective on childhood'. *Sociological Studies of Childhood* 7: 77–101.

James, A. and A. Prout (eds.). 1997. *Constructing and Reconstructing Childhood: Contemporary Issues in the Sociological Study of Childhood*. London: Falmer.

James, D. 2002. ' "To take the information down to the people": Life skills and HIV/AIDS peer educators in the Durban area'. *African Studies* 61(1): 169–92.

Jeffrey, C., P. Jeffery and R. Jeffery. 2008. *Degrees without Freedom: Education, Masculinities and Unemployment in North India*. Stanford: Stanford University Press.

Jewkes, R. and N. Abrahams. 2000. *Violence against Women in South Africa: Rape and Sexual Coercion*. Pretoria: Pretoria: Medical Research Council.

Jewkes, R., K. Dunkle, M.P. Koss, J.B. Levin, M. Ndunae, N. Jamaa and Y. Sikweyiya. 2006. 'Rape perpetration by young, rural South African men: Prevalence, patterns and risk factors'. *Social Science and Medicine* 63: 2 949–961.

Jewkes, R., J. Levin, N. Mbananga and D. Bradshaw. 2002. 'Rape of girls in South Africa'. *The Lancet* 359: 319–20.

Jewkes, R., M. Nduna, J. Levin, N. Jama and K. Dunkle. 2007. 'Evaluation of Stepping Stones: A gender transformative HIV prevention intervention'. Pretoria: Medical Research Council, Gender and Health Research Unit. http://www.popline.org/docs/ 316333, accessed 8 April 2008.

Jones, S. 1993. *Assaulting Childhood: Children's Experiences of Migrancy and Hostel Life in South Africa*. Johannesburg: Wits Press.

Kabeer, N. 1999. 'Resources, agency, achievements: Reflections on the measurement of women's empowerment'. *Development and Change* 30(3): 435–64.

Kahn, M. 2006. 'Matric matters'. In V. Reddy (ed.), *Marking Matric: Colloquium Proceedings*, 127–38. Cape Town: HSRC Press.

Kakuru, D.M. 2006. *The Combat for Gender Equality in Education: Rural Livelihood Pathways in the Context of HIV/AIDS*. Wageningen: Wageningen Academic Publishers.

Kallaway, P. (ed.). 1984. *Apartheid and Education*. Johannesburg: Ravan Press.

———. 2002. *The History of Education under Apartheid, 1948–1994: The Doors of Learning and Culture Shall Be Opened*. New York: Peter Lang.

Kaufman, C.E., T. de Wet and J. Stadler. 2001. 'Adolescent pregnancy and parenthood in South Africa'. *Studies in Family Planning* 32(2): 147–60.

Kelly, M.J. 2002. 'Preventing HIV transmission through education'. *Perspectives in Education* 20(2): 1–12.

Kent, A. 2002. ' "Let's talk about sex, baby!" Negotiating space, performance and sexualities within a compulsory heterosexual school regime in South Africa, in the context of the HIV/AIDS epidemic'. Unpublished Master's dissertation. London: Institute of Education, University of London.

———. 2004. 'Living life on the edge: Examining space and sexualities within a township high school in greater Durban, in the context of the HIV epidemic'. *Transformation: Critical Perspectives on Southern Africa* 54: 59–75.

Kentridge, M. 1990. *An Unofficial War: Inside the Conflict in Pietermaritzburg*. Cape Town: David Philip.

Kenway, J. and L. Fitzclarence. 1997. 'Masculinity, violence and schooling: Challenging "poisonous pedagogies" '. *Gender and Education* 9(1): 117–33.

Kenway, J. and S. Willis, with J. Blackmore and L. Rennie. 1997. *Answering Back: Boys and Girls in School*. Sydney: Allen and Unwin.

Kimmel, M.S. 2001. *The Gendered Society*. New York: Oxford University Press.

Kirumira, E. 2004. 'Interview with Edward Kirumira'. *Transformation* 54: 154–59.

Kitzinger, J. 1988. 'Defending innocence: Ideologies of childhood'. *Feminist Review, Special Issue: Family Secrets, Child Sexual Abuse* 28: 77–87.

————. 1990. '"Who are you kidding?" Children, power and sexual assault'. In A. James and A. Prout (eds.), *Constructing and Reconstructing Childhood: Contemporary Issues in the Sociological Study of Childhood*, 165–88. London: Falmer, 1997.

Kleintjes, S., B. Prince, A. Cloete and A. Davids (eds.). 2005. *Gender Mainstreaming in HIV/ AIDS: Seminar Proceedings*. Cape Town: HSRC Press.

Koggel, C. 2006. *Moral Issues in Global Perspective, Volume I: Moral and Political Theory; Volume II: Human Diversity and Equality; Volume III: Moral Issues*. Peterborough: Broadview Press.

Kotecha, P. 1994. 'The position of Women Teachers', *Agenda* 21: 21–35.

Leach, F. 2002. 'School-based gender violence in Africa: A risk to adolescent health'. *Perspectives in Education* 20(2): 99–112.

Leach, F. and C. Mitchell (eds.). 2006. *Combating Gender Violence in and around Schools: Strategies for Change*. London: Trentham.

Leggett, I. 2005. 'Learning to improve policy for pastoralists in Kenya'. In S. Aikman and E. Unterhalter (eds.), *Beyond Access: Transforming Policy and Practice for Gender Equality in Education*, 128–48. Oxford: Oxfam.

Lemon, A. 2004. 'Redressing school inequalities in the Eastern Cape, South Africa'. *Journal of Southern African Studies* 30(2): 269–90.

Leoschut, L. and P. Burton. 2006. *How Rich the Reward: Results of the 2005 National Youth Victimisation Study*. Monograph Series, No.1. Cape Town: Centre for Justice and Crime Prevention. http://www.cjcp.org.za/Final.RichReward.pdf, 22 January 2008.

Lesko, N. 2007. 'Talking about sex: The discourses of *loveLife* peer educators in South Africa'. *International Journal of Inclusive Education* 11(4): 519–63.

Lister, R. 2004. *Poverty*. Cambridge: Polity Press.

Loila-Nuahn, H. 2004. 'Social support of pregnant adolescents in Durban'. Unpublished M.Ch. dissertation. Durban: University of Natal.

Lugg, R.A. 2007. 'Making different equal? Social practices of policy-making and the National Qualifications Framework in South Africa between 1985 and 2003'. Unpublished Ph.D. thesis. London: Institute of Education, University of London.

Lund, F. 2006. 'Gender and social security in South Africa'. In V. Padayachee (ed.), *The Development Decade? Economic and Social Change in South Africa, 1994–2004*, 160–79. Cape Town: HSRC Press.

————. 2008. *Changing Social Policy: The Child Support Grant in South Africa*. Cape Town: HSRC Press.

Luyt, R. and D. Foster. 2001. 'Hegemonic masculine conceptualisation in gang culture'. *South African Journal of Psychology* 31: 1–11.

Mac an Ghaill, M. 1994. *The Making of Men: Masculinities, Sexualities and Schooling*. Milton Keynes: Open University Press.

Maharaj, R. 2003. 'Frustrated careers? The perceptions of female educators at a Durban primary school'. Unpublished M.Ed. Minor thesis. Durban: University of Natal.

Majeke, S. 2008. 'School-going youth, sexuality and HIV prevention in northern KwaZulu-Natal: A gender perspective'. Ongoing Ph.D. research. Durban: University of KwaZulu-Natal.

Manchester, J. 2004. 'Hope, involvement and vision: Reflections on positive women's activism around HIV'. *Transformation* 54: 85–103.

Mannah, S. 2005. 'The state of mobilisation of women teachers in the South African Teachers' Union'. In L. Chisholm and J. September (eds.), *Gender Equity in South African Education 1994–2004: Perspectives from Research, Government and Unions*, 146–55. Cape Town: HSRC Press.

Marks, M. 2001. *Young Warriors: Youth Politics, Identity and Violence in South Africa.* Johannesburg: Wits Press.

Marks, S. 1987. *'Not Either an Experimental Doll': The Separate Worlds of Three South African Women.* Pietermaritzburg: University of Natal Press.

———. 1994. *Divided Sisterhood: Race, Class and Gender in the South African Nursing Profession.* Johannesburg: Wits Press.

Marriott, D. 1996. 'Reading black masculinities'. In M. Mac an Ghaill (ed.), *Understanding Masculinities*, 185–201. Buckingham: Open University Press.

Martino, W. 1999. '"Cool boys", "party animals", "squids" and "poofters": Interrogating the dynamics and politics of adolescent masculinities in school'. *British Journal of Sociology of Education* 20(2): 239–64.

May, J.D. 1986. *A Study of Income and Expenditure and Other Socio-Economic Structures in Urban KwaZulu: Umlazi.* Durban: KwaZulu Finance and Investment Corporation.

Meintjes, S. 2004. 'The implications of institutionalising gender mainstreaming: Will it mainstream or sidestream gender issues?' In I. Welpe, B. Thege and S. Henderson (eds.), *The Gender Perspective: Innovations in Economy, Organisations and Health within the Southern African Development Community (SADC).* Frankfurt: Peter Lang.

Mitchell, C. 2005. 'Mapping a southern African girlhood in the age of AIDS'. In L. Chisholm and J. September (eds.), *Gender Equity in South African Education 1994–2004: Perspectives from Research, Government and Unions*, 92–112. Cape Town: HSRC Press.

Moletsane, R. 2007. 'South African girlhood in the age of AIDS: Towards girlhood studies'. *Agenda* 72: 155–65.

Moletsane, R., C. Mitchell, A. Smith and L. Chisholm. 2008. *Methodologies for Mapping a Southern African Girlhood in the Age of AIDS.* Rotterdam: Sense Publishers.

Moletsane R., R. Morrell, E. Unterhalter and D. Epstein. 2002. 'Instituting gender equality in schools: Working in an HIV/AIDS environment'. *Perspectives in Education* 20(2): 37–53.

Moller, V., L. Schlemmer, J. Kuzwayo and B. Mbanda. 1978. 'A black township in Durban: A study of needs and problems'. Unpublished paper. Durban: Centre for Applied Social Sciences, University of Natal.

Moorosi, P. 2006a. 'Towards closing the gender gap in education management: A gender analysis of educational management policies in southern Africa'. *Agenda* 69: 58–70.

———. 2006b. 'Policy and practice related constraints to increased female participation in education management in South Africa'. Unpublished Ph.D. thesis. Pietermaritzburg: University of KwaZulu-Natal.

Morley, L. 1999. *Organising Feminisms: The micropolitics of the academy.* Basingstoke: Macmillan.

Moroney, E. 2002. 'Teaching HIV/AIDS education using the life skills approach in two Durban area high schools'. M.Ed. thesis. Durban: University of Natal.

Morrell, R. 1998. 'Gender and education: The place of masculinity in South African schools'. *South African Journal of Education* 18(4): 218–25.

———. 2001a. *Changing Men in Southern Africa.* Pietermaritzburg: University of Natal Press; London: Zed Books.

———. 2001b. *From Boys to Gentlemen: Settler Masculinity in Colonial Natal, 1880–1920.* Pretoria: UNISA Press.

———. 2007a. 'On a knife's edge: Masculinity in black working class schools in post-apartheid education'. In B. Frank and K. Davison (eds.), *Masculinity and Schooling: International Practices and Perspectives*, 35–57. London, Ontario: The Althouse Press.

———. 2007b. 'Do you want to be a father? School-going youth in Durban schools at the turn of the 21st century'. In T. Shefer and K. Ratele (eds.), *From Boys to Men: Social Constructions of Masculinity in Contemporary Society*, 63–81. Cape Town: University of Cape Town Press.

———. 2007c. 'Survey of learner attitudes in two Durban high schools in the year 2000'. http://www.ukzn.ac.za/heard/research/researchSummaries.htm#Learner_Attitudes _in_two_Durban_High_Schools_in_the_Year_2000,_Survey_of, accessed 11 February 2008.

Morrell, R. and D. Epstein. 2008. 'Material effects: Race, class and masculinities among South African teachers'. *Darkmatter Journal* 2. http://www.darkmatter101.org/site/ category/issues/race-matter/.

Morrell, R. and L. Ouzgane (eds.). 2005. *African Masculinities.* New York: Palgrave; Pietermaritzburg: University of KwaZulu-Natal Press.

Morrell, R., E. Unterhalter, L. Moletsane and D. Epstein. 2001. 'HIV and AIDS: Policies, schools and gender identities'. *Indicator South Africa* 18(2): 51–57.

Nair, C.M. 2003. 'Women in management: Perceptions of eight women in the KZN Department of Education'. Unpublished M.Ed. thesis. Durban: University of Natal.

Nattrass, N. 2003. *The Moral Economy of AIDS in South Africa*. Cambridge: Cambridge University Press.

Nayak, A. and M.J. Kehily. 2001. '"Learning to laugh": A study of schoolboy humour in the English secondary school'. In W. Martino and B. Meyenn (eds.), *What About the Boys? Issues of Masculinity in Schools*, 110–23. Buckingham: Open University Press.

Ndabandaba, G.L. 1987. *Crimes of Violence in Black Townships*. Durban: Butterworth.

Nelson Mandela Foundation/HSRC (Human Sciences Research Council). 2002. *Study of HIV/AIDS: South African National HIV Prevalence, Behavioural Risks and Mass Media. Households Survey 2002*. Cape Town: HSRC Publishers.

Niehaus, I. 2000. 'Towards a dubious liberation: Masculinity, sexuality and power in South African lowveld schools, 1953–1999'. *Journal of Southern African Studies* 26(3): 387–407.

NPPHCN (National Progressive Primary Health Care Network)/UNICEF (United Nations Children's Fund). 1997. *Youth Speak Out: A Study on Youth Sexuality*. Braamfontein: NPPHCN/UNICEF.

O'Flynn, S. and D. Epstein. 2005. 'Standardising sexuality: Embodied knowledge, achievement and "standards"'. *Social Semiotics* 15(2): 183–208.

O'Sullivan, L.F., A. Harrison, R. Morrell, A. Monroe-Wise and M. Kubeka. 2006. 'Shifting sexualities: Gender dynamics in the primary sexual relationships of young rural South African women and men'. *Culture, Health and Sexuality* 8(2): 99–113.

Odora Hoppers, C. 2005. 'Between "mainstreaming and transformation": Lessons and challenges for institutional change'. In L. Chisholm and J. September (eds.), *Gender Equity in South African Education 1994–2004: Perspectives from Research, Government and Unions*, 55–73. Cape Town: HSRC Press.

Pandor, N. 2005. 'The hidden face of gender inequality in SA education'. In L. Chisholm and J. September (eds.), *Gender Equity in South African Education 1994–2004: Perspectives from Research, Government and Unions*, 19–24. Cape Town: HSRC Press.

Parker, W. 2003. 'Re-appraising youth prevention in South Africa: The case of *loveLife*'. Paper presented at the South African AIDS Conference. Durban. http://www.cadre. org.za/pdf/Youth%20prevention%20in%20SA.pdf, accessed 8 April 2008.

Parkes, J. 2002. '"Children also have rights, but then who wants to listen to our rights?" Children's perspectives on living with community violence in South Africa'. *Educate* 2(2). http://www.educatejournal.org/index.php?journal=educate&page=article&op =view&path%5B%5D=21&path%5B%5D=17, accessed 20 January 2009.

———. 2007. 'The multiple meanings of violence: Children's talk about life in a violent South African neighbourhood'. *Childhood* 14(4): 401–14.

———. 2008. 'Developing a framework for conceptualising and operationalising research on violence against girls'. Paper presented at the ActionAid Stop Violence Project Research Workshop, Nairobi, September.

————. 2009 (In press). 'Perspectives on children and violence'. In R. Cowen and A.M. Kazamias (eds.), *International Handbook of Comparative Education*. Dordrecht: Springer.

Patton, C. 1994. *Last Served? Gendering the HIV Pandemic*. London: Taylor and Francis.

Peacock, D., B. Khumalo and E. McNab. 2006. 'Men and gender activism in South Africa: Observations, critique and recommendations for the future'. *Agenda* 69: 71–81.

Perry, H. and B. Fleisch. 2006. 'Gender and educational achievement in South Africa'. In V. Reddy (ed.), *Marking Matric: Colloquium Proceedings*, 107–26. Cape Town: HSRC Press.

Pettifor, A.E., H.V. Rees, I. Kleinschmidt, A.E. Steffenson, C. MacPhail, L. Hlongwa-Madikizela, K. Vermaak and N.S. Padian. 2005. 'Young people's sexual health in South Africa: HIV prevalence and sexual behaviors from a nationally representative household survey'. *AIDS* 19: 1 525–534.

Poku, N.K., A. Whiteside and B. Sandkjaer (eds.). 2007. *AIDS and Governance*. London: Ashgate.

Preston-Whyte, E. 1993. 'Women who are not married: Fertility, "illegitimacy", and the nature of households and domestic groups among single African women in Durban'. *South African Journal of Sociology* 24(3): 63–71.

Ramphele, M. 2002. *Steering by the Stars: Being Young in South Africa*. Cape Town: Tafelberg.

Raniga, T. 2006. 'The implementation of the national life skills and HIV/AIDS school policy and programme in the eThekwini region'. Unpublished Ph.D. thesis. Pietermaritzburg: University of KwaZulu-Natal.

Ratele, K. 2001. 'Between "ouens": Everyday makings of black masculinity'. In R. Morrell (ed.), *Changing Men in Southern Africa*, 239–53. Pietermaritzburg: University of Natal Press; London: Zed Books.

Ratele, K., E. Fouten, T. Shefer, A. Strebel, N. Shabalala and R. Buikema. 2007. '"Moffies, jocks and cool guys": Boys' accounts of masculinity and their resistance in context'. In T. Shefer, K. Ratele, A. Strebel, N. Shabalala and R. Buikema (eds.), *From Boys to Men: Social Constructions of Masculinity in Contemporary Society*, 112–27. Cape Town: University of Cape Town Press.

Raynor, J. and K. Wesson. 2006. 'The girls' stipend program in Bangladesh'. *Journal of Education for International Development* 2(2). http://www.equip123.net/JEID/articles/3/Bangladesh.pdf, accessed 22 October 2008.

Reddy, V. 2006. 'The state of mathematics and science education: Schools are not equal'. In S. Buhlungu, J. Daniel, R. Southall and J. Lutchman (eds.), *State of the Nation: South Africa 2005–2006*, 392–416. Cape Town: HSRC Press.

Reid, G. and L. Walker (eds.). 2005. *Men Behaving Differently*. Cape Town: Double Storey.

Renold, E. 2005. *Girls, Boys and Junior Sexualities: Exploring Children's Gender and Sexual Relations in the Primary School*. London: Routledge/Falmer.

Richter, L. 2007. ' "He is my hero": Men and fatherhood in South Africa'. *South African Labour Bulletin* May: 10–13.

Richter, L., A. Dawes and C. Higson Smith (eds.). 2004. *Sexual Abuse of Young Children in Southern Africa*. Cape Town: HSRC Press.

Richter, L. and R. Morrell (eds.). 2006. *Baba: Men and Fatherhood in South Africa*. Cape Town: HSRC Press.

Robeyns, I. 2007. 'When will society be gender just?' In J. Browne (ed.), *The Future of Gender*, 54–74. Cambridge: Cambridge University Press.

Rugalema, G. and V. Khanye. 2004. 'Mainstreaming HIV/AIDS in the education systems in sub-Saharan Africa: Some preliminary insights'. In C. Coombe (ed.), *The HIV/AIDS Challenge to the Education System: A Collection of Essays*, 81–203. Paris: IIEP.

Salo, E. 2003. 'Negotiating gender and personhood in the new South Africa: Adolescent women and gangsters in Manenberg township on the Cape Flats'. *European Journal of Cultural Studies* 6(3): 345–65.

SAP (South African Police Services) Information Management. 2008. http://www.saps. gov.za/statistics/reports/crimestats/2006/_pdf/provinces/rsa_total.pdf, accessed 10 October 2008.

Sathiparsad, R. 2005. ' "It is better to beat her": Male youth in rural KwaZulu-Natal speak on violence in relationships'. *Agenda* 66: 79–88.

Sears, J. 1992. 'Dilemmas and possibilities of sexuality education: Reproducing the body politic'. In J. Sears (ed.), *Sexuality and the Curriculum: The Politics and Practices of Sexuality Education*, 7–33. New York: Teachers College Press.

Sebakwane, S. 1993/4. 'Gender relations in Lebowa secondary schools'. *Perspectives in Education* 15(1): 83–99.

Sen, A. 1979. 'Equality of what?' The Tanner Lecture on Human Values. Stanford University, 22 May.

———. 1992. *Inequality Re-examined*. Oxford: Oxford University Press.

———. 1999. *Development as Freedom*. Oxford: Oxford University Press.

Shefer, T. and K. Ruiters. 1998. 'The masculine construct of heterosex'. *Agenda* 37: 39–45.

Shisana, O., T. Rehle, L. Simbayi, W. Parker, K. Zuma, A. Connolly, S. Jooste and V. Pillay (eds.). 2005. *South African National HIV Prevalence, HIV Incidence, Behaviour and Communication Survey*. Cape Town: HSRC Press.

Silin, J.G. 1995. *Sex, Death and the Education of Children: Our Passion for Ignorance in the Age of AIDS*. New York: Teachers College Press.

Sloth-Nielsen, J. 2007. 'The state of South Africa's prisons'. In S. Buhlungu, J. Daniel, R. Southall and J. Lutchman (eds.), *State of the Nation: South Africa 2007*, 379–401. Cape Town: HSRC Press.

Stewart, F. (ed.). 2008. *Horizontal Inequalities and Conflict: Understanding Group Violence in Multi-ethnic Societies*. London: Palgrave.

Stuart, J. 2006. ' "From our frames": Exploring with teachers the pedagogic possibilities of a visual arts-based approach to HIV and AIDS'. *Journal of Education* 38: 67–88.

Subedar, M. 2003. 'An analysis of the nature and effects of sexual harassment on secondary schoolgirls in South Africa: A case study of four co-educational schools in Pietermaritzburg, KwaZulu-Natal'. Unpublished Ph.D. thesis. Durban: University of Natal.

Swain, J. 2003. 'How young schoolboys become some*body*: The role of the body in the construction of masculinity'. *British Journal of Sociology of Education* 24(3): 299–314.

Tallis, V. 1998. 'AIDS is a crisis for women'. *Agenda* 39: 6–13.

———. 2000. 'Gendering the response to HIV/AIDS: Challenging gender inequality'. *Agenda* 44: 58–66.

Thorne, B. 1993. *Gender Play: Girls and Boys in School*. Buckingham: Open University Press.

Thorpe, M. 2001a. 'Shifting discourse: Teenage masculinity and the challenge for behavioural change'. Paper presented at the AIDS in Context conference, Johannesburg.

———. 2001b. 'An evaluation of the intervention of DramAide's programme: "Mobilising young men to care", in two township schools in Durban'. Commissioned research, Durban, University of Natal.

———. 2005. 'Learning about HIV/AIDS in schools: Does a gender-equality approach make a difference?' In S. Aikman and E. Unterhalter (eds.), *Beyond Access: Transforming Policy and Practice for Gender Equality in Education*, 197–211. Oxford: Oxfam.

Truscott, K. 1994. *Gender in Education*. Johannesburg: Education Policy Unit, University of the Witwatersrand/NECC.

UNAIDS (The Joint United Nations Programme on HIV/AIDS). 2007. *AIDS Epidemic Update*. Geneva: UNAIDS. http://www.unaids.org/en/KnowledgeCentre/HIVData/Epi Update/EpiUpdArchive/2007/default.asp, accessed 22 January 2008.

UNESCO (United Nations Educational, Scientific, and Cultural Organization) Institute for Statistics. 2008. 'Gender parity in education: Not there yet'. http://www.uis. unesco.org/template/pdf/EducGeneral/UISFactsheet_2008_No%201_EN.pdf, accessed 30 January 2009.

Unterhalter, E. 1991. 'The impact of apartheid on women's education in South Africa'. *Review of African Political Economy* 48: 66–75.

———. 1999. 'The schooling of South African girls'. In C. Heward and S. Bunwaree (eds.), *Gender, Education and Development: Beyond Access to Empowerment*, 49–64. London: Zed Books.

———. 2000. 'Remembering and forgetting: Constructions of education gender reform in autobiography and policy texts of the South African transition'. *History of Education* 29(5): 457–72.

————. 2002. 'Gender, race, and different lives: SA women teachers' autobiographies and the analysis of educational change'. In P. Kallaway (ed.), *The History of Education under Apartheid, 1948–1994: The Doors of Learning and Culture Shall Be Opened*, 243–55. Cape Town: Pearson Education.

————. 2005. 'Fragmented frameworks? Researching women, gender, education and development'. In S. Aikman and E. Unterhalter (eds.), *Beyond Access: Transforming Policy and Practice for Gender Equality in Education*, 15–35. Oxford: Oxfam.

————. 2006. 'New times and new vocabularies: Theorising and evaluating gender equality in Commonwealth higher education'. *Women's Studies International Forum* 29(6): 620–28.

————. 2007. *Gender, Schooling and Global Social Justice*. London: Routledge.

Unterhalter, E., S. Aikman and T. Boler. 2008. 'Esssentialism, equality, and empowerment: Concepts of gender and schooling in the HIV and AIDS epidemic'. In S. Aikman, E. Unterhalter and T. Boler (eds.), *Gender Equality, HIV and AIDS: A Challenge for the Education Sector*, 11–32. Oxford: Oxfam.

Urdang, S. 2006. 'The care economy: Gender and the silent AIDS crisis in southern Africa'. *Journal of Southern African Studies* 32(1): 165–77.

US Department of Justice. 2006. 'Crime in the US: Forcible rape'. http://www.fbi.gov/ucr/cius2006/offenses/violent_crime/forcible_rape.html, accessed 10 October 2008.

Valodia, I. 1998. 'Engendering the public sector: An example from the women's budget initiative in South Africa'. *Journal of International Development* 10: 943–55.

Van den Berg, S. 2006. 'The targeting of public spending on school education, 1995–2000'. *Perspectives in Education* 24(2): 49–63.

Van Deventer, I. and P.C. van der Westhuizen. 2000. 'A shift in the way female educators perceive intrinsic barriers to promotion'. *South African Journal of Education* 20(3): 235–41.

Vavrus, F. 2003. *Desire and Decline: Schooling amid Crisis in Tanzania*. New York: Peter Lang.

Verma, R., J. Pulerwitz, V. Mahendra, S. Khandekar, G. Barker, P. Fulpagare and S. Singh. 2003. 'Challenging and changing gender attitudes among young men in Mumbai, India'. *Reproductive Health Matters* 14(28): 135–43.

Vetten, L. and K. Bhana. 2001. *Violence, Vengeance and Gender: A Preliminary Investigation into the Links between Violence against Women and HIV/AIDS in South Africa*. Johannesburg: Centre for the Study of Violence and Reconciliation.

Walker, L. 2005. 'Men behaving differently: South African men since 1994'. *Culture, Health and Sexuality* 7: 225–38.

Walker, L., G. Reid and M. Cornell. 2004. *Waiting to Happen: HIV/AIDS in South Africa*. Cape Town: Double Storey.

Walkerdine, V. 1984. 'Developmental psychology and child-centre pedagogy: The insertion of Piaget into early education'. In J. Henriques, W. Hollway, C. Urwin, C. Venn and V. Walkerdine, *Changing the Subject: Social Regulation and Subjectivity*, 153–202. London: Methuen.

———. 1997. *Daddy's Girl: Young Girls and Popular Culture*. Basingstoke: Macmillan.

Walsh, S. and C. Mitchell. 2006. ' "I'm too young to die": HIV, masculinity, danger and desire in urban South Africa'. *Gender and Development* 14(1): 57–68.

Wedgwood, N. 2005. 'Just one of the boys? A life history case study of a male physical education teacher'. *Gender and Education* 17(2): 189–201.

Wells, H. and L. Polders. 2006. 'Anti-gay hate crimes in South Africa: Prevalence, reporting practices, and experiences of police'. *Agenda* 67: 20–28.

Westwood, S. 1990. 'Racism, black masculinity and the politics of space'. In J. Hearn and D. Morgan (eds.), *Men, Masculinities and Social Theory*, 55–71. London: Unwin Hyman.

Wetherell, M. and N. Edley. 1999. 'Negotiating hegemonic masculinity: Imaginary positions and psycho-discursive practices'. *Feminism and Psychology* 9(3): 335–56.

WHO (World Health Organization). 2002. 'Sexual health'. http://www.who.int/reproductive-health/gender/sexual_health.html, accessed 26 January 2007.

Wolpe, A. 2005. 'Reflections on the gender equity task team'. In L. Chisholm and J. September (eds.), *Gender Equity in South African Education 1994–2004: Perspectives from Research, Government and Unions*, 119–42. Cape Town: HSRC Press.

Wolpe, A., O. Quinlan and L. Martinez. 1997. *Gender Equity in Education: A Report of the Gender Equity Task Team*. Pretoria: Department of Education.

Wood, K. and R. Jewkes. 2001. ' "Dangerous" love: Reflections on violence among Xhosa township youth'. In R. Morrell (ed.), *Changing Men in Southern Africa*, 317–36. Pietermaritzburg: University of Natal Press; London: Zed Books.

Wood, K., F. Maforah and R. Jewkes. 1998. ' "He forced me to love him": Putting violence on adolescent sexual health agendas'. *Social Science and Medicine* 47(2): 233–42.

Zakwe, M. 2005. 'The raising of a Zulu man'. *Agenda* 64: 141–47.

Zulu, L. 1998. 'Role of women in the reconstruction and development of the new democratic South Africa'. *Feminist Studies* 24(1): 147–57.

About the Authors

Robert Morrell was born in Cape Town, along with his twin brother, Christopher. He started his schooling at an elite boys-only school and grew up middle class, in a heterosexual nuclear family. His family loathed apartheid, so he spent his teenage years in Swaziland, where he enjoyed four years of coeducation in a racially mixed school. The regime was laissez faire and he did not do much academic work, preferring to play golf with his brother. He spent his last two years of school miserably unhappy at a South African boarding school. After school he did a year of national service in the South African Defence Force. He then became a student at Rhodes University, where he was involved in left-wing student politics. He remained involved in anti-apartheid politics once he became a lecturer and was deported from Transkei in 1984 for supporting student demonstrations against Kaiser Matanzima, the president of Transkei. Since 1985 he has lived and worked in Durban. He worked in the History Department at the University of Durban-Westville, where he was also the first secretary-treasurer of the Combined Staff Association, which locked horns with the University's vice-chancellor, Professor Greyling. Since 1989 Robert has worked in the Department of Education at the University of Natal (now University of KwaZulu-Natal). His recent publications include *From Boys to Gentlemen: Settler Masculinity in Colonial Natal, 1880–1920* and edited volumes, *Changing Men in Southern Africa, African Masculinities* (with Lahoucine Ouzgane) and *Baba, Men and Fatherhood in South Africa* (with Linda Richter).

Debbie Epstein was born in Pretoria, younger than her siblings by five and eight years respectively. Her parents were liberal, secular Jews, who opposed apartheid from its inception. Her maternal grandparents were strong socialists and Debbie's earliest political memory is of her grandfather showing her pictures of three very important men with beards. She later found out that they were Marx, Lenin and Trotsky. Aged eleven, Debbie went to her first political meeting, where Chief Luthuli was speaking to a white audience and the meeting was disrupted by Afrikaner nationalists. Debbie's father regularly invited

black doctors to their home for dinners and seminars with visiting overseas medical researchers. Debbie began her B.A. at the University of Witwatersrand and became involved in student politics. Halfway through her first year, she left South Africa for the United Kingdom, where she has lived ever since, returning to South Africa for the first time in 1995. She studied at Sussex University and was a teacher for nearly twenty years before becoming a teacher adviser for race equality in Birmingham. Her Ph.D. was on race and racism in white schools. Subsequently she has published widely on sexuality, gender and race and their intersections in educational and other settings. Her books include *Schooling Sexualities* (with Richard Johnson) and *Silenced Sexualities in Schools and Universities* (with Sarah O'Flynn and David Telford). She is now semi-retired, working as a part-time professor at Cardiff University and is in the later stages of training as a psychotherapist.

Elaine Unterhalter was born in Johannesburg. She studied for a B.A. at the University of the Witwatersrand, where she was involved in student politics and became interested in gender inequality and feminism. She went on to teach in rural KwaZulu-Natal at the Charles Johnson Memorial Hospital in Nqutu. After further studies in history at the University of Cambridge, she began a Ph.D. at the School of Oriental and African Studies, University of London. Her focus was a social history of the Nqutu district from 1879 to 1910 and her work highlighted the significance of the schools established by missionaries and questions of gender associated with patterns of migrant labour. These questions became a major theme in her work and in jobs as a researcher and university teacher at three universities in the United Kingdom. She has written about gender, race and class inequalities and their bearing on education, carrying out studies in South Africa, Bangladesh, India and Kenya. She worked with the ANC education committee and women's committee during years of exile in London from the late 1970s onwards. She currently works at the Institute of Education, University of London, and is involved with two large cross-country studies examining the implementation of gender equality policies in schools in South Africa, Kenya, Tanzania and Nigeria. She is author of *Gender, Schooling and Global Social Justice*; co-author with Sheila Aikman of *Beyond Access: Transforming Policy and Practice for Gender Equality in Education*; and co-editor (with Sheila Aikman and Tania Boles) of *Gender Equality, HIV and AIDS: Challenges for the Education Sector*.

Deevia Bhana was born in Durban in 1966. She was one of five daughters (and no sons) in a close-knit Gujerati family. Issues of gender and racial inequalities were very close to her within a broader context of apartheid South Africa. For the first ten years of her life she lived in the area of Mansfield, Durban – which apartheid had defined as mainly 'grey', inhabited by people of mixed racial groups. Her family then had to move to live in areas demarcated as Indian by the Group Areas Act and she was educated in schools defined

under apartheid as Indian. However, she does not see herself as living her life through her colour or racialised group. In 1989, she took on her first job as a school teacher in a secondary school in Durban, where she worked for nine years before taking up an academic job at the University of Durban-Westville. Her Ph.D. study drew her into the world of young children, schooling and the intersection of gender and sexuality. She has since then published in the areas of gender, childhood sexualities and HIV and AIDS.

Relebohile (Lebo) Moletsane was born in Matatiele in the Eastern Cape and went to rural mission boarding schools in the area until she finished high school in 1978. She completed her undergraduate education at the University of Fort Hare in 1982, where she remembers frequent cases of sexual harassment of female students by their male counterparts and incidents of rape of women students, often by groups of men. Such events, then called 'test matches', took place mainly over weekends, especially after social gatherings and parties. After graduating, she went to teach at a high school in what was then the Transkei (now part of the Eastern Cape) and remembers her first principal's cruel use of corporal punishment of boys, in particular, as well as the rampant sexual exploitation of female students. She left to pursue postgraduate studies in the United States and came back in 1997 to teach at the University of Natal (now the University of KwaZulu-Natal) in Durban, where she worked until the end of 2007. She now works for the Human Sciences Research Council as the director of the Gender and Development Unit. Her teaching and research include the areas of curriculum studies and gender and education, including gender-based violence and its links to HIV and AIDS. She is co-author of *Methodologies for Mapping a Southern African Girlhood in the Age of AIDS*.

Index